Drop a pebble in a pond and the results are quite predictable: circular waves flow from the point of impact. Hit a point on a crystalline solid, however, and the expanding waves are highly nonspherical; the elasticity of a crystal is anisotropic.

This book provides a fresh look at the vibrational properties of crystalline solids, elucidated by new imaging techniques. From the megahertz vibrations of ultrasound to the near-terahertz vibrations associated with heat, the underlying elastic anisotropy of the crystal asserts itself. Phonons — the elementary packets of heat — display unique patterns of "caustics" such as the one for CaF_2 shown on the front cover (see Chapter 4).

Written from the viewpoint of an experimentalist, the text contains both the basic theory of wave propagation in solids and its experimental applications. Imaging techniques are described that provide graphic insights into the topics of phonon focusing, lattice dynamics, and ultrasound propagation. Scattering of phonons from interfaces, superlattices, defects, and electrons are treated in detail. The book includes many striking and original illustrations.

This book will be of interest to graduates and researchers in condensed matter physics, materials science, and acoustics.

T0185698

Imaging Phonons

to Kathy

Imaging Phonons

Acoustic Wave Propagation in Solids

JAMES P. WOLFE

*University of Illinois at
Urbana-Champaign*

CAMBRIDGE UNIVERSITY PRESS
Cambridge, New York, Melbourne, Madrid, Cape Town, Singapore, São Paulo

Cambridge University Press
The Edinburgh Building, Cambridge CB2 2RU, UK

Published in the United States of America by Cambridge University Press, New York

www.cambridge.org
Information on this title: www.cambridge.org/9780521620611

© James P. Wolfe 1998

First published 1998
This digitally printed first paperback version 2005

A catalogue record for this publication is available from the British Library

Library of Congress Cataloguing in Publication data
Wolfe, J. P. (James Philip), 1943–
Imaging phonons : acoustic wave propagation in solids / James P.
Wolfe.
p.   cm.
Includes bibliographical references.
ISBN 0-521-62061-9
1. Waves.   2. Solids.   3. Acoustic surface waves.   4. Phonons.
5. Imaging systems.   I. Title.
QC176.8.W3W65   1998
530.4'16 – dc21   97-36117
                                                                CIP

ISBN-13 978-0-521-62061-1 hardback
ISBN-10 0-521-62061-9 hardback

ISBN-13 978-0-521-02208-8 paperback
ISBN-10 0-521-02208-8 paperback

Contents

Preface

Because matter is composed of light and heavy particles – namely, electrons and nucleons – the properties of solid materials naturally divide into two main categories: electronic and vibrational. For the most part, the rapid motion of electrons governs the binding and elasticity of a solid, its conductivity, and its optical and magnetic properties. In contrast, the more sluggish motion of the atomic nuclei determines the vibrational properties of a solid. The descriptions of electronic and nuclear motion in solids are generally developed in terms of wave motion. Such waves in solids are most easily understood for well-ordered systems such as crystals, where the atoms are arranged in regular arrays.

This book is about vibrational waves in crystals – more specifically, those classified as acoustic waves. These are the waves that depend on the elasticity of the medium and whose frequencies extend from zero to a few terahertz (1 THz = 10^{12} Hz). Experimental probes of the motion of acoustic waves in solids traditionally range from ultrasonics to thermal conductivity. The major focus of this book is the propagation of *phonons* – the tiny packets of vibrational energy associated with *heat* and typically observed in the frequency range from 10^{10} to 10^{12} Hz. The propagation of ultrasonic waves in the megahertz range will also be examined here using imaging techniques initially developed for high-frequency phonons.

Phonons govern the thermal properties of nonmetallic solids. At normal temperatures their motion is highly diffusive. However, by cooling a crystal to just a few degrees Kelvin it is possible to observe the unperturbed travel of phonons – the so-called ballistic propagation, or motion without scattering. In such a measurement, the experimenter generates a local concentration of nonequilibrium phonons by either electrical or optical means and watches for their arrival at another point – a time-of-flight experiment. A number of creative variations on this technique have been devised to characterize nonequilibrium phonons and thus learn about the vibrational properties of the medium. This book deals with the processes of generation and detection of acoustic phonons and, more importantly, what happens in between: their ballistic propagation and scattering. It is the scattering of phonons that regulates, for example, the thermal conductivity of a nonmetallic solid.

The ability to *image* the heat flux (i.e., phonons) emanating from a point source in a crystal was first demonstrated in 1978. Remarkable patterns of "caustics" – intense concentrations of heat flux – were observed. The seed for understanding the complex phonon images had been planted a decade before – an effect known as phonon focusing. What makes imaging techniques so important is the inherent anisotropies in the ballistic propagation and scattering of phonons. These methods have now contributed to major areas of phonon physics, including elastic and inelastic scattering, scattering at interfaces, scattering from defects, frequency dispersion, and lattice dynamics. In addition, imaging techniques have been used to study the interaction of phonons with fabricated structures such as semiconductor superlattices and electronic devices. In general, the phonon-imaging methods explore the *angular anisotropies* of a system, in contrast to thermal-imaging tools, which have been developed for *spatial microscopy* of objects.

Basic experimental (and theoretical) ideas are developed for each of the major topics of this book: ballistic propagation (phonon focusing), phonon dispersion (lattice dynamics), bulk scattering (elastic, inelastic, and quasi-diffusive), scattering at interfaces (acoustic mismatch and Kapitza anomaly), and interference of coherent waves (internal diffraction). A wide variety of crystal symmetries are considered. The essential elements of key experiments are described and extensive references are listed, including the title of each paper.

This book is designed to serve as an introduction to the expanding literature on heat pulses and phonon imaging, while providing new perspectives into the broad areas of acoustic-wave propagation and phonon scattering in condensed matter. The aim is toward the needs of beginning graduate students and also scientists whose backgrounds lie in other fields. Currently there is a growing need to understand the dynamics of phonons at low temperatures. For example, the development of phonon-based detectors for high-energy particle detection relies on an understanding of the frequency and spatial distributions of phonons produced by a point disturbance, as well as how these phonons interact with boundaries. In addition, the extension of phonon-imaging techniques to ultrasonic bulk and surface waves promises to open new areas of research and impact materials characterization.

Everyone has his or her own way of understanding a particular concept. Glancing over this book, the reader will see that I am a strong proponent of graphical explanations. It is difficult for most of us to gain a working intuition of fourth-rank tensors, such as the elasticity tensor, but the acoustic slowness surface (replete with lines of zero Gaussian curvature) provides a marvelous visual aid for dealing with acoustic anisotropy. Wherever possible, I have illustrated a concept with a picture. Computer graphics have proven invaluable in exposing the remarkable topologies of the acoustic slowness and wave surfaces. Indeed, phonon images provide unimagined perspectives of these mathematical constructs. Theoretical Monte Carlo simulations and calculations of caustic patterns, which came shortly after the first experiments, have systematically revealed the topologies of phonon focusing for a variety of crystal symmetries. Together, theory and experiments have provided quantitative and graphical understanding of phonon scattering from defects and interfaces.

In the field of physical acoustics, intuition is an acquired ability. Consider the cartoon of a man dropping square stones into the pond (Figure 1).

Figure 1 What is wrong here? The square bricks are producing square waves at a distance, contrary to the result one would obtain by assuming circular Huygens wavelets emanating from each point on the surface of the source. In the real world, the shape of the wavefront at a great distance is relatively independent of the source and takes on the shape of a Huygens wavelet, which in a real pond is circular. But not all media display circular (or spherical) Huygens wavelets (see Figure 2). (Cartoon by I. Riutin, from The Search for the Truth, *by A. B. Migdal.)*

Obviously, this is a very unusual pond because the wavefronts at a great distance are also square. This is not the result that we would predict for a real pond by constructing *circular* Huygens wavelets from each point on the surface of the disturbance. At a distance much larger than the size of the disturbance (e.g., the brick), the wavefronts in a real pond become circular, i.e., they take on the shape of the Huygens wavelet for the medium.

In contrast to a real pond, the Huygens wavelets associated with a point disturbance in a *crystal* are not spherical. This nonintuitive fact is demonstrated by a recent ultrasound imaging experiment, the results of which are displayed in Figure 2. A short ultrasonic pulse is focused to a point near the

Figure 2 Wavelets of ultrasound emanating from a point source in a silicon crystal, as explained in the text. The acoustic wavefront of a shear wave is moving at the speed of sound and displays a highly nonspherical shape, due to the elastic anisotropy of the crystal. Superposed in this image are three snapshots of the expanding ultrasonic wavefront. The delay times after the short excitation pulse are labeled in the figure. This experiment typifies the anisotropic propagation of vibrational waves in most crystals. (The experiment was performed by Matt Hauser at the University of Illinois and is described in Chapter 13.)

center of one face of a silicon crystal. The image contains three time-resolved snapshots of a shear wave arriving at the opposite parallel face, as discussed in Chapter 13. This expanding acoustic wave displays a highly unusual shape. In technical terms, the wave surface has folded sections that give rise to multiple acoustic pulses arriving at a given point. This experiment forecasts the remarkable effects of elastic anisotropy that are treated in detail in this book.

Prologue

Anisotropic heat flow in crystals

A. Historical overview

Anisotropy is a recurring theme in the physics of crystalline solids. The regular arrangement of atoms in crystals produces direction-dependent physical properties. For example, the response to applied electric and magnetic fields usually depends on the direction of the fields with respect to the symmetry axes of the crystal. Properties such as effective mass, electrical conductivity, magnetic susceptibility, and electric polarizability are second-rank tensors because they relate two vectors, such as current and electric field.

Thermal conductivity is also a second-rank tensor, giving the heat flow produced by an imposed temperature gradient. In nonmetallic crystals, heat is conducted by high-frequency elastic waves, or phonons. The velocity of an elastic wave depends on the direction of propagation with respect to the crystal axes because the elasticity of a crystal is anisotropic. Yet, in cubic crystals at room temperature, the thermal conductivity is nearly isotropic. This isotropy results from the fact that, at normal temperatures, phonons scatter frequently, making heat flow a *diffusive* rather than *ballistic* process. At low temperatures, however, the mean free paths of phonons lengthen dramatically, and it is possible to observe the ballistic propagation of heat, i.e., no phonon scattering between source and detector. Ballistic heat flow occurs at or near the velocities of sound and exhibits the anisotropies expected from elasticity theory. Heat flow in the ballistic regime is governed by the fourth-rank elasticity tensor.

In recent years, immense progress has been made in the study of ballistic heat flow and phonon scattering in nonmetallic crystals. This progress is primarily due to the development of phonon-spectroscopy and phonon-imaging techniques. Phonon spectroscopy encompasses a variety of methods that resolve the frequency of nonequilibrium phonons, as described in the reviews cited in Chapter 1. Phonon imaging, on the other hand, reveals the spatial anisotropies in ballistic heat flow. Both techniques are based on detection and analysis of *heat pulses* propagating in a crystalline medium at a few degrees Kelvin. The heat pulses are composed of nonequilibrium phonons with typical frequencies between a few tens of gigahertz (1 GHz = 10^9 Hz) and a terahertz (1 THz = 10^{12} Hz).

The problem of elastic-wave propagation in an anisotropic (e.g., crystalline) medium was essentially solved in the 19th century. The proper description of heat, however, requires the quantization of energy in each vibrational mode, i.e., the concept of phonons. Nevertheless, a phonon with a frequency of a few hundred gigahertz often has nearly the same velocity (magnitude and direction) as a low-frequency acoustic wave with the same wavevector direction and polarization. This is the continuum limit, in which the phonon wavelength is significantly larger than the distance between atoms. Thus in a first approximation ballistic heat flux is described by elasticity theory.

A heat source emits an *incoherent* distribution of phonons with a wide range of frequencies and wavevector directions. The nonequilibrium phonons produced by a pulse of heat generated at a point on the surface of a crystal propagate radially outward from this point source. One can imagine a non-spherical "shell" of thermal energy expanding into the crystal, much like the ultrasound wavefront shown in Figure 2 of the Preface. The radius of this anisotropic shell in a given direction is proportional to the phonon group velocity in that direction. In fact, there are several distinct shells of thermal energy emanating from the point source, corresponding to compression (longitudinal) and shear (transverse) waves of different velocity.*

The heat-pulse technique was first demonstrated for solids by von Gutfeld and Nethercot[1] in 1964. An electrically generated burst of heat on one surface of a cold crystal was detected with a thin-film bolometer on the opposite side of the crystal. Several pulses were observed, having times of flight corresponding to the longitudinal and transverse sound velocities in the crystal. In 1969, Taylor, Maris, and Elbaum[2] noted that for all crystals that they tested there was a large difference in the *intensities* of the longitudinal and transverse pulses, and, furthermore, the ratio of these intensities depended on the propagation direction. They showed that this anisotropy in heat flux could be understood in terms of the basic elastic anisotropy of the crystal. This effect was called phonon focusing.

Phonon focusing is a property of all crystals at low temperatures. The term "focusing" does not imply a bending of particle paths, as in the geometrical-optics sense of the term. Phonon focusing occurs because elastic waves in anisotropic media have a group velocity, $\mathbf{V}$, that in general is not parallel to

---

* The compressional wave arrived at earlier times than those shown in Figure 2 of the Preface (see Chapter 13).

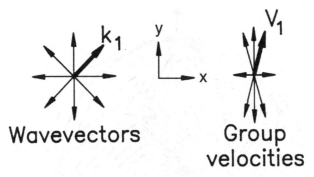

Wavevectors **Group velocities**

Figure 1 Example of how compressional waves with different wavevectors propagate in a medium that is stiffer along y than x. The wave with wavevector k_1 bisecting x and y axes may produce more energy flow along y than x, leading to a group-velocity V_1 pointing more nearly along y. Consequently, a uniform angular distribution of k's gives rise to a nonuniform distribution of V's – an effect called phonon-focusing.

the wavevector **k**; that is, the direction of energy propagation is not normal to the wavefronts.* In particular, waves with quite different **k** direction can have nearly the same **V**, so that the energy flux associated with those waves bunches along certain crystalline directions.

This situation is illustrated schematically in Figure 1. Here one imagines a hypothetical medium that is stiffer to compression along the y direction than along x, leading to a higher compressional-wave velocity along y. The phonon with wavevector k_1 has a group velocity V_1 pointing more nearly along the y axis, as explained in Chapter 2. As a consequence of the noncollinear k and **V**, an isotropic distribution of wavevectors emanating from a point source of heat can produce a highly anisotropic distribution of energy flux.

The discovery of this phonon-focusing effect is relatively recent because it applies to *ballistic* heat propagation, a phenomenon best observed at liquid-helium temperatures. Also, the development of sensitive superconducting phonon detectors and the availability of laboratory computers for both data collection and theoretical predictions have proven essential.

At first, calculations of the phonon-focusing effect were accomplished statistically for a few symmetry directions of a crystal.[3] A random array of phonon k vectors was generated, and the group velocity of each was calculated using elasticity theory. For a fixed arrangement of heat source and detector, the number of phonons intercepting the detector area was counted. In this way, tables of phonon flux along high symmetry directions were compiled. For typical crystals with only approximately 30% angular variation in sound velocity, the angular variation in heat flux was found to be more than a hundred!

---

* The group velocity **V** is related to the acoustic Poynting vector, **P**, by $V = P/U$, where U is the vibrational energy density [Eq. (36), page 49].

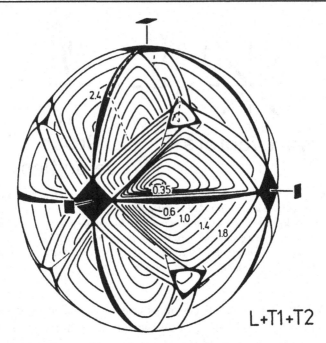

L+T1+T2

Figure 2 Polar plot of phonon intensity emanating from a point source of heat in Ge. All phonon modes (longitudinal and transverse) are included. (Calculated by Rösch and Weis.[5])

The mathematical basis for this phonon enhancement factor was elucidated in a classic article by Maris.[4]

Rösch and Weis[5] extended the statistical computations of Maris and coworkers to nonsymmetry directions and predicted striking angular distributions of the three-dimensional heat flux in Ge, Si, sapphire and quartz. An example of their polar graphs of phonon intensity is shown in Figure 2. These remarkable predictions motivated Eisenmenger to try and "photograph" the distribution of heat flux in Si using the fountain effect of liquid helium. His successful phonon-imaging experiment[6] is depicted in Figure 5 on page 18.

A different phonon-imaging method was developing simultaneously from another quarter. Electron-hole droplets had been discovered in photoexcited germanium at low temperatures.[7] The energetics of this unusual metallic liquid was well understood, but how the micrometer-sized droplets could "diffuse" millimeters from the excitation surface in their 40-μs lifetime remained a mystery. Keldysh[8] postulated that a "phonon wind," produced by the carrier thermalization, could push the droplets these large distances. Indeed, experiments by Bagaev et al.[9] and by Hensel and Dynes[10] demonstrated the motion of electron-hole droplets under the influence of a separate heat source. Subsequently, Greenstein and Wolfe[11] succeeded in imaging the infrared luminescence from a cloud of electron-hole droplets, and they discovered that the cloud was highly anisotropic (see Figure 15 on page 269 and Ref. 24). Using the predictions of Maris and coworkers, they postulated that the anisotropy in the cloud was principally due to the inherent anisotropy of the phonon wind – i.e., phonon focusing.

While studying electron-hole droplets in Ge, Hensel and Dynes[12] found an extreme anisotropy in the nonequilibrium phonon flux emanating from a point source. Using a narrow bolometer detector and a movable laser source, they observed sharp peaks in the heat flux, which they associated with phonon-focusing singularities. These structures were confirmed by statistical calculations similar to those performed by Maris. Independently, Taborek and Goodstein[13] performed calculations that supported the idea of phonon singularities, which they termed phonon caustics. Hensel and Dyne's idea of using a fixed detector and a scannable phonon source is one of the principal features of present-day phonon-imaging systems. At the same time, however, Akimov et al.[14] observed the anisotropy in phonon flux in ruby using a fixed source and a scannable optical beam as a phonon detector.

Spatial images of ballistic heat pulses in a crystal were first achieved by Northrop and Wolfe,[15] who deposited a small bolometer on one side of a Ge crystal and raster scanned a pulsed laser beam across the opposite surface. The heat-pulse intensity for each laser position was collected by a computer and displayed as brightness in an image. The first images showed an unexpected complexity,* which led to new insights into the subtleties of phonon focusing. At that time the statistical calculations were not sufficiently developed to produce a complete phonon image. Instead, an analytical explanation of the results was sought. The three-dimensional constant-frequency surfaces in wavevector space were plotted, and it was quickly realized that there are lines of zero Gaussian curvature on these surfaces that give rise to caustic lines in the phonon images.†

The shapes and angular positions of the phonon caustics depend on the elastic properties of the particular crystal. Systematic calculation of phonon-focusing topologies have been conducted by Every, Tamura, and others.‡ All known cubic crystals have sufficient elastic anisotropy to display singular phonon-focusing patterns. In continuum elasticity theory, the pattern of caustics is independent of phonon frequency. Phonons traveling along a given caustic have a readily predictable wavevector direction (wavenormal) and polarization. These properties have proven valuable in the study of phonon propagation and scattering in crystals. The identification of a sharp focusing structure allows the experimenter to isolate ballistic phonons of specific *mode*, *wavevector direction*, and (with an appropriate detector) *frequency*.

---

* See, for example, Figure 4 on page 16, or the cover of this book.
† This effect is described in detail in Chapter 2.
‡ A study of the caustic patterns in cubic crystals is presented in Chapter 4, and the extension to noncubic and piezoelectric crystals is discussed in Chapter 5. Dispersive wave surfaces are discussed in Chapters 6 and 7.

B. Central themes of this book

Over the past 15 years, phonon-imaging experiments and the corresponding theoretical calculations have contributed to many of the traditional areas of phonon physics. In this book they are covered as follows

Vibrational waves in anisotropic media:
        Experiment   –   Chapters 1, 3, and 13
        Theory   –   Chapters 2, 4, 5, and 13

Lattice dynamics:
        Theory   –   Chapter 6
        Experiment   –   Chapter 7

Phonon scattering:
        Theory   –   Chapter 8
        Experiment   –   Chapters 9 and 10

Interfaces:
        Theory   –   Chapters 11 and 14
        Experiment   –   Chapters 12 and 14

In addition, the imaging techniques have affected the areas of electron–phonon interactions in semiconductors (e.g., electron-hole liquid in Chapter 10 and two-dimensional electron-hole gas in Chapter 15), the propagation of phonons through superlattice structures (Chapter 12), and the development of new techniques such as particle detectors (Chapter 15). The principal aim of this book is to present basic descriptions of propagation and scattering of acoustic phonons from the new perspectives gained by imaging. Earlier reviews of phonon imaging have been written by Northrop and Wolfe,[16] Maris,[17] and Every.[18] A broad viewpoint of the fields of nonequilibrium phonons and lattice dynamics can be acquired from the conference proceedings listed at the end of this Prologue.

Readers anxious for a detailed discussion of why phonons "focus" the way they do may find a short tour through the following sections useful:

Section 2A, Acoustic Waves in an Anisotropic Medium
Section 4D, Origins of the Phonon Caustics
Section 6B, Lattice Dynamics (especially Figure 10)

The section on lattice dynamics shows that a face-centered-cubic lattice with only central-force interactions between nearest neighbors displays a well-defined phonon-focusing pattern (independent of force constant) that is similar to those observed in many crystals. The reader who is familiar with acoustic or ultrasonic techniques might find it useful to begin with Chapter 13.

Acknowledgements

The main purpose of this book is to make the greatly expanded view of phonon propagation that has developed over the past couple of decades more accessible to students and scientists embarking in research on phonons. Every points out in his recent review[18] that "in phonon imaging one has for the first time a tool with which the full grandeur of the theory of crystal acoustics developed by Christoffel, Rayleigh, Musgrave, Federov, Alshits and others can be examined experimentally." On the practical side, insights have been gained into the dynamics of phonons – in particular, their interaction with defects, interfaces, and carriers.

I wish to acknowledge the many contributions of those who made this book possible. Foremost in the theorist ranks, Arthur Every and Shin-ichiro Tamura have provided numerous creative explanations and predictions that continue to complement and drive experimental work. The variety of physics covered in this book is due in large part to their strong collaborations with experimentalists, including those in my own research group. I also greatly appreciate their willingness to read parts of this manuscript and suggest improvements.

Humphrey Maris, besides introducing pivotal ideas in this field, has sparked a number of new research directions covered in this book. An example was his good-natured accusation (in front of a large audience) that the Illinois group was faking the sharp caustics in *photoexcited* semiconductors. He reasoned that the usual theory of quasidiffusion would mask the observation of caustics. Only recently has a feasible resolution to this anomaly been forwarded, as discussed in Chapter 10. Theorists Y. B. Levinson and S. Esipov contributed important insights into these areas.

The real heroes are the graduate students around the globe who performed many of the experiments described in this book. Greg Northrop's thesis work, entitled "Phonon Imaging," has formed the basis for many experimental and theoretical developments in the field of phonon imaging over the last decade. Also at the University of Illinois at Urbana-Champaign, Michael Greenstein, Michael Tamor, Gregory Koos, Donna Hurley, Steven Kirch, Saad Hebboul, Mark Ramsbey, Jeffrey Shields, Madeleine Msall, Rob Vines, and Matt Hauser have made unique research contributions to phonon physics, as evidenced in the following chapters.

I want to especially acknowledge Werner Dietsche, whose year-long collaboration with my group at Illinois in the early stages of phonon imaging (1980–81) provided us with tunnel-junction technology and many valuable perspectives. Helmut Kinder and his students (especially Christian Höss) in München provided the atmosphere for a productive sabbatical year (1988–89), during which my work on this book was initiated. I thank the Alexander von Humboldt Foundation for making this possible. The method of phonon imaging could not have been developed without seminal advances in detector technology and phonon spectroscopy, pioneered in large part by well-known scientists in Germany, and I have benefited greatly from numerous interactions with them and their students.

The Illinois work described herein was funded by the National Science Foundation through Materials Research Laboratory grants. Many critical interactions – for example, with A. C. Anderson, D. van Harlingen, and A. V. Granato – were catalyzed by the Illinois Materials Research Laboratory and the Physics Department.

Particularly important to the completion of this book were Madeleine Msall and Matt Hauser. Madeleine has provided many valuable editorial suggestions and helped immensely with the final corrections. Matt has greatly assisted in the figure production. I also thank those students and colleagues who gave me suggestions on various parts of the manuscript, especially Janet Tate, Saad Hebboul, Mark Ramsbey, Jeff Shields, James Kim, Howard Yoon, Malcolm Carroll, Rob Vines, and Doug Wake.

References

1. R. J. von Gutfeld and A. H. Nethercot, Heat pulses in quartz and sapphire at low temperature, *Phys. Rev. Lett.* **12**, 641 (1964).
2. B. Taylor, H. J. Maris, and C. Elbaum, Phonon focusing in solids, *Phys. Rev. Lett.* **23**, 416 (1969).
3. B. Taylor, H. J. Maris, and C. Elbaum, Focusing of phonons in crystalline solids due to elastic anisotropy, *Phys. Rev. B* **3**, 1462 (1971).
4. H. J. Maris, Enhancement of heat pulses in crystals due to elastic anisotropy, *J. Acoust. Soc. Am.* **50**, 812 (1971).
5. F. Rösch and O. Weis, Geometric propagation of acoustic phonons in monocrystals within anisotropic continuum acoustics, Parts I and II, *Z. Physik B* **25**, 101 (1976).
6. W. Eisenmenger, Phonon detection by the fountain pressure in superfluid helium films, in *Phonon Scattering in Condensed Matter*, H. J. Maris, ed. (Plenum, New York, 1980).
7. J. C. Hensel, T. G. Phillips, T. M. Rice, and G. A. Thomas, The electron-hole liquid in semiconductors, in *Solid State Physics*, Vol. 32, H. Ehrenreich, F. Seitz, and D. Turnbull, eds. (Academic Press, NY, 1977).
8. L. V. Keldysh, Phonon wind and dimensions of electron-hole drops in semiconductors, *JETP Lett.* **23**, 86 (1976).
9. V. S. Bagaev, T. I. Galkina, and N. N. Sibeldin, Interaction of EHD with deformation field, ultrasound and nonequilibrium phonons, in *Phonon Scattering in Condensed Matter*, W. Eisenmenger, K. Lassmann, and S. Dottinger, eds. (Springer-Verlag, Berlin, 1984).
10. J. C. Hensel and R. C. Dynes, Interactions of electron-hole drops with ballistic phonons in heat pulses: the phonon wind, *Phys. Rev. Lett.* **43**, 1033 (1977).
11. M. Greenstein and J. P. Wolfe, Anisotropy in the shape of the electron-hole drop cloud in germanium, *Phys. Rev. B* **26**, 5604 (1982).
12. J. C. Hensel and R. C. Dynes, Observation of singular behavior in the focusing of ballistic phonons in Ge, *Phys. Rev. Lett.* **43**, 1033 (1979).
13. P. Taborek and D. Goodstein, Phonon focusing catastrophes, *Sol. State Comm.* **33**, 1191 (1981); *Phys. Rev. B* **22**, 1850 (1980).

14. A. V. Akimov, S. A. Basun, A. A. Kaplyanskii, V. A. Rachin, and R. A. Titov, Direct observation of focusing of acoustic phonons in ruby crystals, *JETP Lett.* **25**, 461 (1977).
15. G. A. Northrop and J. P. Wolfe, Ballistic phonon imaging in solids – a new look at phonon focusing, *Phys. Rev. Lett.* **43**, 1424 (1979); *Phys. Rev. B* **22**, 6196 (1980).
16. G. A. Northrop and J. P. Wolfe, Phonon imaging: theory and applications, in *Nonequilibrium Phonon Dynamics*, W. E. Bron, ed. (Plenum, New York, 1985).
17. H. J. Maris, Phonon focusing, in *Nonequilibrium Phonons in Nonmetallic Crystals*, W. Eisenmenger and A. A. Kaplyanskii, eds. (North-Holland, Amsterdam, 1986).
18. A. G. Every, Thermal phonon imaging, in *Proceedings of Karpacz Winter School* (1993).

Selected works and conference proceedings on nonequilibrium phonons

19. R. J. von Gutfeld, Heat pulse transmission, in *Physical Acoustics, Vol. 5*, R. P. Mason, ed. (Academic, New York, 1968).
20. A. K. McCurdy, H. J. Maris, and C. Elbaum, *Phys. Rev. B* **2**, 4077 (1970).
21. J. Philip and K. S. Viswanathan, Phonon magnification in cubic crystals, *Phys. Rev. B* **17**, 4969 (1978).
22. M. Lax and V. Narayanamurti, Phonon magnification and the Gaussian curvature of the slowness surface in anisotropic media: detector shape effects with application to GaAs, *Phys. Rev. B* **22**, 4876 (1980).
23. J. P. Wolfe, Ballistic heat pulses in crystals, *Phys. Today* **33**, 44 (December 1980).
24. J. P. Wolfe, Transport of degenerate electron-hole plasmas in Si and Ge, *J. Lumin.* **30**, 82 (1985).
25. H. J. Maris, ed., *Phonon Scattering in Condensed Matter* (Plenum, New York, 1980).
26. W. E. Bron, ed., *International Conference on Phonon Physics, J. Phys.* (Paris) **42**, C6 (1981).
27. W. Eisenmenger, K. Lassmann, and S. Dottinger, eds., *Phonon Scattering in Condensed Matter* (Springer-Verlag, Berlin, 1984).
28. A. C. Anderson and J. P. Wolfe, eds., *Phonon Scattering in Condensed Matter V* (Springer-Verlag, Berlin, 1986).
29. W. E. Bron, ed., *Nonequilibrium Phonon Dynamics* (Plenum, New York, 1985).
30. W. Eisenmenger and A. A. Kaplyanskii, eds., *Nonequilibrium Phonons in Nonmetallic Crystals* (North-Holland, Amsterdam, 1986).
31. S. Hunklinger, W. Ludwig, and G. Weiss, eds., *Phonons 89* (World Scientific, Singapore, 1990).
32. M. Meissner and R. O. Pohl, eds., *Phonon Scattering in Condensed Matter VII* (Springer-Verlag, Berlin, Heidelberg, 1993).
33. G. P. Srivastava, *The Physics of Phonons* (Adam Hilger, Bristol, 1990).
34. T. Paszkiewicz, ed., *Physics of Phonons*, Lecture Notes in Physics, Vol. 285 (Springer-Verlag, Berlin, 1987).
35. J. P. Wolfe, Acoustic Wavefronts in crystalline solids, *Phys. Today* **48**, 34 (September 1995).
36. T. Nakayama, S. Tamura, and T. Yagi, eds., *Phonons 95, Physica B* **219–220** (North-Holland, Amsterdam, 1996).

▌1

Ballistic heat pulses and phonon imaging – A first look

Time, space, and energy. These are the parameters of a heat-pulse experiment. The principal subject of investigation is *nonequilibrium acoustic phonons* in a crystalline solid. A phonon is the elementary quantum of vibrational energy of the atoms in the crystal. Historically, the thermal properties of solids have been measured by static or steady-state methods such as heat capacity and thermal conductivity.[1] It is not surprising, then, that time- and space-resolved measurements of nonequilibrium phonons have provided new perspectives on the thermal properties of condensed matter. The introduction of the heat-pulse method in 1964 by von Gutfeld and Nethercot[2,3] has certainly revolutionized the way we view and understand phonons.

The basic method is quite simple: A nonmetallic crystal is cooled to a few degrees above absolute zero. A heat pulse is produced by passing a short burst of current through a metal strip deposited on the crystal or, alternatively, by optically exciting a metal film that has been deposited on one surface of the crystal. The electrically or optically excited film acts as a Planckian emitter of heat. Figure 1(a) schematically depicts the spatial distribution of heat energy (high-frequency acoustic phonons) at some instant of time after the excitation pulse. Several nonspherical "shells" of thermal energy, corresponding to longitudinal and transverse modes, propagate away from the heat source. The rise in temperature on the opposite face of the crystal is monitored with a superconducting metal film detector, labeled D. The heat capacity of this deposited film is extremely small, allowing rapid variations in temperature to be measured.

If the crystal is sufficiently free of defects, several sharp pulses, delayed in time after the generator pulse, are detected. The times of flight of these heat pulses generally correspond to the velocities of sound (compression and shear waves) in the crystal. This is a remarkable result, considering that the phonons in a heat pulse typically have frequencies of a few hundred gigahertz, compared

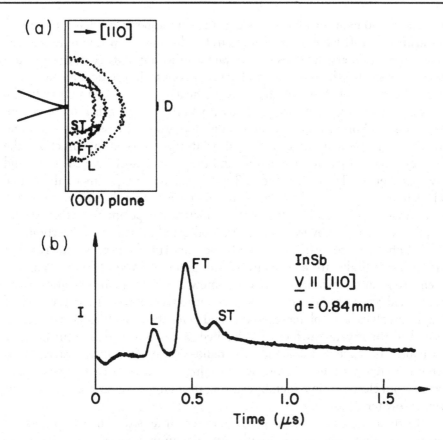

Figure 1 (a) Schematic drawing of the expanding shells of heat-pulse energy produced by a single laser pulse absorbed at the metallized surface of a crystal. Longitudinal (L), slow transverse (ST), and fast transverse (FT) modes propagate at different, direction-dependent velocities. See also Figure 2 in the Preface; there the FT mode was not excited by the ultrasound, and the L wavefront had already past. (b) Heat-pulse signals in InSb detected by a PbTl tunnel junction (430-GHz onset) at a crystal temperature of 1.6 K. V = group velocity, d = crystal thickness. The small negative signal near t = 0 is due to optical excitation near the detector by scattered laser light. (From Hebboul and Wolfe.[4])

to the megahertz frequencies of ultrasound waves $(1 \text{ GHz} = 10^3 \text{ MHz} = 10^9 \text{ Hz})$. Figure 1(b) shows the detected heat pulses for a semiconductor crystal InSb, cooled to a temperature of 1.6 K.[4]

The principal significances of these observations are:

(1) At sufficiently low temperatures, heat propagation in high-purity non-metallic crystals is *ballistic*, i.e., the phonons travel without scattering. This rapid transport is in stark contrast to the sluggish propagation of heat at normal temperatures. Viewed in terms of the quantum vibrations of the crystal,

phonon–phonon scattering is so strong at room temperature that the thermal propagation is diffusive over all measurable distances. At low temperature, however, the cold crystal does not contain a sufficient density of equilibrium phonons to scatter the few nonequilibrium phonons in the heat pulse.

(2) The times of flight of the arriving heat pulses serve to identify the particular modes of propagation, each with a different polarization direction. The acoustic phonon branch contains one longitudinal (compressional) and two transverse (shear) modes. For an arbitrary propagation direction, the transverse modes are nondegenerate and they are usually labeled as fast and slow transverse (FT and ST) modes. Thus a heat-pulse experiment allows one to isolate phonons of a particular mode and propagation direction. This may be contrasted to a standard thermal conductivity experiment, which senses the phonon transport averaged over all modes and propagation directions.

What physical properties are actually measured in a heat-pulse experiment? The times of flight are a measure of the phonon *group* velocities along a given propagation direction. This is somewhat different information from that gained by the radio-frequency pulse-echo technique. The latter method usually involves large planar detectors and gives the acoustic *phase* velocities normal to the crystal surface. (A detailed discussion of this point is given in Chapter 13, page 333.) In addition, radio-frequency acoustic waves travel freely through metallic crystals, whereas high-frequency phonons strongly scatter from electrons in a metal, resulting in submicrometer mean free paths even at low temperatures.

Unfortunately, comparisons of the absolute intensities of heat pulses for two nonmetallic crystals (say, with different impurity concentrations) are very difficult. This is because superconducting phonon detectors are difficult to reproduce precisely, and their sensitivity often depends strongly on the bath temperature and electrical bias. On the other hand, for a given detector and propagation direction, the *ratio* of intensities of the various phonon modes can be accurately determined. Such observations led to the discovery of phonon focusing, described below. Measurements of the ratios of heat-pulse intensities have also proven useful in studying the phonon scattering from certain impurities in semiconductors. For example, the Cr^{3+} impurity in GaAs is found to strongly scatter the transverse phonon modes, and comparisons along various symmetry directions have revealed the polarization selection rules. However, so far only a small number of such studies have been made.[5-7]

More generally, heat pulses are the basis for the powerful methods of phonon spectroscopy and phonon imaging. This evolution was made possible by the development of novel phonon generators and detectors.

Phonon spectroscopy is the measurement of phonon interactions as a function of frequency. This involves the generation or detection of a tunable, quasi-monochromatic band of phonons. At present, the most widely applicable technique involves the use of superconducting tunnel junctions, which can be deposited on nearly any crystal. In addition, optical methods for phonon generation and detection have proven valuable in systems containing appropriate probe impurities. Important applications of phonon spectroscopy include the study of impurities in semiconductors and insulators. Informative reviews of phonon-spectroscopy techniques are listed at the end of this chapter.[8-10]

Phonon imaging deals with the *directional* properties of high-frequency phonons. In addition to employing frequency-selective detectors, imaging methods involve continuous scanning over the propagation direction of the phonons. As shown below, this allows selection of phonon mode and polarization direction. In principle, spectroscopy and imaging techniques can be combined, but currently known frequency-tunable detectors are specialized and are usually specific to particular crystal systems. Thus it is meaningful to distinguish phonon-spectroscopy and phonon-imaging techniques, although spectroscopy experiments frequently have spatial resolution and imaging experiments often have frequency selection. The distinction lies in the scanning parameter: frequency or space.

Why would one want to scan over space, i.e., phonon propagation direction? To begin to answer this question, let us look again at the heat pulses in InSb, shown in Figure 2. The deposited copper film is irradiated with a focused laser beam, which provides the heat pulse. The position of the laser spot is movable. Figure 2(a) shows the heat pulses for two selected laser positions. The two traces appear identical in all respects but one: the upper trace (A) contains a dominant FT pulse, which is absent in the lower trace (B). A subtraction of the two traces shows only the FT pulse, plus a small tail of scattered phonons.

Now here is the surprise: The displacement of the laser beam between point A and point B is only approximately 0.1 mm. This corresponds to a displacement angle of only approximately 5 degrees for the 1-mm-thick crystal! One must conclude that the FT energy flux is highly directional. To map out the precise angular dependence of the intensity, we can "sit" at the arrival time of the FT pulse (using a boxcar integrator; see page 63) and slowly scan the position of the laser spot along x.* The resulting line scan of the FT intensity is plotted in Figure 2(b).

The remarkable effect plotted here is called phonon focusing[†] – the tendency for the ballistic heat flux emitted from a point source to concentrate along certain directions of the crystal. The use of the term "focusing" does not

---

* The novel idea of scanning the *source* of the phonons was introduced by Hensel and Dynes.[11] In principle, the variation in angular density of the phonons could be measured by a small movable detector; however, in practice, the usual thin-film detectors are quite firmly attached to the crystal.

[†] Another descriptive word is "channeling"; however, this term has been given a specific meaning with regard to phonon scattering (Chapter 9, page 223).

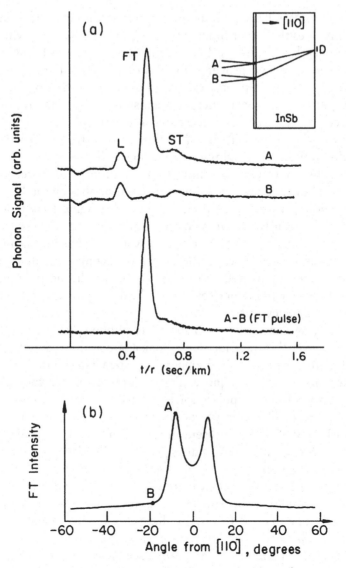

Figure 2 (a) Heat pulses in InSb produced by exciting a 2500 Å copper film with a focused laser beam. As the beam is translated slightly a dramatic change in the FT pulse intensity is seen, due to phonon focusing. A subtraction of the two traces shows only the FT pulse with a long tail attributed to phonons scattered in the bulk. The detector is a PbTl tunnel junction (430-GHz onset). (b) Peak intensity of the FT heat pulse as the laser beam is continuously translated in the (110) plane. The maxima correspond to caustics in phonon flux. T = 1.6 K. (From Hebboul and Wolfe.[4])

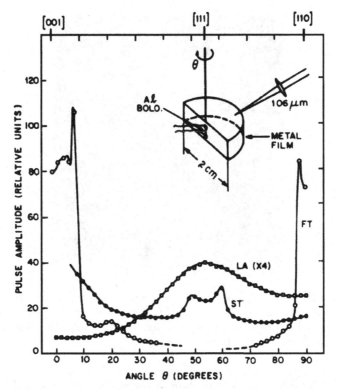

Figure 3 Intensities of longitudinal and transverse modes as a function of propagation directions in the (110) plane of Ge. (From Hensel and Dynes.[11])

imply any bending of the phonon trajectories. All the ballistic phonons travel in straight lines away from the point source of heat. The anisotropy in flux occurs because the number of phonons emitted from the heat source along one crystalline direction differs from that emitted along another. This concept is illustrated in Figure 1 on page 3.

The *patterns* of phonon flux from a point source are very anisotropic and complex for all real crystals that have been examined. The anisotropy in ballistic flux is not relegated to the FT mode alone. Figure 3 displays the earliest angular-scanning experiment, which was performed by Hensel and Dynes[11] on a Ge crystal. The detector in their experiment was a long, thin strip of aluminum that was held at its superconducting transition temperature. The cylindrically cut crystal is rotated as shown to allow sampling over a wide range of angles in the {110} plane. The line scans for both ST and FT heat pulses show sharp structures, similar to the InSb result described above. The longitudinal (L) mode, on the other hand, displays a relatively slowly varying intensity with propagation angle.

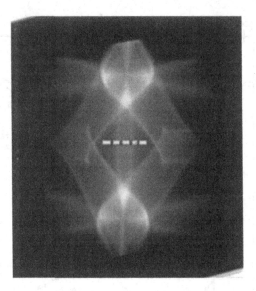

Figure 4 Phonon image of InSb at T = 1.8 K obtained by raster scanning the laser heat source across the (110) face of the crystal. Bright regions correspond to directions of large heat flux. The complex pattern is due to phonon focusing. (From Hebboul and Wolfe.[4])

To get the complete picture of phonon focusing, one must collect phonon intensities in the two-dimensional space of the propagation angles (θ, ϕ). To achieve adequate angular resolution, this requires a small phonon detector, a sharply focused laser beam, and a means of accurately raster scanning the position of the laser beam. Also, a computer is invaluable for collecting the two-dimensional array of intensities. The data is usually displayed as a grey-tone image, where brightness is proportional to the heat flux. This procedure was first accomplished with a Ge crystal by Northrop and Wolfe (Ref. 15 on page 9).

Consider the line scan of FT heat-pulse intensity shown in Figure 2(b). A phonon image is composed of a collection of these horizontal line scans for many vertical positions of the laser beam. Typically, a single line scan contains 256 intensity measurements along the x direction. An image contains 256 such lines, 65,536 data points in all. The computer controls the scanning of mirrors, which deflect the laser beam to a particular point on the crystal. For many undoped semiconductor crystals, the phonon signals are strong, and a respectable image can be collected in a minute or two. An image recorded by Hebboul and Wolfe[4] for InSb is shown in Figure 4. Similar, though not identical, patterns are obtained for GaAs, Si, Ge, LiF, and a number of other crystals.*

---

* As described in Chapter 4, these are crystals in the negative-Δ class, as determined by their elastic constants. For dispersive effects,[12] see Chapter 7.

The white dashed line across the center of this image corresponds to the line scan of Figure 2(b). This scan cuts across an intense vertical ridge, which is due to phonon focusing of the FT phonons. Actually, the image of Figure 4 was taken with a rather broad time selection in order to include both of the transverse modes, as seen in the time traces of Figure 2(a). The box-shaped structures near the top and bottom of the image and the ramps radiating diagonally are due to the ST mode. The center of this image corresponds to a [110] direction in the crystal. The ST boxes are centered on ⟨100⟩ directions, and the threefold cusp structures at the left and right are centered on ⟨111⟩ directions. The topological origins of these structures will be discussed in Chapter 2.

A completely different way of observing the two-dimensional phonon-focusing pattern was demonstrated by Eisenmenger.[13] A diagram of his experiment is shown in Figure 5. A small resistive film was evaporated at the center of a disk of Si 2 cm in diameter and 5 mm thick. The sample was immersed in superfluid He at approximately 2 K, with the heater film located on the bottom surface. The level of the He was adjusted to partially immerse the crystal so that the upper surface was covered with a thin film of superfluid He.

When current was applied to the fixed heater film, a continuous stream of nonequilibrium phonons was emitted, principally along the strong focusing directions of Si. Where the concentrated phonon beams intersected the upper surface of the crystal a thickening of the He film occurred, which could be visually observed when the surface was illuminated. The thickening of the He film is due to the local rise in temperature, which raises the film pressure and pulls more liquid from the He bath. This unique property of superfluid helium is responsible for the well-known fountain effect. The threefold focusing structure observed in Eisenmenger's experiment is centered around the [111] axis of the Si disk.

A third type of phonon-imaging experiment is closely related to the laser-scanning technique previously described. Instead of a laser beam, however, a focused electron beam is used to produce the heat pulse. The crystal is mounted on a cold stage inside an electron-beam microscope, and a tiny superconducting bolometer is used as the phonon detector. This experiment was developed by Eichele et al.,[14] and one of their phonon images of a quartz (SiO_2) crystal is shown in Figure 6. Again, the sharp focusing structures are due to transverse phonons. The method of display is a little different in this case: the image is composed of closely spaced line scans, so that the bolometer signal is added to the y deflection rather than modulating the brightness.

Interesting variations on these techniques have been developed and will be described in later chapters. For example, Schreyer et al.[15] produced a spatially selective detector by locally sensitizing a large-area superconducting tunnel junction with a scannable laser beam. In principle, this permits the

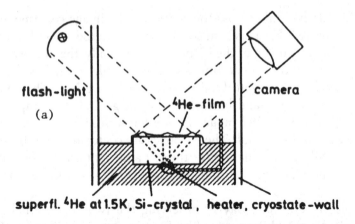

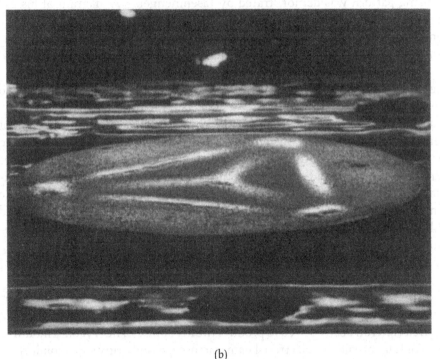

(b)

Figure 5 (a) Experimental arrangement used by Eisenmenger, to observe phonon focusing in Si by the fountain effect of superfluid ^{4}He. (b) Photograph of the Si surface showing the thickening of the helium film where the heat flux is greatest. The heat is continuously emitted from a resistive film on the center of the immersed Si surface. (From Eisenmenger.[13])

experimenter to adjust the phonon frequency by using a tunable phonon source, such as a superconducting tunnel junction. Other forms of detectors used in phonon imaging include semiconductor devices[16] and optical detection.[17,18]

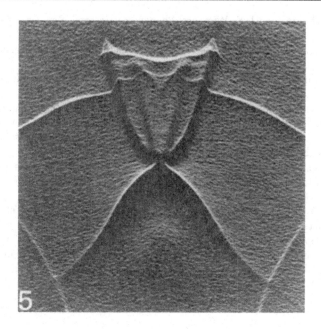

Figure 6 Phonon image of y-cut quartz taken with an Al bolometer detector and a scanned electron beam as the heat-pulse source. (From Eichele et al.[14])

References

1. See, for example, R. Berman, *Thermal Conduction in Solids* (Clarendon, Oxford, 1976).

2. R. J. von Gutfeld and A. H. Nethercot, Heat pulses in quartz and sapphire at low temperatures, *Phys. Rev. Lett.* **12**, 641 (1964).

3. R. J. von Gutfeld, Heat pulse transmission, in *Physical Acoustics, Vol. V*, W. P. Mason, ed. (Academic, New York, 1968).

4. S. E. Hebboul and J. P. Wolfe, Lattice dynamics of InSb from phonon imaging, *Z. Phys. B* **73**, 437 (1989).

5. V. Narayanamurti, M. A. Chin, and R. A. Logan, Direct determination of symmetry of Cr ions in semi-insulating GaAs substrates through anisotropic ballistic-phonon propagation and attenuation, *Appl. Phys. Lett.* **33**, 481 (1978).

6. M. Hamdache, P. J. King, D. T. Murphy, and V. W. Rampton, Phonon spectroscopy of chromium doped GaAs using super conducting tunnel junctions, *J. Phys. C* **15**, 5559–80 (1982).

7. J. C. Culbertson, U. Strom, and S. A. Wolf, Deep level symmetry studies using ballistic-phonon transmission in undoped semi-insulating GaAs, *Phys. Rev. B* **36**, 2962 (1987); Z. Xin, F. F. Ouali, L. J. Challis, B. Salce, and T. S. Cheng, Phonon scattering by impurities in semi-insulation GaAs wafers, *4th International Conference on Phonon Physics* (Hokkaido, 1995), Physica B.

8. W. E. Bron, Spectroscopy of high-frequency phonons, *Rep. Prog. Phys.* **43**, 301 (1980).

9. H. Kinder, Monochromatic phonon generation by superconducting tunnel junctions, in *Nonequilibrium Phonon Dynamics*, W. E. Bron, ed. (Plenum, New York, 1985).

10. M. N. Wybourne and J. K. Wigmore, Phonon spectroscopy, *Rep. Prog. Phys.* **51**, 923 (1988).

11. J. C. Hensel and R. C. Dynes, Observation of singular behavior in the focusing of ballistic phonons in Ge, *Phys. Rev. Lett.* **43**, 1033 (1979).

12. G. A. Northrop, S. E. Hebboul, and J. P. Wolfe, Lattice dynamics from phonon imaging, *Phys. Rev. Lett.* **55**, 95 (1985).

13. W. Eisenmenger, Phonon detection by the fountain pressure in superfluid helium films, in *Phonon Scattering in Condensed Matter*, H. J. Maris, ed. (Plenum, New York, 1979).

14. R. Eichele, R. P. Huebener, and H. Seifert, Phonon focusing in quartz and sapphire imaged by electron beam scanning, *Z. Phys. B* **48**, 89 (1982).

15. H. Schreyer, W. Dietsche, and H. Kinder, Laser induced nonequilibrium superconductivity – a spatially resolving phonon detector, *Proceedings of 17th Int. Conf. on Low Temp. Phys.*, U. Eckern et al., eds. (North Holland, Amsterdam, 1984), p. 665.

16. H. Karl, W. Dietsche, A. Fischer, and K. Ploog, Imaging of phonon-drag effect in GaAs-AlGaAs heterostructures, *Phys. Rev. Lett.* **61**, 2360 (1988).

17. A. V. Akimov et al. (Ref. 14 of Prologue).

18. M. T. Ramsbey, I. Szafranek, G. Stillman, and J. P. Wolfe, Optical detection and imaging of nonequilibrium phonons in GaAs using excitonic photoluminescence, *Phys. Rev. B* **49**, 16427 (1994).

2

Phonon focusing

For a crystal containing N atoms, there are $3N$ normal modes of vibration. These normal modes are generally described in terms of plane waves of the form $\cos(\mathbf{k} \cdot \mathbf{r} - \omega t)$, where $\mathbf{k}$ is the wavevector with magnitude $2\pi/\lambda$ and $\omega \equiv 2\pi\nu$ is the angular frequency of the wave with wavelength λ and frequency ν. The boundary conditions of the crystal restrict the possible values of $\mathbf{k}$, leading to a uniform grid of discrete wavevectors in k space, bounded by the Brillouin zone of the particular lattice. This quantization of wavevectors, $\mathbf{k}$, is simply a result of classical mechanics.

Models of lattice dynamics seek to determine the vibrational frequency for a given $\mathbf{k}$ by considering the elastic forces between atoms. As an example,[1] the dispersion relation, $\omega(\mathbf{k})$, for Ge along the [111] axis is reproduced in Figure 1(a). The acoustic and optical branches are shown. Given this dispersion relation, the density of modes per unit frequency, $D(\omega)$, can be calculated. The density of modes – essentially a "vibrational spectrum" of the crystal – is basic to the thermal and electrical properties of the particular material.

Now quantum mechanics comes into play. If we think of a vibrational mode with wavevector $\mathbf{k}$ and (angular) frequency ω as a plane wave spread uniformly over the entire crystal, then, according to the quantum mechanics of the harmonic oscillator, the energy (and, hence, amplitude) of this wave can have only discrete values. The quantum number, $n = 1, 2, 3\ldots$, determines the allowed vibrational energies of this mode, $E_n = \hbar\omega(n + 1/2)$, with $\hbar = $ (Planck's constant)$/2\pi$. One may interpret n as the number of phonons in the mode of wavevector $\mathbf{k}$ and frequency ω. Thus the wave is composed of a zero-point energy $\hbar\omega/2$ and n phonons of energy $\hbar\omega$. The thermal-average number of phonons in a given mode of frequency ω is given by the Planck distribution:

$$\bar{n} = 1/[\exp(\hbar\omega/k_B T) - 1], \tag{1}$$

21

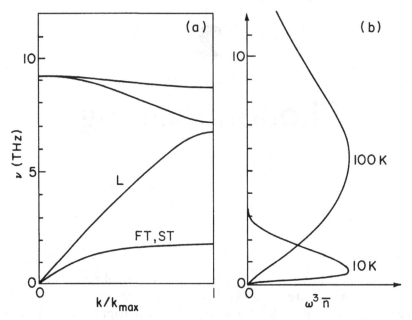

Figure 1 (a) Approximate phonon dispersion relation for **k** ∥ [111] in Ge (adapted from Ref. 1). $k_{max} = \sqrt{3}\pi/a$ with a = lattice constant. Longitudinal and transverse acoustic branches are labeled L and T. The fast and slow transverse waves are degenerate along this direction. (b) Two normalized Planck spectra, assuming the continuum-limit density of states, $D(\omega) = C\omega^2$, where $\omega = 2\pi\nu$. The quadratic density of states is valid only for the linear portions of the dispersion curves, so the actual phonon distributions are more complicated than shown here.

where k_B is the Boltzmann constant and T is the lattice temperature. This quantization of vibrational energy becomes noticeable when only a few phonons occupy a mode; i.e., when the phonon energy $\hbar\omega$ is comparable to or larger than $k_B T$. These are the phonons we usually associate with heat. For lattice temperatures of a few degrees Kelvin, the relevant phonon frequencies, $\nu = \omega/2\pi$, are in the range of 10^{11} to over 10^{12} Hz (i.e., approximately 100 GHz to several tetrahertz).

The number of thermally excited phonons per unit frequency is obtained by multiplying $\bar{n}$ by the density of states, $D(\omega)$. In the long-wavelength (continuum) approximation, $D(\omega) = C\omega^2$. The vibrational energy per unit frequency, $(\hbar\omega)D(\omega)\bar{n}$, then goes as $\omega^3\bar{n}$, a Planck spectrum. The peak of this Planck spectrum is at $\nu = 2.8 k_B T/h$, which for $T = 10\,\mathrm{K}$ is approximately 600 GHz. To give an idea of which phonon modes are occupied at a given temperature, Figure 1(b) shows a Planck spectrum plotted adjacent to the dispersion curves of Ge. The actual number of phonons per unit frequency, $\bar{n}D(\omega)$, is more complicated than the Planck spectrum because the true density

Table 1. *Conversion factors for typical acoustic phonon energies*

Energy $h\nu$ (meV)	Frequency ν (GHz)	Wavenumber ν/c (cm^{-1})	Wavelength[a] λ_{typ} (Å)	Planck temperature[b] $h\nu/2.8k_B$ (K)
4	967	32.2	50	16.5
1	242	8.1	200	4.1
0.25	60	2.0	800	1.0

[a] Assuming $\lambda_{typ} = v_{typ}/\nu$, using $v_{typ} \cong 5$ km/s.
[b] Temperature of the Planck distribution that has a peak at energy $h\nu$.

of modes[2] is more complex than that given by the continuum approximation, which assumes a linear dispersion relation, $\omega = vk$. However, at temperatures below 10 K, the Planck spectrum is often a fair approximation of the actual phonon spectrum (e.g., for the L mode in Ge at 10 K in Figure 1). Typical acoustic-phonon energies and the corresponding frequencies are given in Table 1.

For vibrational wavelengths, $\lambda = 2\pi/k$, much larger than the atomic spacing a (i.e., in the continuum limit, $k \ll \pi/a$), the dispersion relation is linear. Most heat-pulse experiments are conducted in this nondispersive frequency regime. This is why the times of flight of the heat pulses described in Chapter 1 correspond to the known sound velocities in the crystal. The striking phonon-focusing effects can be understood in terms of continuum elasticity theory. We begin, therefore, with a graphical description of acoustic waves in a crystal.

A.   Acoustic waves in an anisotropic medium

In an elastically anisotropic medium, the restoring force on an atom depends on the direction of its displacement from equilibrium. Consequently, the velocity of a displacement wave depends on its propagation direction and polarization.[3] This is true for all crystals.

A vibrational wave carries energy. One of the nonintuitive results of elasticity theory in an anisotropic medium is that generally the vibrational energy does not flow along the wavevector direction! This important concept can be illustrated using the Huygens principle, i.e., by constructing the wavefront from an extended disturbance as a superposition of waves from point sources along the boundary of the disturbance.

If we drop a pebble into an isotropic pond, a circular wavefront is generated, as indicated in Figure 2(a). If we drop a stick into the same pond, the wavefront is easily constructed, as shown in the right portion of the figure.

(a) Isotropic Medium

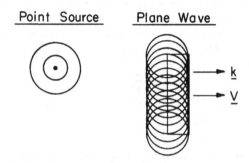

(b) Anisotropic Medium

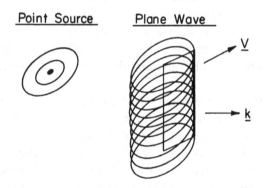

*Figure 2 (a) Wavefronts emanating from a point source and a line source in an isotropic medium. (b) Wavefronts emanating from point and line sources in an anisotropic medium. In this case, the wavevector **k**, which is perpendicular to the wavefront, is not parallel to the energy flow, which is along the group-velocity direction, **V**.*

For propagation distances smaller than the length of the stick, we observe that (neglecting edge effects) the vibrational energy is radiating outward in a direction perpendicular to the stick. This energy flux propagates along the group-velocity vector, **V**.

If instead of *dropping* the stick we use it to drive a sinusoidal oscillation, we can generate a plane wave with a wavevector **k** perpendicular to both the stick and the wavefronts. It seems quite natural that energy is propagating outward along the direction of the wavevector.

Now imagine the same experiment in an anisotropic pond, an imaginary medium that has a faster wavefront velocity in one direction than another. Dropping a pebble into this pond produces a wave surface that displays the anisotropy of the medium, as shown in Figure 2(b). Dropping a stick at an arbitrary angle produces the wavefront shown. In this case the direction of the vibrational energy flux is no longer perpendicular to the stick. The energy flux is traveling more nearly along the highest-velocity direction of the medium. A closer analysis shows that the group-velocity direction is determined by the point on the wave surface (the ellipse) that has a tangent parallel to the stick.

For an oscillating stick of infinite length, the wavevector is perpendicular to the stick, just as in the isotropic case. This is because the wavevector is always orthogonal to the plane wavefronts, by virtue of its definition:

$$u = u_0 \cos(\mathbf{k} \cdot \mathbf{r} - \omega t),$$

where u is the wave amplitude and $\mathbf{r}$ is the radial distance from an origin. The scalar phase velocity is defined by $v = \omega/k$ and corresponds to the velocity of the plane wavefronts. In the anisotropic case, v is clearly not equal to the magnitude of the group-velocity vector, $\mathbf{V}$. And $\mathbf{k}$ is not collinear with $\mathbf{V}$, except when $\mathbf{k}$ is along a symmetry axis.

A useful relationship between $\mathbf{k}$ and $\mathbf{V}$ can be deduced from Figure 2(b). If we set the origin of $\mathbf{r}$ somewhere on the source of the line disturbance, then $\mathbf{k} \cdot \mathbf{r} = \mathbf{k} \cdot \mathbf{V}t =$ constant describes a line of constant phase. Because $\mathbf{k} \cdot \mathbf{r} = \omega t$ for a moving front (e.g., peak amplitude) of the plane wave,

$$\omega = \mathbf{V} \cdot \mathbf{k}, \tag{2}$$

which will be proven rigorously later in this chapter. This expression is reminiscent of the dispersion relation, $\omega = vk$ (also valid here), but involves the group velocity rather than the phase velocity. Together, these relations imply that $V = v/\cos \zeta$, where ζ is the angle between $\mathbf{V}$ and $\mathbf{k}$.

An acousto-optic experiment[4] vividly demonstrates the noncollinearity of $\mathbf{k}$ and $\mathbf{V}$. Figure 3 shows a radio-frequency acoustic wave propagating in a crystal of quartz. A standing longitudinal wave is generated by a transducer bonded to the lower region of the left surface. This surface is perpendicular to the y axis of the quartz crystal. The picture is produced by sending monochromatic light through the crystal and observing the scattered components of the light by the Schlieren technique. The acoustic wave modulates the dielectric constant of this piezoelectric crystal and has sufficient intensity to deflect the laser light. The phase fronts of the plane wave are seen as the closely spaced vertical lines that are parallel to the generator surface, implying that $\mathbf{k}$ is normal to the surface. The energy flux, on the other hand, is at an oblique angle to the surface. This is an example of an oblique wave, well known in the field of physical acoustics.

For a continuous medium, the wave surface is independent of frequency. That is, the velocity in a given direction is the same for all frequencies. The wavevector, of course, depends linearly on the driving frequency: $k = \omega/v$.

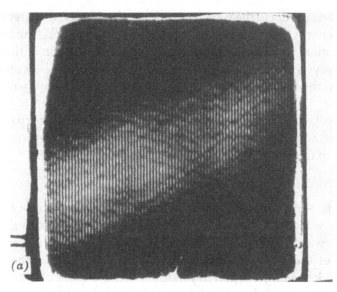

Figure 3 *Propagation of an oblique acoustic wave with* **k** *parallel to the Y crystal axis in quartz. The energy flux from the acoustic transducer attached near the bottom of the left surface is clearly not parallel to* **k**, *which is perpendicular to the vertical wavefronts seen in the photo. (From Staudte and Cook.[4])*

It is very useful to construct the locus of wavevectors for a given frequency – i.e., a constant-frequency surface in k space. This surface has the shape of a "slowness surface" because its radial dimension, $|\mathbf{k}| = \omega/v$, in a particular direction is proportional to the inverse of the phase velocity in that direction. It is often convenient to define a slowness vector, $\mathbf{s} = \mathbf{k}/\omega$, which is parallel to the **k** vector. The radial dimension of the slowness surface,

$$s = |\mathbf{k}|/\omega = 1/v,$$

is independent of frequency in the continuum limit. Because the slowness surface and the constant-ω surface have the same shape in this limit, the terms are often used interchangeably.

For the simple ellipsoidal wave surface of the anisotropic pond imagined above, the shape of the slowness surface can be easily determined, as indicated in Figure 4. The wavevector corresponding to a given **V** is perpendicular to the wave surface, as indicated in Figure 2(b), and has a length given by $k = 2\pi/\lambda$. For a given frequency, the wavevector corresponding to the fast axis of the wave surface is smaller than that corresponding to the slow axis. This leads to an ellipsoidal slowness surface with a long axis rotated 90 degrees from that of the hypothetical wave surface. In three dimensions a prolate-spheroid slowness surface produces an oblate-spheroid wave surface, and vice versa.[5]

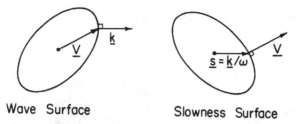

Wave Surface   Slowness Surface

Figure 4 Geometrical relation between the wave (or constant-t) surface and the slowness (or constant-ω) surface.

It is also important to realize that for a given wavevector **k**, pointing from the origin to a constant-ω surface, the corresponding group velocity **V** is perpendicular to the surface at that point. This is the familiar result in physics that the group velocity is the gradient of the constant-frequency surface in k space, $\mathbf{V} = \nabla_k \omega(\mathbf{k}) \equiv d\omega/d\mathbf{k}$. This concept is illustrated in Figure 5. Consider a group of two wavevectors pointing to adjacent points on the constant-ω (slowness) surface. These waves have slightly different wavelengths and **k** directions. When superposed, the waves interfere to form stationary lines of zero intensity that are perpendicular to the slowness surface. (A more complete discussion is given in Chapter 13.) Because energy does not flow across these null lines, their direction is also the direction of energy flux, or group velocity.* These relationships among **k**, **V**, and energy flux have a firm mathematical basis, which will be developed later in this chapter.

In real crystals wave surfaces and slowness surfaces are not simple ellipsoids, as in the above example. They must reflect the symmetry of the particular crystal. We shall see below that a phonon-imaging experiment reveals the shapes of the wave surfaces (i.e., group-velocity surfaces) of the crystal. There is also a type of experiment that directly samples wavevector space. This acoustic-optic technique, developed by Schaeffer and Bergman,[6] measures cross sections of the three-dimensional slowness surface.

In this technique a collimated beam of light is directed into an optically transparent crystal. The crystal is excited by a radio-frequency acoustic beam, which bounces obliquely off the crystal surfaces, exciting an almost continuous distribution of acoustic wavevector directions. In effect, the crystal is a multimode cavity. Through elasto-optic coupling, the incident photons with wavevector **K** scatter nearly elastically off these acoustic waves. By momentum conservation, the wavevector **K**′ of a scattered photon is given by $\mathbf{K}' = \mathbf{K} + \mathbf{k}$, where **k** is an acoustic wavevector, and $k \ll K$ for radio-frequency acoustic frequencies. Energy conservation dictates that $c(K' - K) = vk$, which means

---

* Wave packets are formed from a distribution of frequencies. The packet travels along **V**. Note also that **V** is well defined for a single sinusoidal wave. The energy flux for this wave is given by the Poynting vector $\mathbf{P} = U\mathbf{V}$, where U is the energy density in the wave.

$$\underline{V} = \frac{\partial \omega}{\partial \underline{k}} \quad \therefore \quad (\underline{V} \cdot \Delta \underline{k}) = \Delta \omega = 0$$

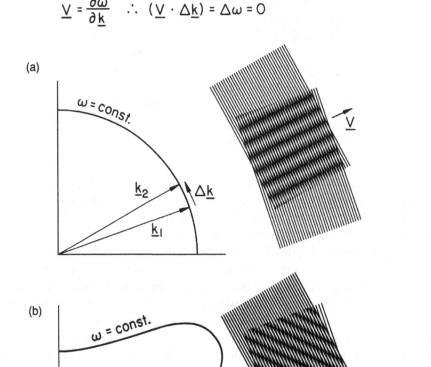

Figure 5 *Interference of two waves with the same frequency but different wavevector directions. Energy flows in the direction of* **V**, *which is parallel to the interference nodes and normal to the (ω = constant) slowness surface. (a) Isotropic medium, (b) anisotropic medium.*

that K' nearly equals K because the velocity of light, c, is much larger than the velocity of sound, v. Consequently, a photon can only scatter from acoustic wavevectors that are nearly normal to the incident beam, and the scattering angle (in radians) is approximately equal to $k/K \propto 1/v$.

Thus, the acoustically deflected photons, striking a screen behind the crystal, map out a slice of the constant-ω surfaces of the acoustic waves in the plane perpendicular to the incident light beam. The results for several crystals are shown in Figure 6. It can be seen that the sound velocity (or slowness, $1/v$) typically varies by approximately 30% with direction.

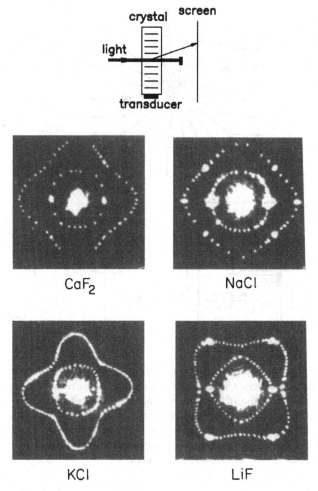

Figure 6 *Diffraction images formed by sending a laser beam through several types of cubic crystals. The incident beam is along [100]. The loci of diffracted points define cross sections of the acoustic slowness surfaces in the crystals. (From Schaeffer and Bergmann.[6])*

B. Phonons

How do these nonspherical slowness surfaces affect the propagation of heat pulses? An ideal heat pulse contains a Planckian frequency distribution of phonons, which implies an isotropic distribution of wavevectors. Of course, the angular distribution of wavevectors may be modified somewhat by interfaces and bulk scattering, but initially we may consider the distribution in k space to be nearly isotropic.

An isotropic distribution of wavevectors does not imply an isotropic distribution of energy flux. This key idea is illustrated in Figure 7(a) for a simple

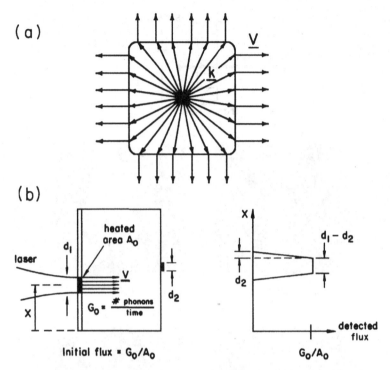

Figure 7 (a) Hypothetical slowness surface for a crystal. A uniform distribution of wavevectors produces a nonuniform distribution of energy flux – the effect known as phonon focusing. (Adapted from Elbaum.[27]) (b) Schematic diagram showing how the mathematically infinite flux along (100) directions in (a) becomes finite when a phonon source and detector of finite sizes are used. The detected signal becomes arbitrarily sharp as the source and detector dimensions are reduced.

hypothetical slowness surface. The energy flux is strongly directed perpendicular to the flat surfaces of the slowness surface. This is the basic idea of phonon focusing introduced by Taylor, Maris, and Elbaum[7] and developed by Maris.[8] More precisely, the energy flux along a group velocity normal to the slowness surface is inversely proportional to the curvature at that point.

A cross section of a real slowness surface of Ge is shown in Figure 8(a). This surface for the ST mode of Ge was calculated from the measured elastic constants, tabulated on page 94.[9] This surface leads to a strong concentration of flux (i.e., strong focusing) along directions close to the ⟨100⟩ axes of the crystal.

To help interpret the pattern of flux in real space, we show in Figure 8(b) the corresponding cross section of the group-velocity surface (wave surface). This is a collection of normals to the slowness surface, with magnitudes given by the gradient of $\omega(\mathbf{k})$ in $\mathbf{k}$ space. The origin of the folds in the wave surface can be traced to inflection points on the slowness surface (e.g., corresponding

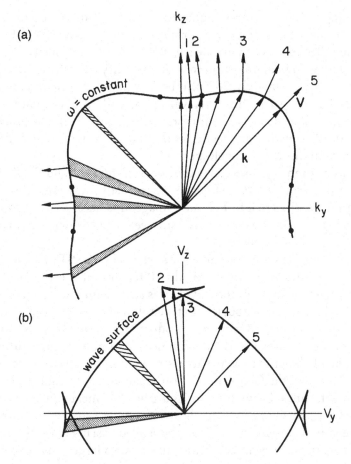

Figure 8 (a) Intersections of a constant-frequency surface with the (001) plane for the slow transverse (ST) mode in Ge. The radius of this surface along $\mathbf{k}$ is $|\mathbf{k}| = \omega/v$, which is proportional to the slowness, $|\mathbf{s}| = |\mathbf{k}/\omega| = 1/v$, where v is the anisotropic phase velocity. (b) Corresponding slice of the wave surface, showing folds (e.g., at the group velocity labeled 2) that correspond to the zero-curvature inflection points in the slowness surface. The shaded and cross-hatched regions show how a constant solid angle in real space (such as that subtended by a phonon detector) corresponds to widely varying solid angles in k-space – and, thus, widely varying phonon flux – as $\mathbf{V}$ is scanned. (From Northrop and Wolfe[9]; see also Musgrave.[5])

to the group velocity labeled 2). At these points, the curvature (inverse radius) of the slowness surface changes from positive to negative. Exactly at these inflection points, the curvature vanishes.

Because phonon flux is inversely proportional to the curvature of the slowness surface, a vanishing curvature implies a mathematically infinite (but integrable) flux from a point heat source along the corresponding group-velocity direction. Equivalently, sharp singularities in flux (caustics) occur along the directions of the folds in the wave surface. A very large number of k vectors around the inflection point have nearly the same group velocity. Thus even with a moderate angular variation in sound velocity over the entire slowness surface (here, approximately 30%), immense angular variations in phonon flux occur. The *measured* intensity variations will depend upon the size of the heat source and detector, as illustrated in Figure 7(b). In practice, the sharpness of the caustic can depend on the crystal perfection and ultimately upon diffraction limits for the phonons.[10]

The peaks in intensity observed in the line scans for InSb and Ge in Chapter 1 (pages 14 and 15) correspond to folds in the respective wave surfaces. In order to explain the features of a full phonon image, it is necessary to examine the three-dimensional slowness and wave surfaces. This exercise involves some interesting topological structures.

Figure 9 shows the slowness and wave surfaces for the ST mode in Ge.[9] The inflection *points* described above for the (001) slice of the slowness surface are parts of "parabolic" *lines* that divide this surface into different topological regions. The local shape of a surface in three-dimensional space is described by the Gaussian curvature, which is the product of the two principal curvatures at a point on the surface, i.e., the maximum and minimum curvatures at that point. If one of the principal curvatures vanishes, the surface element has a parabolic shape and the Gaussian curvature vanishes (see Figure 21).

The parabolic lines [heavy lines in Figure 9(a)] delineate the locus of points where the Gaussian curvature vanishes. These lines separate regions of positive and negative Gaussian curvature. The shaded area in Figure 9(a) is a saddle region of negative Gaussian curvature. It separates large convex regions and smaller concave regions, all with positive curvature.

The parabolic lines on the slowness surface correspond to folds on the wave surface, which is represented in Figure 9(b). One can imagine the wave surface extending into the crystal with its origin at the point source of heat. This group-velocity surface, multiplied by the time duration after the excitation pulse, represents the expanding distribution of ballistic thermal energy in the crystal. Where the folds in this expanding wave surface intersect the detection surface, a singular flux, or caustic, can be observed. The lines of intersection of the caustics with a crystal surface form a caustic map, some sections of which are plotted on the right side of Figure 10. The left side of this figure shows the parabolic lines on the slowness surface from which these caustic patterns originate.

Figure 10(a) corresponds to the ⟨100⟩ ST-box structure similar to that observed in the image of GaAs (Figure 4 on page 16). The heaviest lines show the relationships between one branch of the parabolic lines and the corresponding real-space caustic. The concave cloverleaf region around the [100] axis, as shown in Figure 10(b), gives rise to the complex inner-box

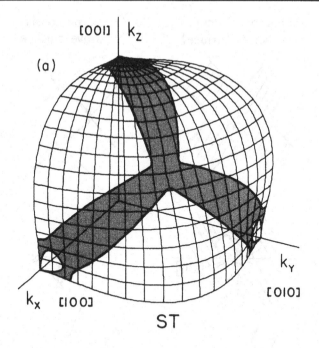

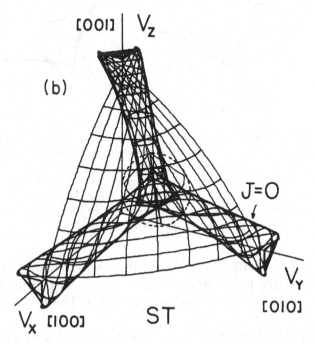

Figure 9 (a) Slowness surface for the ST mode in Ge. The heavy lines are parabolic lines of zero Gaussian curvature, which separate regions of convex, saddle, and concave curvature on the surface. The saddle region is shaded. (b) Corresponding group-velocity surface. (From Northrop and Wolfe.[9])

Parabolic lines
(slowness surface)

Phonon caustics
(wave surface)

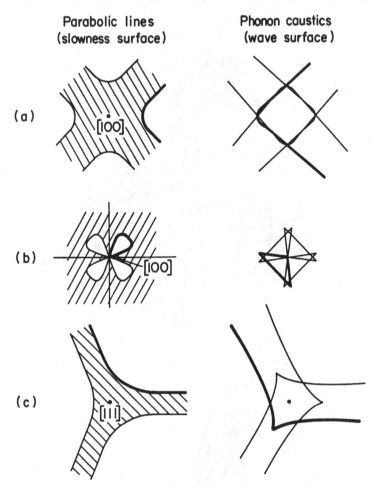

(a)

(b)

(c)

Figure 10 At left, sections of the ST slowness surface of Ge showing lines of zero Gaussian curvature. The structures (a) and (b) are superposed along the ⟨100⟩ axes in Figure 9(a). (a) Shows the lines separating convex and saddle (shaded) curvatures; (b) Shows the lines separating saddle (shaded) and concave curvatures; (c) shows lines separating saddle and convex regions near [111]. On the right are projections of the folds in the real-space wave surface. Heavy lines correspond left to right.

structure around the ⟨100⟩ axis. Finally, the ramps radiating outward from ⟨100⟩ directions (see Figure 9) intersect in a three-cusp structure centered on the [111] direction, as shown in Figure 10(c). The FT caustics are not shown.

An experimental phonon image of Ge is shown in Figure 11. This crystal has wave surfaces similar to those of GaAs and InSb. Both inner- and outer-box structures are observable in the image. Accompanying this image is a caustic map derived from the elastic constants of Ge. FT caustics are

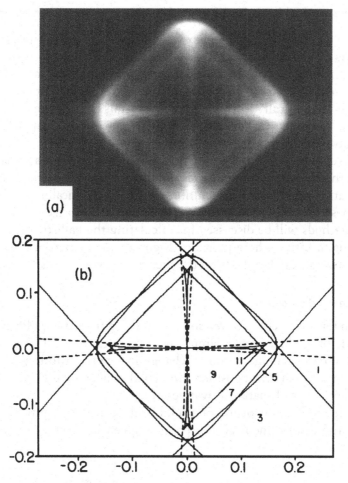

Figure 11 (a) Image of the phonon-focusing pattern in Ge, taken with an Al bolometer at T = 2 K. The center of the image corresponds to the [001] direction, and the angular range is 28° from left to right. The angular resolution is 0.4 degrees of arc. (b) Caustic map of Ge calculated by continuum elasticity theory. The numbers indicate how many distinct sheets of the ST wave surface there are for a given propagation direction, i.e., the number of distinct heat pulses that could be detected with sufficient temporal resolution. Axes are measured in radians. (From Northrop and Wolfe.[9])

indicated by dashed lines, and the numbers indicate the number of overlapping wavesheets.* The agreement between experiment and the results of continuum elasticity theory is good, although the experimental inner-box structure is discernibly smaller than its theoretical counterpart. (This discrepancy is discussed in Chapter 7, page 170.)

* Detailed pictures of the ST wave surface are given in Figure 25 and in Figure 6(b) on page 103.

C. Mathematical basis for phonon focusing

The reason for these highly complex wave surfaces in real crystals is that the elasticity is described by a fourth-rank tensor, which relates two second-rank tensors – stress and strain. In contrast, the second-rank thermal-conductivity tensor relates two vectors – heat flow and temperature gradient. Thermal conductivity is theoretically isotropic in cubic crystals because the diffusive motion of the phonons washes out the acoustic anisotropy. For ballistic heat pulses, however, the acoustic anisotropy remains.

In this section, we introduce the notation and formalism associated with elasticity theory. After defining stress and strain tensors, we write down the wave equation and discuss its solution, which is graphically represented as the slowness surface. An expression for the group velocity will be derived. Finally, methods will be discussed for calculating the ballistic heat flux. The approach is intuitive where possible. Rigorous mathematical proofs can be found in several standard texts on the theory of elastic waves.[5,11-13]

i) Stresses and strains

The strain tensor describes how a solid is distorted from its equilibrium shape as a result of applied forces. Simply put, the strain ε_{xx} of a uniform rod under compression is its fractional change in length. More generally, the displacement of an atom at position $\mathbf{r}$ in a solid is given by $\mathbf{u}(\mathbf{r})$, and the strain is defined in terms of spatial derivatives of $\mathbf{u}(\mathbf{r})$.

A simple uniform distortion of a solid is illustrated in Figure 12. For this example, the displacement at any point is given in terms of x, y and the

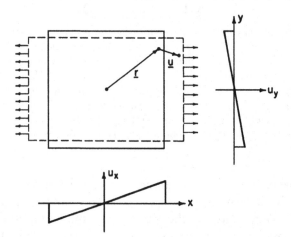

Figure 12 Displacements u_x and u_y associated with a uniform uniaxial strain. The strain tensor ε has the nonzero elements $\varepsilon_{xx} = \partial u_x/\partial x$ and $\varepsilon_{yy} = \partial u_y/\partial y$.

constant derivatives, $\partial u_x/\partial x$ and $\partial u_y/\partial y$. These derivatives are elements of a dimensionless strain tensor, which defines the fractional distortions of a unit cell in the crystal. If we subdivide the volume shown into smaller cells, each cell has exactly the same fractional distortion; i.e., the strain is uniform. Strain is defined in terms of derivatives because it is the fractional distortion of the cell surrounding an atom that determines the force on the atom, not the total displacement $\mathbf{u}$.

The particular strain shown in the figure could be produced by pulling uniformly on the left and right surfaces. If no forces are applied to the other surfaces, there will be some elastic shrinkage perpendicular to the horizontal forces, as shown. The ratio of $\partial u_y/\partial y$ to $\partial u_x/\partial x$ is known as Poisson's ratio and is in the range -0.2 to -0.3 for most solids. Elasticity theory gives the relative values of these and other derivatives when arbitrary stresses are applied to a solid. (At this stage we generalize the notation: the subscripts l, $m = 1, 2, 3$ are used for the cartesian axes x_1, x_2, x_3 representing x, y, z.)

How is the strain tensor defined? In practice, the elements of the second-rank strain tensor are not simply $\partial u_l/\partial x_m$. Instead, the symmetric combination

$$\varepsilon_{lm} = \frac{1}{2}(\partial u_l/\partial x_m + \partial u_m/\partial x_l) \tag{3}$$

is chosen. With this definition the elements of the strain tensor have a simple form for pure rotations or pure distortions, as depicted in Figure 13. Changes in volume (compression or dilation) and shape (shear) of the solid element are possible. One can also define the antisymmetric combination,

$$w_{lm} = \frac{1}{2}(\partial u_l/\partial x_m - \partial u_m/\partial x_l). \tag{4}$$

We see from the figure that $\partial u_1/\partial x_2$ and w_{12} are nonzero for rotation of the solid. Pure rotation of the crystal is described by a uniform w, which produces no restoring forces. Its diagonal components are zero and its off-diagonal components are used to form the axial vector operator known as *curl* or *rot*. The strain tensor ε, on the other hand, describes pure distortions – that is, displacements for which there is a restoring force.

It is noteworthy that the pure distortion depicted in Figure 13 is physically similar to the distortion shown in Figure 12, with the coordinate axes rotated by 45 degrees. The strain tensor for this uniaxial stress contains both dilation and shear components. Rotational transformation of a second-rank tensor is briefly considered in Figure 14, discussed below.

Of course, the distortions need not be uniform across the solid. The strain tensor is determined by the *local* spatial derivatives of displacement. In an elastic wave, for example, the strain varies from point to point, so that ε is a function of position and time, $\varepsilon(\mathbf{r}, t)$.

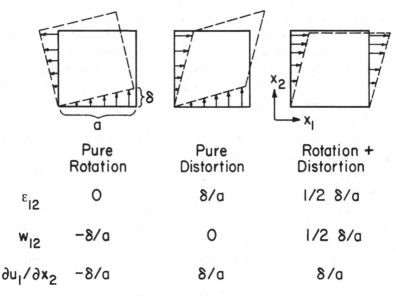

	Pure Rotation	Pure Distortion	Rotation + Distortion
ε_{12}	0	δ/a	$1/2\ \delta/a$
w_{12}	$-\delta/a$	0	$1/2\ \delta/a$
$\partial u_1/\partial x_2$	$-\delta/a$	δ/a	δ/a

Figure 13 Examples of uniform rotation and distortion of a solid. Elements of the strain and rotation tensors are determined from Eqs. (3) and (4). In each case the displacement u_1 is given by $\partial u_1/\partial x_2 = \varepsilon_{12} + w_{12}$, but only the strain tensor is required to describe elastic waves.

It is possible to define the strain tensor in vector form, rather than the component form given in Eq. (3). This formalism is well developed in Auld's text.[11] In terms of the operator $\nabla = \hat{x}\partial/\partial x_1 + \hat{y}\partial/\partial x_2 + \hat{z}\partial/\partial x_3$, the gradient of a vector is the outer product (or dyad) $\nabla\mathbf{u}$, which is a second-rank tensor with the elements $(\nabla\mathbf{u})_{lm} = \partial u_l/\partial x_m$. The strain tensor is thus given by

$$\varepsilon = \frac{1}{2}(\nabla\mathbf{u} + \widetilde{\nabla\mathbf{u}}) \equiv \nabla_s\mathbf{u}, \tag{5}$$

where the second term represents the transposed matrix (interchanged rows and columns). The subscript s indicates the symmetric operation defined by Eq. (5).

Now we consider the forces on a solid, or a small element of the solid. Stress describes the force per unit area on a surface, i.e., a type of pressure. Pressure exerted on the surface of the solid is transmitted through the solid, so the stress is locally defined inside the solid. The force vector applied to a surface, however, need not be directed normal to the surface; there can be both compressional and shear components. The (shear) stress component σ_{yx} is the y component of the force acting on the x plane (per unit area), $\sigma_{yx} = F_y/A_x$, as in Figure 14(a).

If we isolate a cubic volume element in the solid with sides parallel to the coordinate axes (x, y, z), the net force on this element at a given time is the

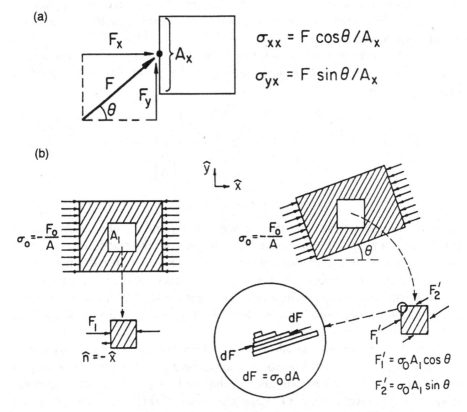

Figure 14 (a) Definition of the stress-tensor elements in terms of the force, F, applied to a surface. (b) Two examples of finding the stress tensor in uniformly stressed solids. A small cube is chosen with axes $\hat{x}$, $\hat{y}$ parallel to its faces of area A_1. The total force on each side of the cube is determined by geometrical considerations. The elements of the stress tensor are found using the definition of the traction force, $F/A_1 = \sigma \cdot \hat{n}$. Stress components involving z are zero. σ_0 is defined to be negative for compression. The second case shows how σ transforms under rotation: In general, for an axial rotation given by $x_i' = a_{ij} x_j$, one has $\sigma_{ij}' = a_{ik} a_{jl} \sigma_{kl}$. The circular inset indicates how the total force on a surface can be considered as the sum of forces dF on thin volume elements that are aligned along the applied force direction:

Case 1	Case 2

$$\mathbf{F}_1 = -\sigma_0 A_1 \hat{x}$$
$$\mathbf{F}_1/A_1 = -\sigma \cdot \hat{x}$$
$$= -\sigma_{xx}\hat{x} - \sigma_{yx}\hat{y}$$
Therefore

$$\sigma_{xx} = \sigma_0$$
$$\sigma_{yx} = 0$$
$$\sigma_{yy} = 0$$

$$\mathbf{F}_1' = -\sigma_0 A_1 \cos\theta (\hat{x}\cos\theta + \hat{y}\sin\theta)$$
$$\mathbf{F}_1'/A_1 = -\sigma_{xx}'\hat{x} - \sigma_{yx}'\hat{y}$$

$$\sigma_{xx}' = \sigma_0 \cos^2\theta$$
$$\sigma_{yx}' = \sigma_0 \cos\theta\sin\theta$$
$$\sigma_{yy}' = \sigma_0 \sin^2\theta \ (\text{from } \mathbf{F}_2')$$

vector sum of the forces acting on all six sides. These forces are applied by the surrounding solid. The stress tensor contains the detailed information about these forces.

The vector force per unit area acting on a surface element is known as the traction force. For a surface with normal in the $+x$ direction, the traction force is given by

$$\boldsymbol{\sigma}_x \equiv \sigma \cdot \hat{x} \quad \text{(a vector)}$$

$$= \begin{bmatrix} \sigma_{xx} & \sigma_{xy} & \sigma_{xz} \\ \sigma_{yx} & \sigma_{yy} & \sigma_{yz} \\ \sigma_{zx} & \sigma_{zy} & \sigma_{zz} \end{bmatrix} \begin{bmatrix} 1 \\ 0 \\ 0 \end{bmatrix}$$

$$= \sigma_{xx}\hat{x} + \sigma_{yx}\hat{y} + \sigma_{zx}\hat{z}. \tag{6}$$

σ_{xx} is the normal pressure on the surface, and σ_{yx} and σ_{zx} are components of the traction force parallel to the surface. In general, the traction force acting on an elemental area with normal $\hat{n}$ is $\sigma_n = \sigma \cdot \hat{n}$, where the dot indicates a sum over the second subscript of σ_{ij} (see also, Appendix II). Simple examples of these concepts are given in Figure 14, which considers uniform stresses. For an elastic wave, the stress tensor is defined locally and is a function of position and time, $\sigma(\mathbf{r}, t)$.

For both static strains and vibrational motion, the stress tensor is symmetric, $\sigma_{ij} = \sigma_{ji}$. This important property arises because there are no net torques on an infinitesimal volume element in the solid. A mathematical argument may be found in Auld's text.[11] Despite arguments of this sort, there is an interesting debate in the literature whether, under some conditions, acoustic waves can have some torsional character. The most recent experiments,[14] however, argue against such internal torques, at least for long-wavelength waves.

ii) The elasticity tensor

The two second-rank tensors, stress σ and strain ε, are related to each other by a fourth-rank elasticity tensor, in a generalized version of Hooke's law:

$$\sigma = C:\varepsilon. \tag{7}$$

This relationship can also be written in component form,

$$\sigma_{ij} = C_{ijlm}\varepsilon_{lm}, \tag{8}$$

where summation over repeated indices is assumed. The double dot in Eq. (7) indicates a sum over two of the four indices C_{ijlm}. C is also called the stiffness tensor. This expression represents nine equations, each with nine terms.

Fortunately, the elastic stiffness constants, C_{ijlm}, are not all independent. Because $\sigma_{ij} = \sigma_{ji}$ due to vanishing torques, and $\varepsilon_{ij} = \varepsilon_{ji}$ by definition, C_{ijlm} is invariant to interchange of i and j or l and m. That is, $C_{ijlm} = C_{jilm} = C_{ijml}$. A further argument made from energy considerations shows that, in addition,

$C_{ijlm} = C_{lmij}$. These symmetries reduce the number of independent stiffness constants in any medium from 81 to 21. The symmetry of the crystal lattice can reduce the number further. In cubic crystals, for example, there are only three independent stiffness constants!

The symmetries discussed above lead to a simpler expression for C with only two subscripts, known as the Voigt contraction.[15] We write

$$C_{IJ} = C_{ijlm}, \tag{9}$$

with the abbreviated subscripts I and J given by

I or J	1	2	3	4	5	6
ij or lm	xx	yy	zz	yz	xz	xy.

Hooke's law is thus written in simpler form,

$$\sigma_I = C_{IJ}\varepsilon_J, \tag{10a}$$

or

$$\sigma = C \cdot \varepsilon, \tag{10b}$$

where the single dot indicates a sum over one index. Further discussion of this abbreviated notation is given in Appendix II.

The above symmetry, $C_{ijlm} = C_{ijml}$, also implies that, when ε is multiplied by C, one may use the simple expression

$$\varepsilon = \nabla \mathbf{u}, \tag{11}$$

or $\varepsilon_{lm} = \partial u_l / \partial x_m$, because both terms in parentheses in Eq. (5) yield the same product.

iii) The wave equation

Consider a small cubic volume element of dimensions δx, δy, and δz. Now allow σ and ε to vary in space and time. According to Eq. (6), the traction force on the surface element with normal $\hat{x}$ at position x, y, z is $\sigma_x(x, y, z)$, and the total force on this surface element is $\sigma_x(x, y, z)\delta y \delta z$. The total force applied to the six surfaces of this volume element by the surrounding solid is

$$\begin{aligned}
\delta \mathbf{F}_{\text{tot}} &= \sigma_x(x, y, z)\delta y \delta z + \sigma_y(x, y, z)\delta x \delta z + \sigma_z(x, y, z)\delta x \delta y \\
&\quad - \sigma_x(x - \delta x, y, z)\delta y \delta z - \sigma_y(x, y - \delta y, z)\delta x \delta z \\
&\quad - \sigma_z(x, y, z - \delta z)\delta x \delta y \\
&= [(\partial \sigma_x / \partial x) + (\partial \sigma_y / \partial y) + (\partial \sigma_z / \partial z)]\delta x \delta y \delta z. \tag{12}
\end{aligned}$$

By Newton's second law this net force is equal to the mass of the element, $\rho \cdot (\delta x \delta y \delta z)$, times its acceleration. Dividing through by volume, this gives

$$\rho \partial^2 \mathbf{u} / \partial t^2 = (\partial \sigma_x / \partial x) + (\partial \sigma_y / \partial y) + (\partial \sigma_z / \partial z). \tag{13}$$

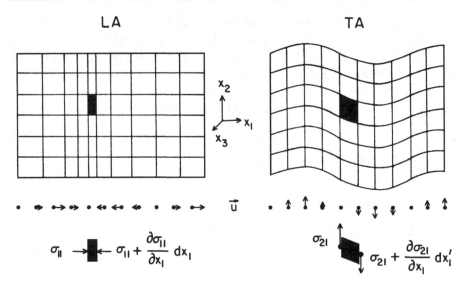

Figure 15 Schematic representation of longitudinal and transverse waves in a solid medium. An infinitesimal volume element is displaced a distance **u** *from its equilibrium position. For the longitudinal wave the net force on a volume element $dx_1 dx_2 dx_3$ is $(\partial \sigma_{11}/\partial x_1)dx_1 dx_2 dx_3$, where σ_{11} is the compressive stress. Because the mass of this element is $\rho dx_1 dx_2 dx_3$, the equation of motion is $\rho \ddot{u}_1 = \partial \sigma_{11}/\partial x_1$. (From Wolfe.[16])*

It is convenient to write this vector equation in component form. Again, $(x, y, z) = (x_1, x_2, x_3)$. Thus the component form of Newton's law is

$$\rho \partial^2 u_i / \partial t^2 = \partial \sigma_{ij} / \partial x_j, \tag{14}$$

where summation over repeated indices is implied. A simple illustration[16] of this equation for longitudinal and transverse waves is given in Figure 15. Using Hooke's law and the definition of the strain tensor, Eq. (11), we have the elastic wave equation,

$$\rho \partial^2 u_i / \partial t^2 = (\partial / \partial x_j) C_{ijlm} \varepsilon_{lm}$$
$$= C_{ijlm}(\partial^2 u_l / \partial x_j \partial x_m). \tag{15}$$

We assume a plane-wave solution to this equation,

$$\mathbf{u} = u_0 \mathbf{e} \exp[i(\mathbf{k} \cdot \mathbf{r} - \omega t)], \tag{16}$$

where the wavevector $\mathbf{k}$ is given and the frequency ω and polarization $\mathbf{e}$ of the wave are to be determined. Plugging this into Eq. (15), and using summation notation, as usual, we have,

$$\rho \omega^2 e_i = C_{ijlm} k_j k_m e_l. \tag{17}$$

This is a set of three equations for the components of the unit polarization vector **e**, known as the Christoffel equation and commonly written as

$$[D_{il} - v^2\delta_{il}]e_l = 0, \tag{18}$$

where

$$D_{il} = C_{ijlm}n_j n_m/\rho \tag{19}$$

is known as the Christoffel matrix.[17] The symbols n_i are direction cosines of the wavevector, i.e., components of the wave normal $\mathbf{n} = \mathbf{k}/k$. δ_{il} is the Kronecker delta. The scalar $v = \omega/k$ is called the phase velocity. Physically, it represents the speed of the wavefronts (the planes of constant phase), so one can also associate the vector $v\mathbf{n}$ with the phase velocity.

For a given wave normal $\mathbf{n}$, or wavevector direction (θ_k, ϕ_k), there are three eigenvalues v_α given by the roots of the characteristic equation,

$$|D_{il} - v^2\delta_{il}| = 0. \tag{20}$$

Because D_{il} depends only on the *direction* of $\mathbf{k} = (k, \theta_k, \phi_k)$, the phase velocity is likewise independent of k, and the frequency of the wave is given by

$$\omega = v(\theta_k, \phi_k)k. \tag{21}$$

For a given wavevector direction, ω is proportional to k. This linear dispersion relation is a characteristic of continuum elasticity theory.

Given the eigenvalues v_α the corresponding eigenvectors $\mathbf{e}(\alpha; \theta_k, \phi_k)$ can be found by solving Eq. (18). The subscript α identifies the phonon mode. These three modes are usually labeled as the fast transverse (FT), slow transverse (ST), and longitudinal (L) modes. The eigenvectors for a given $\mathbf{k}$ are always orthogonal to each other. However, for a general k vector direction, a particular mode is not purely longitudinal ($\mathbf{k} \parallel \mathbf{e}$) or transverse ($\mathbf{k} \perp \mathbf{e}$) in character. Hence, it is more correct to call the modes quasi-longitudinal or quasi-transverse, but for compactness we adopt the labels L, FT, and ST.

A general closed-form solution for the eigenvalues, v_α, of the Christoffel equation has been derived by Every.[18,19] This solution is reproduced in Appendix I. In practice, the solution of this problem is most easily found numerically, without recourse to the analytic solutions. For a given wavevector direction, the eigenvalues v_α and eigenvectors $\mathbf{e}(\alpha; \theta_k, \phi_k)$ of the Christoffel matrix D_{il} are determined by diagonalization of this matrix with standard computer software.

iv) Slowness surface

The three eigenvalues of the Christoffel equation are the phase velocities $v_\alpha(\theta_k, \phi_k)$. We seek the most useful way to display this result graphically. The slowness vector $\mathbf{s} = \mathbf{k}/\omega = \mathbf{n}/v$ is a useful vector with direction along $\mathbf{k}$

and magnitude $s(\alpha; \theta_k, \phi_k) = 1/v_\alpha(\theta_k, \phi_k)$. This definition allows us to write the Christoffel equation [Eq. (18)] in a frequency-independent form:

$$(C_{ijlm}s_j s_m - \rho\delta_{il})e_l = 0. \tag{22}$$

A radial plot of $s(\alpha; \theta_k, \phi_k)$ gives a slowness surface for the mode α, which has the shape of a constant-frequency surface in $\mathbf{k}$ space: $k(\theta_k, \phi_k) = \omega_0/v_\alpha$ (θ_k, ϕ_k). More precisely, one says that there are three *sheets*, $\alpha = $ L, FT, ST, of the slowness surface of a given crystal. The mode index, α, appears subscripted on the scalar v_α, but parenthetically in the vectors $\mathbf{s}(\alpha; \theta_k, \phi_k)$ and $\mathbf{e}(\alpha; \theta_k, \phi_k)$, to avoid confusion between the vector components and the mode index.

In this book, we define the inner, middle, and outer sheets of the slowness surface as L, FT, and ST, respectively, even though the wave polarizations are not purely longitudinal or transverse. Sometimes it will be convenient to choose a numerical label for α at the beginning of a calculation. Usually, but not always, $\alpha = 1$ will denote the FT mode, $\alpha = 2$ the ST mode, and $\alpha = 3$ the L mode.

What are some general features of slowness surfaces? The three sheets of the slowness surface may touch each other in places but they do not pass through each other.* It can happen that the group velocity along a particular propagation direction for an ST wave is actually faster than that for the FT wave along the same direction! But these ST and FT waves actually have k vectors in different directions. For a given wavevector direction, the ST phase velocity is always slower or equal to the FT phase velocity. This is the justification for defining ST and FT modes.

The slowness surfaces for the different modes may touch each other either *tangentially* or *conically* at a point.† An example of both cases is shown in Figure 16(a). This is a cross section of the slowness surfaces for Ge in the (110) plane. Along [100] the ST and FT surfaces meet tangentially. Along [111] it appears that the two surfaces cross each other; however, they actually just meet conically at a point. That is, in a cross section taken slightly out of the (110) plane, the two surfaces do not meet. A representation of this conical gramophone connection along [111] is shown in the figure.

One of the interesting aspects of this conical point is that there is a cone of directions, centered around the [111] axis, inside of which there are no FT group velocities. This can be seen by constructing the normals ($\mathbf{V}$ directions) to the FT slowness surface adjacent to the conic point. The ST mode makes up for this deficiency, however, by having (at least) two group velocity values for a given real-space direction inside this cone. This situation is illustrated in the (110) cross section of the group-velocity surfaces plotted in Figure 16(b). The ST and FT branches meet perfectly at the boundary of the FT-forbidden cone. This effect is known as internal conical refraction because phonons entering

---

* Except for hexagonal crystals, as treated in Chapter 5.
† In hexagonal crystals, degeneracy along a line is also possible.

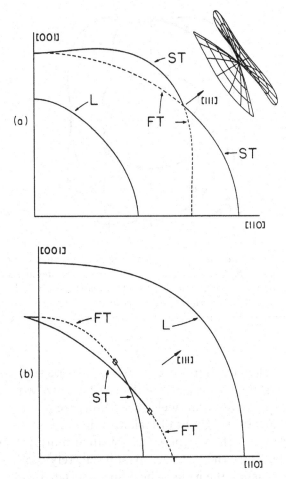

*Figure 16 (a) Intersection of the three sheets of the Ge slowness surface with the (1̄10) plane, showing the conic contact between FT and ST sheets along [111]. Slightly out of this plane the two sheets do not touch each other. The inset shows the three-dimensional nature of this conic region. (b) Intersection of the group-velocity surfaces with the (1̄10) plane. For the ST mode, only branches from the (1̄10) plane in **k**-space are shown. The small diamonds mark the continuous transition from ST to FT and correspond to **k** ∥ [111] in (a). (From Northrop and Wolfe.[9])*

the crystal from another medium with wavevectors normal to the (111) surface will be refracted into a cone. The wave-surface topology associated with conic points has been studied by Every[19] and by Hurley and Wolfe,[20] and a further discussion will be presented in Chapter 4.

As discussed above, the three-dimensional transverse slowness surfaces can have convex, concave, and saddle topology. The boundaries of these regions have vanishing Gaussian curvature, which leads to the phonon caustics. The

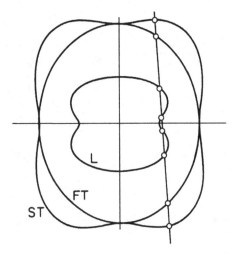

Figure 17 If the L slowness surface had a region of negative curvature, it would be possible to draw a line that intersected it four times and thus had eight total intersections with all sheets. A maximum of six intersections are possible because the surface is generated by a sixth-order equation (v^6). Therefore, the L surface is entirely convex and produces no caustics. (From Maris.[21])

inner sheet of the slowness surface (L), however, has no such parabolic lines. The geometrical argument presented in Figure 17 shows why this is so.[21,22] Since the characteristic equation yields a sixth-degree polynomial equation, there can be at most six solutions of v along a line in k space. If there were inflections in the inner (L) slowness surface, then more than six solutions would be possible. Thus, the L surface must be entirely convex. Although it does not display caustics, the ballistic heat flux associated with the L mode *is* anisotropic because the L sheet of the slowness surface is not spherical.

A particularly interesting example of the interrelation of L, FT, and ST sheets occurs for TeO_2, shown in Figure 4 on page 127. In this highly anisotropic tetragonal crystal, the longitudinal and transverse surfaces touch each other, and conic points occur along nonsymmetry directions. The inner (L) sheet touches the middle (FT) sheet conically in several wavevector directions, and the polarization of the L mode has more transverse than longitudinal character in some directions. Cross sections of these surfaces in the symmetry planes (Figure 3, page 126) suggest that the different sheets pass through each other, but the three-dimensional representations show otherwise.

v) Group velocity and energy propagation

The concept of group velocity is most familiar in the context of phonon dispersion in one dimension. As illustrated in Figure 18, two large-k acoustic

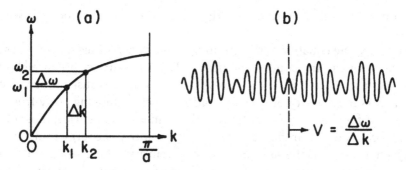

Figure 18 Concept of group velocity in one dimension. Two waves with different ω, k and phase velocity, ω/k, superpose to form packets moving at the group velocity, $\Delta\omega/\Delta k$. For example, $\cos(k_1 x - \omega_1 t) + \cos(k_2 x - \omega_2 t) = 2\cos(\bar{k}x - \bar{\omega}t)\cos[(\Delta k/2)x - (\Delta\omega/2)t]$, where $\bar{\omega} = (\omega_1 + \omega_2)/2, \Delta\omega = \omega_2 - \omega_1$, etc. The latter factor is the modulation envelope with velocity $\Delta\omega/\Delta k$.

waves of different frequency and phase velocity combine to produce a modulated wave. The velocity of the modulation envelope is defined as the group velocity, i.e., the velocity of a packet. If the wavevectors are chosen from the nonlinear region of the dispersion curve the group velocity is not equal to the phase velocity. Because energy does not transfer across the nodes in the envelope, the group velocity is also the velocity of energy propagation.

It is straightforward to generalize this one-dimensional example to three dimensions.[11] The result is

$$\mathbf{V} = d\omega(\mathbf{k})/d\mathbf{k} = \nabla_k \omega(\mathbf{k}). \tag{23}$$

This is the gradient of $\omega(\mathbf{k})$ in k space, which is normal to the constant-ω surfaces. For a distribution of phonon frequencies, a wave envelope is formed with velocity given by the gradient of ω.

The distinction between phase and group velocity is certainly necessary when the dispersion relation is not linear, which in the one-dimensional example means that group and phase velocities are not equal, $\omega/k \neq \Delta\omega/\Delta k$. Due to elastic anisotropy, however, the concept of group velocity is necessary even in the case of linear dispersion. Just as two waves with the same direction and different frequencies interfere with each other, so will two waves with the same frequency and different wavevector directions. Figure 5 illustrated that two waves with the same frequency but slightly different $\mathbf{k}$ directions produce an interference pattern. The node lines are fixed in space, and there is no energy flow perpendicular to them. These node lines turn out to be perpendicular to an elemental segment of slowness surface containing the two wavevectors. By Eq. (23), this is also the direction of the group velocity; that is, energy flow is along the direction of the group velocity. If we choose a distribution of frequencies for this wavevector direction, packets of vibrational energy which travel with velocity **V** will be formed.

Physically, the equality of the group and energy velocities can also be seen from the "stone in the pond" experiment. The wave surface produced by the single impulse is certainly the group-velocity surface, because group velocity is defined as the velocity of the modulation envelope moving away from the source. But for this single impulse (involving many frequencies) most of the energy is contained near the wavefront, so the velocity of energy transmission in a given direction is the same as the group velocity. The mathematical basis for this equality is not so obvious. How does one compute the energy flow of a plane acoustic wave?

Before going through this mental exercise, let us first derive a practical expression for the group velocity in terms of the elastic constants, the given **k** vector, and the solution to the wave equation, ω and **e**. This is accomplished by dotting the vector form of the wave equation [Eq. (17)], with **e** and differentiating with respect to k_n (summation notation dictates that $e_i e_i = 1$),

$$\rho \omega^2 = e_i C_{ijlm} k_j k_m e_l \tag{24}$$

$$2\rho \omega \, \delta\omega/\delta k_n = e_i C_{inlm} k_m e_l + e_i C_{ijln} k_j e_l + \cdots$$

$$= 2 e_i C_{ijln} k_j e_l. \tag{25}$$

The dummy variables m in the first term were replaced with j. Then i and l were interchanged because $e_i e_l = e_l e_i$, and $C_{ijln} = C_{lnij}$ was used. Additional terms on the right side of this equation can be combined in the form $\mathbf{e} \cdot (\partial \mathbf{e}/\partial \mathbf{k})$, by virtue of Eq. (17). This expression is zero because **e** has a constant magnitude (unity) as **k** is varied. So, the desired result for the components of the group velocity is

$$V_n = \partial \omega/\partial k_n = e_i C_{ijln} k_j e_l/\rho \omega. \tag{26}$$

Once **e** and ω are found for a given **k**, this expression gives **V**. The explicit form of the components V_n for cubic symmetry is given in Eq. (12) of Appendix II.

An intuitive feel for this rather complicated expression can be gained by considering the energy flow of an acoustic wave. The acoustic energy crossing an imaginary surface in the crystal is described in terms of an acoustic Poynting vector, analogous to the Poynting vector in electromagnetic theory. The acoustic Poynting vector is defined in terms of higher-order tensors.

The power (energy per second) transmitted to an object equals the applied force, **F**, times the velocity, **v**, of the object:

$$\mathcal{P} = \mathbf{v} \cdot \mathbf{F}. \tag{27}$$

For the acoustic problem, the power per area transmitted to a section of the crystal is

$$\mathcal{P}/A = \mathbf{v} \cdot (\mathbf{F}/A), \tag{28}$$

where $\mathbf{v}$ is the velocity of the atoms on the surface of that section. Because we are dealing with a force distributed over the surface, the force per area is generalized to the stress tensor, σ. The acoustic Poynting vector for real displacements is thus defined as

$$\mathbf{P} = -\overline{\mathbf{v} \cdot \sigma}. \tag{29}$$

where the overbar indicates a time average over one cycle. For complex waves, $\exp[i(\mathbf{k} \cdot \mathbf{r} - \omega t)]$, the appropriate form is $\mathbf{P} = -\mathrm{Re}(\mathbf{v} \cdot \sigma^*)/2$. $\mathbf{P}$ has units of energy per area per second. The negative sign reflects the fact that $\sigma \cdot \mathbf{n}$ is defined as the traction force of the crystal at a surface with normal $\mathbf{n}$, which points outward from the surface being stressed. $\mathbf{P}$ is defined locally throughout the crystal. Note that the magnitude of the local velocity $\mathbf{v}$ of the medium is not the phase velocity of the wave.

Assuming a real elastic wave with polarization $\mathbf{e}$, the local displacement and velocity are

$$\mathbf{u} = u_0 \mathbf{e} \cos(\mathbf{k} \cdot \mathbf{r} - \omega t), \tag{30}$$

$$\mathbf{v} = \omega u_0 \mathbf{e} \sin(\mathbf{k} \cdot \mathbf{r} - \omega t). \tag{31}$$

The strain associated with this displacement is

$$\varepsilon = \nabla \mathbf{u}$$
$$= -\mathbf{k} \mathbf{e} \, u_0 \sin(\mathbf{k} \cdot \mathbf{r} - \omega t), \tag{32}$$

where the outer product is a second-rank tensor with elements $(\mathbf{ke})_{lm} = k_l e_m$. The stress associated with this displacement is given by

$$\sigma = C \cdot \varepsilon$$
$$= -C \cdot \mathbf{k} \mathbf{e} \, u_0 \sin(\mathbf{k} \cdot \mathbf{r} - \omega t). \tag{33}$$

The important point here is that the stress tensor and the velocity oscillate in phase. In contrast, the displacement of the surface is not in phase with the velocity and stress. The restoring force of the crystal depends not on the absolute displacement but on the differential displacement (microscopically, the *relative* displacements of the atoms). Figure 19 illustrates this situation for a simple longitudinal wave traveling along a symmetry axis.

The equal phase of stress and velocity means that there is a net energy flow for the traveling wave that is given by

$$\mathbf{P} = \mathbf{e} \cdot C \cdot \mathbf{k} \mathbf{e} \, \omega u_0^2 \, \overline{\sin^2(\mathbf{k} \cdot \mathbf{r} - \omega t)} \tag{34}$$

$$= \mathbf{e} \cdot C \cdot \mathbf{k} \mathbf{e} \, \omega u_0^2 / 2. \tag{35}$$

We now define the velocity of energy flux as

$$\mathbf{V} = \mathbf{P}/U, \tag{36}$$

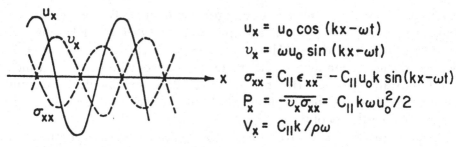

$$u_x = u_0 \cos (kx - \omega t)$$
$$v_x = \omega u_0 \sin (kx - \omega t)$$
$$\sigma_{xx} = C_{||} \epsilon_{xx} = -C_{||} u_0 k \sin(kx - \omega t)$$
$$P_x = -\overline{v_x \sigma_{xx}} = C_{||} k \omega u_0^2 / 2$$
$$V_x = C_{||} k / \rho \omega$$

*Figure 19 Illustration of the principal relations involved in the energy flow of a plane acoustic wave. A longitudinal wave with **k** along [100] in a cubic crystal is assumed. In this symmetric case, the group velocity is along [100] and equals the phase velocity, $v = \omega / k = (C_{11}/\rho)^{1/2}$.*

where $U = (1/2)\rho v_0^2$ is the local energy density, with $v_0 = \omega u_0$. The final result is,

$$\mathbf{V} = (1/\rho\omega)\mathbf{e} \cdot C \cdot \mathbf{k}\mathbf{e}. \tag{37}$$

This is the vector form of the group-velocity equation that we derived above [Eq. (26)]. Energy flux, as defined by Eqs. (29) and (36), is along the group-velocity direction.

It bears repeating that the group-velocity surface, or wave surface, gives the shape of a ballistic heat-pulse distribution in the crystal, assuming a short temporal pulse emanating from a point source. Along the direction of **V**, the position of the heat pulse at time t is given by,

$$\mathbf{r} = \mathbf{V}t. \tag{38}$$

At the beginning of this chapter, we gave a geometrical argument (Figure 2) that the wavevector is perpendicular to this surface. This fact can now be put on a more mathematical basis. Dotting **V** with **s**, we obtain

$$\mathbf{s} \cdot \mathbf{V} = s_n V_n = e_i s_n C_{ijln} s_j e_l / \rho = e_i \cdot e_i,$$

using Eq. (22), so

$$\mathbf{s} \cdot \mathbf{V} = 1, \tag{39}$$

or $\mathbf{k} \cdot \mathbf{V} = \omega$. Now, dotting Eq. (38) with the slowness vector,

$$\mathbf{s} \cdot \mathbf{r} = \mathbf{s} \cdot \mathbf{V}t = t. \tag{40}$$

One may view the expanding wavefront as a constant-t surface in real space. Of course, the arrival time at position **r** depends on the direction of propagation; that is, $t = t(\mathbf{r})$. By the definition of the gradient,

$$dt = (dt/d\mathbf{r}) \cdot d\mathbf{r}. \qquad (41)$$

For $d\mathbf{r}$ along the constant-time surface, $dt = 0$, implying that the gradient $dt/d\mathbf{r}$ is perpendicular to the surface. Now, $\mathbf{s}$ in Eq. (40) does not depend on $\mathbf{r}$, so, from this equation, we can identify $\mathbf{s} = dt/d\mathbf{r}$ as a vector that is perpendicular to the wave surface.* Consequently, $\mathbf{k} \equiv \mathbf{s}/\omega$ is perpendicular to the wave surface at the $\mathbf{V}$ that corresponds to $\mathbf{k}$. [See also Figure 4 on page 338.]

The condition $\mathbf{s} \cdot \mathbf{V} = 1$ derived in Eq. (39) means that the slowness and wave surfaces are polar reciprocals of each other. The slowness surface is by far the simplest of the two. It has no folds and, as indicated in Figure 17, can be intersected by a straight line at only six points or fewer. The wave surface is physically more meaningful in a heat-pulse experiment, but it is far more complicated. The mathematical relation of these polar-reciprocal surfaces is discussed by Duff[22] and by Every,[19] who points out that a straight line can intersect the wave surface at up to 98 points for cubic symmetry (150 points in general). Fortunately, the wave-surface topology for high symmetry crystals is not so severe and can be graphically visualized, as shown later in this chapter and in Chapter 4.

Figure 20 summarizes the basic relations we have developed. In the continuum elasticity theory, the constant-frequency surfaces in k space, when divided by ω, reduce to a single slowness surface. (Only one of its three sheets is shown.) The normal to this surface at a given $\mathbf{s}$ is the group velocity $\mathbf{V}$. Likewise, the constant-time surfaces in real space, when divided by t, reduce to a single group-velocity surface. The normal to this surface is the wavevector $\mathbf{k}$, or slowness $\mathbf{s}$.

vi) Ballistic heat flux

The ballistic heat flux seen in a phonon image is inversely proportional to the local curvature of the slowness surface. How do we quantify the angular variation in intensity of the heat pulses that are observed? Maris[8] defined the enhancement factor, A, which is the ratio of the energy flux for a given propagation direction in the crystal to that of an isotropic medium. For a given infinitesimal solid angle, $\Delta\Omega_k$, of wavevectors there is a corresponding solid angle, $\Delta\Omega_v$, of group-velocity vectors. The enhancement factor is simply

$$A = \Delta\Omega_k/\Delta\Omega_v, \qquad (42)$$

which equals unity for an isotropic medium.

A graphic example of the enhancement factor is found in Figure 8. Two equal $\Delta\Omega_v$ sections are shown as the cross-hatched and shaded regions on the wave-surface diagram. The corresponding $\Delta\Omega_k$ are plotted on the slowness-surface diagram. For the cross-hatched case ($\mathbf{V} \parallel [110]$), $\Delta\Omega_k/\Delta\Omega_v$ is less

---

* I thank H. Kinder for pointing out this argument based on the time domain, in contrast to the frequency domain discussed earlier.

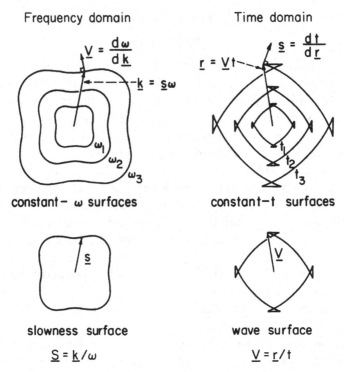

Frequency domain   Time domain

constant − ω surfaces   constant−t surfaces

slowness surface   wave surface

$$\underline{S} = \underline{k}/\omega$$ $$\underline{V} = \underline{r}/t$$

Figure 20 Summary of the basic relationships among **k**, ω, **s**, **V**, **r**, *and* t *as described in the text. Wavevector space is on the left and real space is on the right.*

than unity, indicating a defocusing or weak heat-pulse intensity for this propagation direction. The shaded $\Delta\Omega_v$ crosses three branches of the wave surface, corresponding to three different **k** directions. The total $\Delta\Omega_k/\Delta\Omega_v$ for this propagation direction **V** near [100] is much greater than unity, indicating a large enhancement of phonon flux.

The enhancement factor is directly related to the Gaussian curvature of the slowness surface, as previously discussed. The Gaussian curvature K is the product of the two principal curvatures (inverse radii, K_1 and K_2) at a point on the surface, i.e., $K = K_1 K_2$. Figure 21(a) defines these curvatures and illustrates various types of surface elements. For the parabolic element, one of the principal curvatures vanishes, and, therefore, $K = 0$. As shown in Figure 21(b), the slowness surface is parabolic at the borders between two opposite-curvature regions – here, convex and saddle. These parabolic lines produce the folds in the wave surface, as shown in Figure 21(c).

A cusp in the wave surface is produced by a depression in the slowness surface, as shown in Figure 22. The cusp corresponds to the point on the slowness surface where the direction of nonvanishing principal curvature is normal to

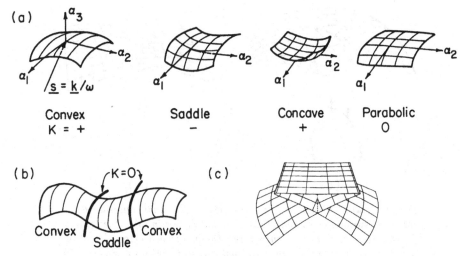

Figure 21 (a) Surface elements of the slowness surface. Orthogonal axes can be defined such that α_3 is perpendicular to the surface and $\alpha_3 = a\alpha_1^2 + b\alpha_2^2$ is the equation of the surface element. Principal curvatures are $K_1 = d^2\alpha_3/d\alpha_1^2 = 2a$ and $K_2 = d^2\alpha_3/d\alpha_2^2 = 2b$. The Gaussian curvature is $K = K_1 K_2 = 4ab$. Infinite heat flux occurs when one or both principal curvatures vanish. (b) Example of parabolic ($K = 0$) lines separating saddle and convex regions. (c) Corresponding folds on the wave surface.

the parabolic line – that is, where the direction of vanishing curvature is parallel to the parabolic line. At this point the direction of the group velocity is stationary with respect to small displacements along the parabolic line.

The algebraic relationship between the enhancement factor and the Gaussian curvature is shown geometrically in Figure 23. The result is

$$A^{-1} = |s^3 V K|, \tag{43}$$

where s and V are the magnitudes of slowness and group velocity. For an isotropic medium, the slowness surface is a sphere with curvature $K = 1/s^2$, and $s \cdot V = 1$, so $A = 1$. A small Gaussian curvature along a given $\mathbf{k} = \omega\mathbf{s}$ means a large flux enhancement for that wavevector direction. If either of the principal curvatures vanishes, K equals zero, and the enhancement is theoretically infinite.

A convenient way of calculating the Gaussian curvature is given in Ref. 23. It is directly related to the second-order derivatives of the phase velocity,

$$K = (v^4/V^4)(\beta_{22}\beta_{33} + \beta_{33}\beta_{11} + \beta_{11}\beta_{22} - \beta_{12}^2 - \beta_{23}^2 - \beta_{31}^2), \tag{44}$$

where $\beta_{ij} = \partial^2 v/\partial n_i \partial n_j = \partial V_j/\partial n_i$, using $V_j = \partial v/\partial n_j$. For further discussions, see Maris,[21] Every and McCurdy,[23] and Lax and Narayanamurti.[24]

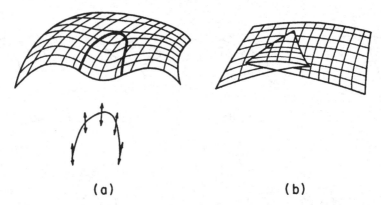

(a)   (b)

Figure 22 (a) A depression in the slowness surface that causes a cusp in the wave surface, as shown in (b). The heavy line in (a) is the locus of points with zero Gaussian curvature. Double-headed arrows show the direction of the nonvanishing principal curvature. Vanishing curvature is perpendicular to the arrows. The cusp occurs when the direction of vanishing curvature is along the parabolic line.

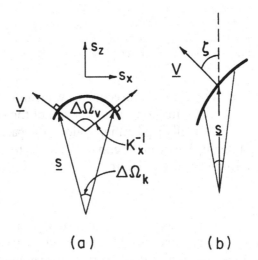

(a)   (b)

Figure 23 (a) Symmetric section of a slowness surface (heavy line) showing schematically the corresponding k space and V space solid angles. The enhancement factor is $A = \Delta\Omega_k/\Delta\Omega_v = |(ds_x ds_y/s^2)/(ds_x ds_y \times K_x K_y)| = |1/s^2 K|$. (b) For an arbitrary section of the slowness surface, $A = |\cos\zeta/s^2 K| = |s \cdot V/s^3 VK| = |1/s^3 VK|$.

Equivalently, the enhancement factor relates the area of a surface element on the slowness surface to that of the corresponding surface element on the wave surface. When one two-dimensional space is mapped onto another two-dimensional space – here, (θ_k, ϕ_k) onto (θ_v, ϕ_v) – the ratio of corresponding elemental areas in these spaces is the Jacobian of the transformation. Since we are interested in solid angles, it is best to consider $\cos\theta$ and ϕ as the variables, rather than θ and ϕ. That is,

$$d\Omega_v = d(\cos\theta_v)d\phi_v = J d(\cos\theta_k)d\phi_k = J d\Omega_k. \tag{44}$$

By writing out the explicit form of $\mathbf{V}$ in Eq. (26), Northrop and Wolfe[9] showed that the relation between the $\mathbf{k}$ and $\mathbf{V}$ angles can be written in the following form:

$$\cos\theta_v = f(\cos\theta_k, \phi_k) \tag{45}$$

$$\phi_v = g(\cos\theta_k, \phi_k). \tag{46}$$

Setting $\eta = \cos\theta_k$ and $\phi = \phi_k$, the Jacobian of this transformation is

$$J(\eta, \phi) = \frac{\partial f}{\partial\eta}\frac{\partial g}{\partial\phi} - \frac{\partial f}{\partial\phi}\frac{\partial g}{\partial\eta}. \tag{47}$$

The enhancement factor is simply

$$A = J^{-1}. \tag{48}$$

The algebra in determining the functions f and g and the Jacobian is quite involved. Fortunately, Northrop[25] has made this calculation of $\mathbf{V}$ and J available in the form of a computer program. The inputs are (θ_k, ϕ_k) and the elastic constants.

Thus far the principal utility of the enhancement factor lies in computing the positions of the parabolic, or caustic, lines. This is necessary to produce caustic maps such as those in Figure 10. For this task, the computer is sent on a root-finding search to determine the $\mathbf{k}$ directions where $J = 0$, and the corresponding loci of $\mathbf{V}$ directions. The resulting caustic map is the intersection of the wave-surface folds with the detection surface. An example is given in Figure 11(b).

In practice, the theoretical heat-flux intensities are determined by Monte Carlo calculations. The difficulty in using the enhancement factor for computing heat flux is that one usually wants to know the flux for a given real-space direction. But the input to A is a $\mathbf{k}$-space direction. There is no one-to-one correspondence between $\mathbf{k}$ and $\mathbf{V}$ directions. That is, for a given $\mathbf{V}$ direction, there may be many $\mathbf{k}$ vectors contributing to the flux. One can easily see this in the plot of Figure 8(b). Indeed, when one also considers group velocities mapped into this [100] plane from k vectors outside of the symmetry plane, a large number of intersections with the wave surface are found. For Ge or Si, as many as 11 different pulses are predicted in some directions, and these separate pulses are not usually resolved in an experiment. The myriad of folds in the group-velocity surface are apparent in the caustic maps of Figures 10 and 11.

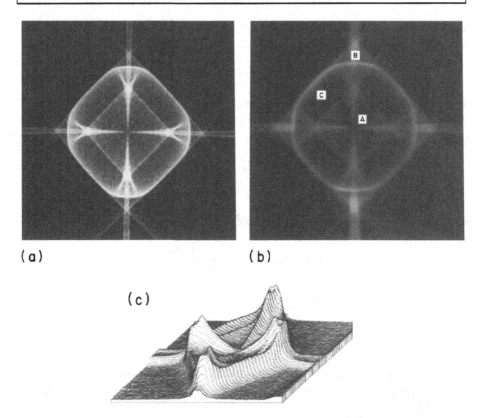

(a)   (b)

(c)

Figure 24 (a) Monte Carlo simulation of heat flux in Si. The center of the image is the [100] propagation direction. (b) Experimental image of Si taken by Shields and Wolfe.[26] The angular width of the image is 23.5 degrees. The labels A, B, and C correspond to points on sections of the wave surface shown in Figure 25. (c) Pseudo–three-dimensional plot of the heat-pulse intensities corresponding to (b).

A Monte Carlo simulation of the heat flux is obtained by (1) generating a number of randomly oriented wavevectors, (2) calculating the group-velocity direction for each by the method described above, and (3) plotting the intersection of these trajectories with the detection surface of the crystal. Each detected phonon is plotted as a bright point in a dark plane (or vice versa). The result[26] for Si is shown in Figure 24(a).

For comparison, a high-resolution experimental image of Si is shown in Figure 24(b). This image is produced with a 2-cm cube of Si, with a $10 \times 10 \, \mu m^2$ detector, and with a laser spot focused to a 15-μm diameter. The resulting angular resolution is approximately 0.06 degrees. The detector is a PbTl tunnel junction sensitive to phonons above 400 GHz. The agreement between experiment and theory is very good. A pseudo–three-

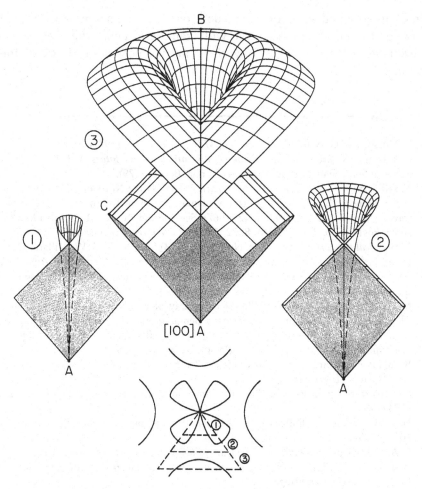

Figure 25 Sections of the ST wave surface in Si corresponding to the clover-leaf region of the slowness surface, shown at the bottom. Labels A, B, and C correspond to those in the phonon image of Figure 24(b). For further details, see Chapter 4.

dimensional representation of the experimental heat flux is plotted in Figure 24(c).

As previously discussed, the caustics in the phonon image are directly related to the folds in the group-velocity surface. Obviously, this is a very convoluted surface. Small sections of this wave surface for Si, corresponding to triangular sections on the slowness surface (near the cloverleaf), are drawn in Figure 25. The folds labeled A, B, and C correspond to the labels of the caustics in the phonon image of Figure 24(b). Additional details about this wave structure are discussed in Chapter 4.

In Chapters 4 and 5 the formalism developed in this chapter will be applied to specific crystal symmetries. We will examine in detail the topologies of the acoustic surfaces and show how they vary with changing elastic stiffness constants.

References

1. G. Nilsson and G. Nelin, *Phys. Rev. B* **6**, 3777 (1972); **3**, 364 (1971).
2. H. Bilz and W. Kress, *Phonon Dispersion Relations in Insulators*, Vol. 10 of Springer Series in Solid-State Sciences (Springer-Verlag, Berlin, 1979).
3. C. Kittel, *Introduction to Solid State Physics*, 3rd ed. (Wiley, New York, 1966), Chapter 4.
4. J. H. Staudte and B. D. Cook, Visualization of quasilongitudinal and quasitransverse elastic waves, *J. Acoust. Soc. Am.* **41**, 1547 (1967). See also Ref. 10, p. 231. See also Merkulov and Yakovlev, *Sov. Phys. Acoust.* **8**, 72 (1962); R. W. Green, *Ultrasonic Investigation of Mechanical Properties* (Academic, New York, 1973), pp. 36–37; G. W. Willard, Criteria for normal and abnormal ultrasonic light diffraction effects, *J. Acoust. Soc. Am.* **21**, 101 (1949); D. T. Pierce and R. L. Byer, Experiments on the interaction of light and sound for the advanced laboratory, *Am. J. Phys.* **41**, 314 (1973).
5. M. J. P. Musgrave, *Crystal Acoustics* (Holden-Day, San Francisco, 1970).
6. Cl. Schaefer and L. Bergmann, Über neue Beugungserscheinungen and Schwingenden Kristallen, *Naturwissenschaften* **41**, 685 (1934); L. Bergmann, *Ultrasonics and their Scientific and Technical Applications* (Wiley, New York, 1938).
7. B. Taylor, H. J. Maris, and C. Elbaum (Refs. 2 and 3 of Prologue).
8. H. J. Maris (Ref. 4 of Prologue).
9. G. A. Northrop and J. P. Wolfe, Ballistic phonon imaging in germanium, *Phys. Rev. B* **22**, 6196 (1980).
10. H. J. Maris, Effect of finite phonon wavelength on phonon focusing, *Phys. Rev. B* **28**, 7033 (1983).
11. B. A. Auld, *Acoustic Waves and Fields in Solids, Vol. I* (Wiley, New York, 1973).
12. A. E. H. Love, *Treatise on Mathematical Theory of Elasticity*, 4th ed. (Cambridge University Press, Cambridge, 1927).
13. F. I. Fedorov, *Theory of Elastic Waves in Crystals* (Plenum, New York, 1968).
14. K. D. Swartz and A. V. Granato, Experimental test of the Laval-Raman-Viswanathan theory of elasticity, *J. Acoust. Soc. Am.* **38**, 824 (1965).
15. J. F. Nye, *Physical Properties of Crystals* (Oxford University Press, London, 1957).
16. J. P. Wolfe, Ballistic heat pulses in crystals, *Phys. Today* **33**, 44 (December 1980).
17. The D_{ij} tensor is closely related to the dynamical matrix $\mathcal{D}_{ij}$ in the microscopic theory of lattice dynamics. As seen in Eq. (20) of Chapter 6, $\mathcal{D}_{ij}$ has the eigenvalues ω^2, so $\mathcal{D}_{ij} = k^2 D_{ij}$ in the continuum limit.
18. A. G. Every, General closed-form expressions for acoustic waves in elastically anisotropic solids, *Phys. Rev. B* **22**, 1746 (1980). See also Barnet, *J. Phys. F* **3**, 1083 (1973).
19. A. G. Every, Ballistic phonons and the shape of the ray surface in cubic crystals, *Phys. Rev. B* **24**, 3456 (1981).
20. D. C. Hurley and J. P. Wolfe, Phonon focusing in cubic crystals, *Phys. Rev. B* **32**, 2568 (1985).
21. H. J. Maris, Phonon focusing, in *Nonequilibrium Phonons in Nonmetallic Crystals*, W. Eisenmenger and A. A. Kaplyanskii, eds. (North-Holland, Amsterdam, 1986).
22. G. F. D. Duff, *Philos. Trans. Royal Soc. London* **252**, 249 (1960).

23. A. G. Every and A. K. McCurdy, Phonon focusing in piezoelectric crystals, *Phys. Rev. B* **36**, 1432 (1987).

24. M. Lax and V. Narayanamurti, Phonon magnification and the Gaussian curvature of the slowness surface in anisotropic media: detector shape effects with application to GaAs, *Phys. Rev. B* **22**, 1 (1980).

25. G. A. Northrop, Acoustic phonon anisotropy: phonon focusing, *Comp. Phys. Comm.* **28**, 103 (1982).

26. J. A. Shields and J. P. Wolfe, Measurement of a phonon hot spot in photoexcited Si, *Z. Phys. B* **75**, 11 (1989).

27. C. Elbaum, Anisotropic heat conduction in cubic crystals, *International Conference on Phonon Scattering in Solids*, Paris (1972).

3

Generation and detection of phonons – Experimental aspects

The physics of nonequilibrium phonons in crystals, as with most scientific phenomena, has been unveiled by a combination of experimental and theoretical work. We have seen that the propagation of acoustic phonons in elastically anisotropic media contains a rich spatial complexity. The advent of imaging techniques has permitted visual insights into elasticity theory and, as shown in later chapters, into the propagation and scattering of high-frequency phonons.

To study phonons one must be able to create and detect them. How does one jiggle the atoms in the crystal and detect their collective motion? Because the nuclei (or atomic cores) are charged, one possibility is the application of oscillating electric fields such as electromagnetic waves. Examples of this include ultrasound generation in piezoelectric materials and phonon generation by absorption of infrared light. Alternatively, because electronic motion produces time-varying local fields, an effective way of creating nonequilibrium phonons is to electrically or optically excite the electrons in the crystal. It comes as no surprise, then, that an efficient means of creating and detecting nonequilibrium phonons is to employ materials with high electron density – namely, metals. Conversely, the macroscopic motion of phonons is studied in materials with low densities of free carriers – namely, insulators and semiconductors. The common approach, therefore, is to attach metallic generators and detectors to the nonmetallic crystals of interest.

This chapter begins with an overview of the measurement techniques involved in a phonon-imaging experiment. The discussion then concentrates on the commonly used methods of phonon detection: superconducting bolometers and tunnel junctions. Finally, we consider the most common method by which phonons are generated in a heat-pulse experiment: electrical or optical heating of a metal film that has been deposited onto the sample surface. The physical mechanisms by which nonequilibrium phonons are *directly* generated in nonmetallic media (e.g., photoexcitation of semiconductors) are left for Chapter 10, after the fundamental processes of phonon scattering and downconversion have been treated.

A. Tools of the trade

How does one record the spatial pattern of ballistic heat flux emitted from a point source in a crystal? The usual experimental equipment includes (1) a cryostat for cooling the crystal; (2) a sensitive superconducting detector of small dimensions fabricated directly on the sample; (3) a laser or electron beam to produce the heat pulses, and associated optics to focus and scan the beam; (4) electronics to amplify and resolve the detected heat-pulse signals by their times of flight across the crystal; and (5) a computer to control the beam scanning and to store and display the phonon intensities as a two-dimensional array. The digitally recorded image [e.g., Figure 24(b) on page 56] is much more than a pretty picture; it contains a large array of intensities that may be processed and analyzed in a variety of ways.

Many of the experiments described in this book are conducted with the crystal directly immersed in a helium bath. The bath is usually cooled below the superfluid λ temperature of helium, $T = 2.17$ K, in order to provide temperature stability for the detector and to eliminate helium bubbles in the optical path. (The cooling is accomplished by vacuum pumping the liquid to a pressure of a few Torr.) One consequence of this experimental arrangement is that the superfluid helium bath provides a good heat sink for the crystal (see Chapter 10). Indeed, a significant portion of the phonons generated in the heat source by surface excitation will be emitted into the helium bath.

A schematic diagram of a phonon-imaging experiment is given in Figure 1. In this scheme, a pulsed laser beam is used to produce the phonon source. (An alternative technique utilizes electron beams, as mentioned in Chapter 1.) Depending on the energy desired per pulse and the required pulse length, various modulation methods may be employed, as indicated in Table 1. The optical arrangement in Figure 1(a) represents one of several possible configurations for focusing the laser beam. Lens L1 provides a sharp focal spot that is refocused approximately 1:1 onto the crystal with lens L2. The purpose of this transfer lens is to provide sufficient working distance to insert deflection mirrors into the optical path while retaining the small focal spot of lens L1. (In the diffraction limit, the spot size scales in proportion to the focal length of lens L1.) The mirrors are attached to galvanometer motors, which produce a rotation of up to $\pm 30°$ proportional to an applied current. Typical components are listed in Appendix A of Reference 1.

This optical arrangement can easily provide a spot size of approximately 30 μm, scannable over approximately 2×2 cm^2 in the crystal plane. For

Table 1. *Characteristics of typical laser sources used for phonon imaging experiments*

Type of laser	Type of modulation	Typical pulse length	Maximum pulse energy of power	Pulse repetition rate
Ar ion	Cavity dumped	10 ns	100 nJ	0–100 kHz
Ar ion	External acousto-optic	20 ns to cw	1 W	0–100 kHz
Nd:YAG (cw pump)	Q-switched	200 ns	100 μJ	0–1 kHz
HeNe	Mechanical chopper	1 ms	10 mW	0–1 kHz

cw = continuous wave.

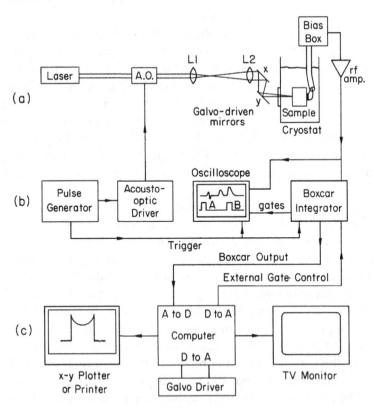

Figure 1 *Schematic diagram of a phonon-imaging apparatus. A.O. = acousto-optic modulator, A to D = analog-to-digital. In some experiments, the boxcar integrator may be replaced with a transient recorder, digital oscilloscope, or signal averaging board for the computer.*

higher resolution, lens L2 may be replaced by a standard macro camera lens, positioned between the scanning mirrors and the sample and preferably mounted directly on the cryostat to minimize the effect of vibrations. In this configuration, a 10-μm spot scannable over approximately 5×5 mm^2 is possible. The use of a microscope objective in the cryostat can improve the spatial resolution further.[1] In all cases, the angular positions of the mirrors are controlled by the computer, which can be programmed to produce single-line scans or an x–y raster pattern.

The detected heat-pulse signals are amplified and fed into a boxcar integrator, shown in Figure 1(b). This useful device provides an output voltage that is proportional to the average phonon signal during a selected time window, t to $t + \Delta t$, after the laser pulse at $t = 0$. To accomplish this feat, the boxcar receives a trigger pulse at $t = 0$, samples the phonon signal during a time gate Δt that begins at a preset delay time t, and averages this signal over many pulse cycles. Alternatively, a digital oscilloscope or transient recorder may be used in place of the boxcar. These multichannel instruments can average and store the entire heat pulse for a given propagation direction (laser-spot position), but recording this information for a large array of propagation directions requires massive data storage.

In recording an image, the laser is pulsed at a high repetition rate and the laser spot is moved relatively slowly across the crystal surface, so the boxcar output follows the time-selected phonon flux as the propagation angle is scanned. The image represents a submicrosecond snapshot in time, but it is acquired by sampling over millions of repetitive laser pulses. An acceptable image may be collected in perhaps 15 s, but higher signal-to-noise ratios are attained by increasing the output time constant of the boxcar and scanning more slowly (or by digitally averaging many scans). As indicated in Figure 1(c), the computer records the boxcar output for an array of laser-beam positions. Line scans can be displayed on an x–y plotter (or printer), and two-dimensional arrays may be displayed as gray-scale images on a TV monitor (by means of a video graphics board) or as pseudo–three-dimensional graphs on the x–y plotter.

Increasing the width of the time gate improves the signal collection efficiency but decreases the time resolution. Often a wide gate width is chosen in order to encompass the arrival time of all, or a subset of, the phonon modes (L, FT, ST) and, hence, collect a time-integrated phonon image. If, on the other hand, a narrow fixed gate is chosen, there will be a small range of propagation distances, centered on $R = t_g/V$, where ballistic signals are detected. In this formula, t_g is the selected delay between the peak of the laser pulse and the center of the time gate, and V is a representative phonon velocity. If V were isotropic, the image would consist of a narrow circle of high intensity, corresponding to the intersection of a spherical shell of radius R with the crystal surface opposite the phonon point source. As we know, the velocity is not isotropic [see Figure 9(b) on page 33], so a more complex pattern is obtained.

Figure 2 is a series of time-resolved images of Si taken with increasing sampling times, t_g, and a gate width of only 50 ns. The detector and metal-heater film are on opposing $\langle 110 \rangle$ faces for this $\simeq$3-mm-thick crystal, and a laser pulse of approximately 15 ns is used. For increasing t_g the detected phonons of a given mode appear at increasing distances from the excitation point.[2] First to appear are the longitudinal phonons, which are most strongly focused near $\langle 111 \rangle$ directions, and next are the FT phonons at the center of the image, [110]. At longer delay times, the ST mode appears. If a broad gate is chosen, phonon signals for all the modes and propagation distances are sampled, and the time-averaged image shown in Figure 2(f) is obtained.

An alternative technique is to select phonon velocity rather than arrival time, which depends on the particular sample geometry. This is accomplished by continuously adjusting the gate delay time to be proportional to the distance between source and detector. As illustrated in Figure 3, for a given x–y position and selected (average) velocity, V, the ballistic arrival time, $t_b = R/V$, is determined by the computer and a voltage proportional to t_b is supplied to the boxcar to continuously control the gate delay such that $t_g = t_b$.

Examples of velocity-selected images of Si are shown in Figure 3. In contrast to Figure 2, the entire pattern of FT ridges appears in one image, and, likewise, the ST structures are more complete. The ST box near $\langle 100 \rangle$ has nearly the same velocity as the FT ridges in the {100} planes. The image with the smallest selected velocity picks out the ST ramps that converge near the $\langle 111 \rangle$ directions, where the group velocities are the slowest.

For a given source position, the boxcar gate delay may be slowly scanned from zero to any desirable time and the boxcar output plotted on an x–y recorder. To optimize the time resolution, a narrow gate width is chosen, and the result is a time trace of the ballistic heat pulses, like one of the many plotted in this book. A digital oscilloscope with nanosecond sampling times (or a transient recorder) is an alternative way of recording time traces.*

Recent advances in multichannel data acquisition and desktop computer speed have permitted the experimenter to record a series of time-resolved images during a single x–y raster scan. For example, at each laser position a time trace may be acquired with a high-speed digital oscilloscope and stored in the computer. An array of these time traces forms an x–y–t data cube (see also Chapter 13). The data can be played out as a time trace for a selected

---

* A boxcar integrator can take advantage of higher pulse-repetition rates than a multichannel recorder, so if the experiment allows for high repetition rates, the overall data collection efficiency of the single and multichannel instruments may be comparable. As data-handling speeds improve, the multichannel devices will undoubtedly have the advantage.

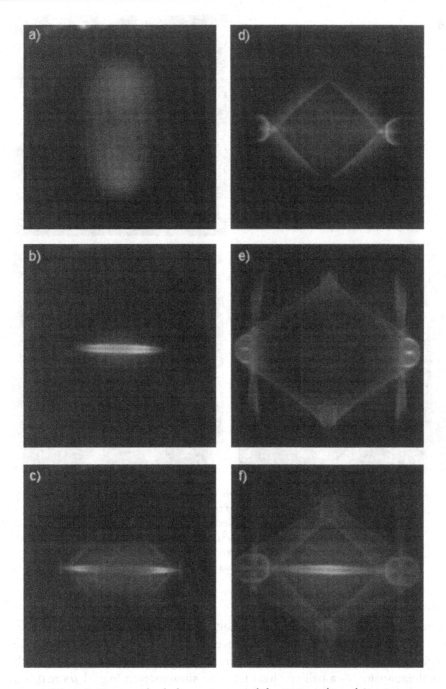

Figure 2 Time-resolved phonon images of the (110) surface of Si using a gate width of 50 ns and fixed gates at the times of (a) 350 ns, (b) 500 ns, (c) 550 ns, (d) 600 ns, and (e) 700 ns. Image (f) is taken with a broad gate (550 ns) to include all modes and arrival times. The crystal thickness is about 3 mm, and T = 1.8 K. (Data from Shields.[2])

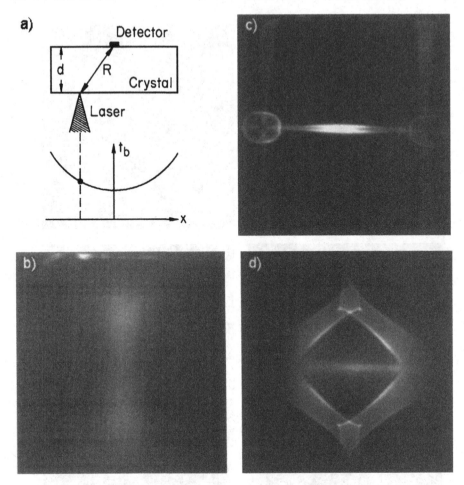

Figure 3 (a) Principle behind constant-velocity scanning. As the laser is moved to position (x, y), the center of the boxcar time gate, t_g, is adjusted to correspond to the ballistic propagation time, t_b, assuming a velocity V. That is, $t_g = t_b = R/V = (d^2 + x^2 + y^2)^{1/2}/V$. Velocity-selected images of (110) Si with a gate width of 30 ns correspond to the velocities of (b) the L mode near [111], (c) the FT and ST modes near [100], and (d) the ST mode near [111]. (Data from Shields.[2])

propagation direction or as a phonon image for a selected time interval (e.g., for constant time or constant velocity). Once recorded, the time-resolved images may be played back successively as a "movie" that enables one to observe the spatial expansion of a ballistic heat pulse in slow motion (e.g., 1 μs real time corresponds to 30 s viewing time).

Several frames of such a movie, produced by Hauser from a 2-hour experimental run are shown in Figure 4(a)–(d). The scanning surface of the 2-mm-thick Si crystal is (100). First one sees the nearly isotropic expansion

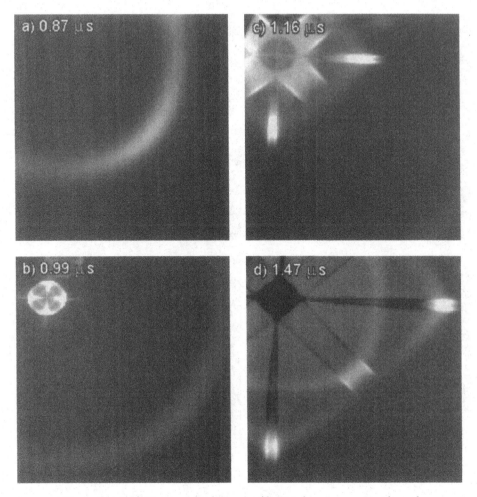

Figure 4 (a)–(d) Time-resolved images of heat pulses propagating through a 5-mm-thick Si crystal and detected on a (100) face. T = 1.8 K. The cavity-dumped Ar⁺ laser pulse width is 15 ns, and the peak absorbed power in the evaporated Cu film is about 1 W. A granular Al bolometer with $2 \times 5 \ \mu m^2$ dimensions is used. The x–y–t data cube has pixel dimensions $128 \times 128 \times 256$. The dark remnants of the intense caustics are an experimental artifact (see also Nb images in Chapter 15). (Data from Wolfe and Hauser.[54]) (e)–(h) The ST box at successive time intervals obtained by boxcar acquisition with 100-ns gate width for a 1-cm-thick Si crystal: (e) 3.4 μs, (f) 3.45 μs, (g) 3.55 μs, (h) 3.75 μs. (Data from Shields.[2])

of the L mode. Then, the ST and FT modes appear and propagate with a wavefront given by the shape of their respective wave surfaces, similar to that shown in Figure 8(b) on page 31 for the ST mode.

A high-resolution view of the time evolution of the ST box is displayed in Figure 4(e)–(h). These data were taken by Shields with a 1-cm-thick crystal

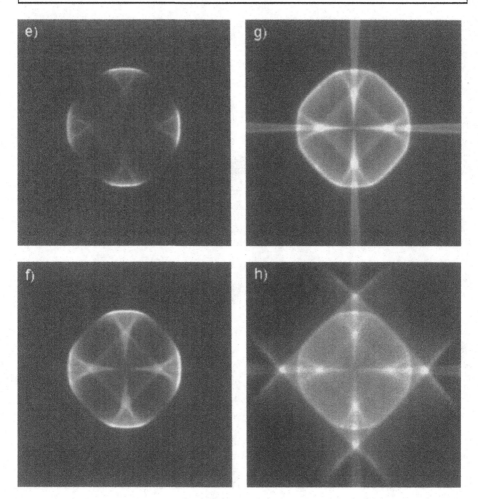

Figure 4 (cont.)

and standard boxcar acquisition of single images. The corners of the ST box appear earliest, indicating that these phonons are faster than any FT phonon.* As the box fills in, the FT mode appears. Finally, the ST ramps are launched from the ST box towards the ⟨111⟩ directions.

A global picture of the phonon focusing pattern in a crystal can be obtained by tilting the crystal in the cryostat, such that three sides can be excited within a single x–y raster scan.[3] The scan range is chosen to extend beyond the boundaries of the crystal, so that a three-dimensional perspective is obtained. Figure 5(a) is the phonon image of a cube of Ge, with the detector centered on the back left face (001). Figure 5(b) is the same crystal with the detector

---

* As pointed out in Chapter 2, for a given k-space direction, the ST phase velocity is always slower than or equal to the FT phase velocity. That is, the ST slowness surface lies entirely outside the FT slowness surface, justifying the slow and fast designations.

Figure 5 Wide-angle phonon images of Ge obtained by a single raster scan over an obliquely oriented crystal. A pulsed Nd:YAG laser directly excites the 1-cm cube immersed in liquid helium at 2 K. In (a) the detecting bolometer is located at the center of the back left face, (001). In (b) the detecting bolometer is located at the center of the back right face, (110). Bright lines are phonon-focusing caustics. (From Northrop and Wolfe.[3])

centered on the back right face (110). The observed caustic patterns may be readily identified with the folds in the group-velocity surfaces shown in Figures 9 and 10 on pages 33 and 34.

B. Phonon detection

With few exceptions (one discussed below), heat-pulse experiments utilize single-element detectors. This is because detectors with sufficient sensitivity, rapid response time, small size, and desired frequency range are difficult enough to fabricate one at a time, much less as arrays of identical devices. Furthermore, a single-element detector is adaptable to submicrosecond time resolution and requires only one set of detection electronics. As indicated in Chapter 1, generally a superconducting detector is deposited onto the crystal at a fixed position, and the position of the phonon source is scanned in one or two dimensions.

For imaging purposes, detectors with dimensions of 10 to 30 μm are commonly used, although devices as small as 2 μm and as large as 250 μm have been employed. A small size is required for angular resolution of phonon-focusing structures. For example, the phonon image of Si shown in Figure 24(b) on page 56 was taken with a 10×10 μm^2 detector, a 15-μm laser spot size, and a phonon path length of approximately 2 cm. Thus, the angular resolution is roughly $(20\,\mu\text{m})/(2\,\text{cm}) = 10^{-3}$ radians $= 0.06°$. If one assumes a relatively high instantaneous power of 1 W emitted from the heated spot, the average intensity at the detector (ignoring phonon focusing and losses to the bath) is approximately $1\,\text{W}/[4\pi(2\,\text{cm})^2] = 2 \times 10^{-2}$ W/cm^2, which implies that the detector receives only $(2 \times 10^{-2}\,\text{W/cm}^2) \times (10\,\mu\text{m})^2 = 2 \times 10^{-8}$ W during the pulse. If the pulse length is 10 ns, then the average energy absorbed by the detector is approximately $(2 \times 10^{-8}\,\text{W}) \times (10\,\text{ns}) = 2 \times 10^{-16}$ J $= 0.2$ fJ per pulse! These are rather stringent size and sensitivity requirements on the phonon detector. Fortunately, the phonon signals can be enhanced by pulsing at a high repetition rate (say, 100 kHz) and time averaging the signals with a boxcar integrator or multichannel analyzer.

In addition to small size and high sensitivity, thin-film superconducting detectors (bolometers and tunnel junctions) have extremely small heat capacities, which implies very short thermal relaxation times when they are in good thermal contact with a cold substrate or bath. The time response of a detector ultimately depends on the relaxation times of excited electrons (quasiparticles) in the superconductor, which may be nanoseconds to microseconds.[4,5] Typically, in a heat-pulse experiment one requires time responses of the detector plus associated electronics in the range of 50 ns or less, which is achievable with these films. The obvious drawback of the devices is that they operate only at low temperatures. Of course, ballistic-phonon propagation

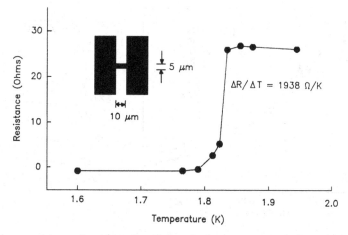

Figure 6 *Resistivity of a granular Al bolometer used for phonon imaging. The bolometer is operated at T = 1.83 K. Inset: the sensitive area of the detector is the 5 × 10 μm² bridge between the two large contact pads. Electrical connections are made using indium pressed onto the 1 × 2 mm² pads. (Data from Msall.[6])*

over macroscopic distances occurs only at low temperatures. So, the tool nicely fits the need.

i) Superconducting bolometers

The simplest superconducting phonon detector is a strip of evaporated metal cooled precisely to its superconducting transition temperature, as shown for granular Al in Figure 6.[6] This device requires only a single evaporation process, but its operation demands accurate stabilization of the crystal temperature. This is generally accomplished by immersing the sample in a liquid-helium bath and regulating the vapor pressure. Under such conditions, the slight rise in temperature caused by a heat pulse produces a significant change in resistivity. Typically, a constant bias current is applied and the change in the voltage drop across the device is detected. The detector characteristic shown in Figure 6 is obtained by evaporating Al in a partial pressure of oxygen, to produce a granular film that has a convenient transition temperature above that of pure Al. This type of detector was used in the first heat-pulse experiments by von Gutfeld and Nethercot[7] and has been employed in many such experiments since that time.

Tiny (2 × 2 μm²) bolometer detectors have been fabricated by photolithography,[8] but practical applications have generally employed devices 10 × 10 μm² or larger. Until now, the phonon sources have proven to be larger than the smallest detectors. Attempts have been made to produce micrometer-sized phonon sources by sharply focused electron beams in an electron microscope,[9] but it turns out that in practice the phonon source size

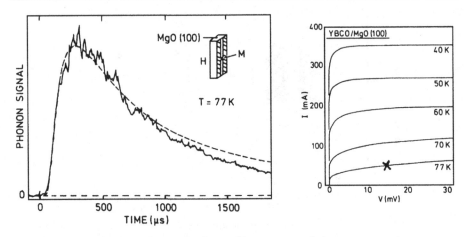

Figure 7 Diffusive heat pulse detected by a YBaCuO bolometer on MgO at 77 K. The detector is a 0.5 × 0.1 mm microbridge (M) made by laser ablation of a 180-nm YBaCuO film on a 1-mm-thick (100) MgO substrate. The I–V characteristic is typical of flux-flow resistivity. (From Berberich.[13])

(due to the electron thermalization length) for a typical 26-keV electron beam is in the range of 25–60 μm (depending on modulation frequency).[8] On the other hand, heat sources produced by optical excitation can have a size less than 20 μm,[10] and smaller sources are likely to be achievable by tighter focusing of the laser beam. So the ability to make bolometers of approximately 1-μm dimension may prove useful in future experiments.

Bolometers composed of type II superconductors, such as NbN[11] and high-T_c ceramics,[12] have also been used for phonon detection, but they operate in a mode somewhat different from that described above. Typically, the temperature of the sample is adjusted well below the superconducting transition temperature of the film. When a sufficient bias current is applied, the device exhibits a finite resistivity, thought to be due to flux creep. Because of the high transition temperatures of YBaCuO (T_c = 90 K), this material has proven to be particularly useful for temperature-dependent studies of heat pulses. Berberich[13] observed the highly diffusive propagation of heat in MgO at 77 K, as shown in Figure 7. Also plotted is the I-V characteristic of his YBaCuO bolometer.

The shape of the heat pulse in Figure 7 is in fairly good agreement with that predicted by the diffusion constant, $D = \kappa/C$, where κ and C are the measured thermal conductivity and specific heat, respectively, as indicated by the dashed line in the figure. In another study, Obry et al.[14] have observed the transition from ballistic to diffusive heat flow in a Si crystal as the temperature is raised from 18 K to 32 K. Again, fairly good agreement is obtained with the scattering times derived from thermal conductivity data. This evolution from ballistic to diffusive transport regimes was qualitatively observed in sapphire and NaF by von Gutfeld,[15] who used a normal-metal In film, which also possesses a temperature-dependent resistivity.

The resistance of a bolometer above its superconducting transition can be controlled by the width, length, and thickness of the film, as well as deposition conditions (such as oxygen pressure in the case of granular Al). Generally, a total resistance of 50–100 Ω at the superconducting transition is desirable, in order to match the impedance of standard electrical coax and the signal preamplifier. For the device shown in Figure 6, the detector's sensitive area is the small bridge ($5 \times 10 \ \mu m^2$) between two larger contact pads. To produce the film, the Al is sputter deposited on a photolithographically patterned sample. Alternatively, larger detectors can be deposited through a metallic mask in physical contact with the sample. In both cases the Al deposition is performed in a mixed argon–oxygen atmosphere in order to control the oxygen content of the films and hence the transition temperature. For phonon detection, the crystal temperature is adjusted just below the superconducting transition, and a bias current is used to raise the weak link (the bridge that carries the highest current density) to the transition temperature, T_c. In fact, crystal temperatures considerably below T_c can be studied by using larger bias currents to raise the local temperature of the bolometer to T_c.

The superconducting bolometer is a broadband phonon detector. Phonons that are absorbed in the metal film act to raise its temperature and, therefore, change its resistance. The mean free path of phonons in the metal, however, is larger for low frequencies, so phonons of sufficiently long wavelength are not absorbed by the film. (They may be directly transmitted into the helium bath or reflected back into the crystal.) This low-frequency cutoff in detector sensitivity generally occurs well below 100 GHz; it depends on the particular metal and the film thickness. Also, the sensitivity and frequency response of the bolometer (as well as of other thin-film detectors) depend on the quality of the interface between the sample and the film, although no systematic studies of this have been made. It is well known that the thermal boundary resistance of ordinary crystal-metal interfaces depends on temperature and, thus, on phonon frequency.

ii) Superconducting tunnel junctions

A more sophisticated phonon detector is the superconducting tunnel junction. According to the celebrated theory of Bardeen, Cooper, and Schrieffer (BCS), as a metal is cooled below some transition temperature, T_c, an energy gap, 2Δ, opens up at the Fermi level, producing superconductivity. This phenomenon is illustrated schematically on the left side of Figure 8. In ordinary metals, the gap comes about because the phonons in the metal produce a slight attraction between two electrons of opposite wavevector and spin, forming Cooper pairs. At zero temperature, all the electrons are correlated into pairs, but, as the temperature is raised, the pairs are thermally dissociated and the gap decreases.

Table 2. *Typical values of superconducting gap energy and frequency near $T = 0$*

	$2\Delta_o$ (meV)	v (GHz)	Ref.
Al	0.3	73	18, 19
In	1.1	260	17
Sn	1.2, 1.3	290, 320	17, 19
Pb	2.8	680	17
$Pb_{0.7}Bi_{0.3}$	3.5	860	17
Nb	3.0	730	17

Note: 1 meV = 242 GHz. Values are rounded to two significant figures.

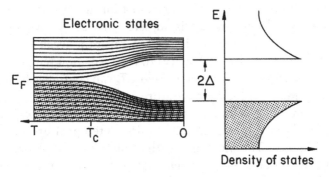

Figure 8 Schematic drawing of the opening of a superconducting gap at the Fermi level of a metal as the temperature is lowered below a critical temperature T_c. The shaded region indicates filled electron states. At the right is a sketch of the resulting density of states at low temperature.

According to the elementary BCS theory,[16] at $T \simeq T_c$,

$$\Delta(T) = 1.74\Delta_o(1 - T/T_c)^{1/2}, \tag{1}$$

where the energy gap at zero temperature is given by $2\Delta_o$. The zero-temperature energy gaps and critical temperatures for several superconductors are given in Table 2.[17-19]

It is convenient to describe this situation in the so-called semiconductor picture, illustrated on the right side of Figure 8. (See the reviews by Eisenmenger[18] and Kinder[19] for more complete descriptions.) Plotted is the density of electronic states versus energy, at low temperature. Near the gap, the density of states displays integrable singularities of the form $(E - E_o)^{1/2}$. At $T = 0$, the lower band is completely filled (i.e., all electrons are paired). As the temperature is raised, thermal phonons with energy hv greater than 2Δ become more abundant, and they can break up the Cooper pairs to produce quasiparticles – that is, unpaired electrons. The equilibrium density of quasiparticles

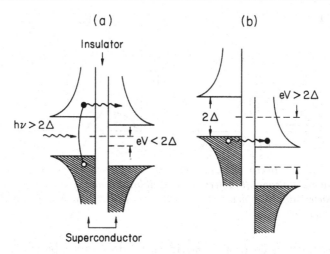

(a)

(b)

Insulator

$h\nu > 2\Delta$

$eV < 2\Delta$

2Δ

$eV > 2\Delta$

Superconductor

Figure 9 *Behavior of a superconductor/insulator/superconductor junction as the applied voltage across the junction is increased. (a) When $V < 2\Delta/e$, phonons with energy $h\nu$ greater than 2Δ can excite quasiparticles across the gap, which can tunnel through the barrier and cause current flow. (b) When $V > 2\Delta/e$, quasiparticles can tunnel directly across the barrier without phonon assistance, producing a sharp rise in the tunneling current, as shown in Figure 10.*

in the superconductor increases exponentially with temperature,

$$n_Q \propto \exp(-\Delta/k_B T), \qquad (2)$$

due to the exponentially decreasing number of phonons with energies above the superconducting gap.

Cooper pairs can also be broken up by nonequilibrium phonons with energies above 2Δ. To utilize this effect as a detector of nonequilibrium phonons, one must devise a way to collect the excited quasiparticles. Recall that semiconductor detectors collect photoexcited electrons and holes that migrate across a p–n junction. In the phonon case, a two-terminal device is created by depositing two superconducting films, with an insulating barrier between them, and then applying a voltage V between the films. At $T = 0$ and $V < 2\Delta/e$, no current flows through the insulating barrier. At nonzero temperature, or in the presence of a pair-breaking phonon flux, quasiparticles are created that can tunnel across the barrier and produce a current, as indicated in Figure 9(a). This is the basis for the superconducting tunnel-junction detector, first demonstrated by Eisenmenger and Dayem.[20] Details of its operation can be found in the review by Eisenmenger.[18] (Other uses of superconducting tunnel junctions include microwave mixers, ultrasensitive magnetometers, fast switches, voltage standards, and particle or photon detectors.)

If the voltage across the junction is raised above $2\Delta/e$, as shown in Figure 9(b), it is energetically possible for an electron to tunnel from the

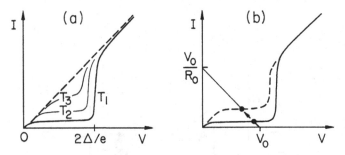

Figure 10 (a) Effect of temperature on the characteristics of a superconducting tunnel junction. (b) Effect of nonequilibrium phonons on the junction characteristics. The detected signal (arrow) follows the load line of the external circuit.

filled band on one side of the barrier into the empty band on the other side, creating a quasiparticle on each side of the barrier. This effect produces a sharp rise in junction current and the onset of ohmic behavior, as indicated schematically in Figure 10(a). At applied voltages less than $2\Delta/e$, the current is approximately constant, given by the equilibrium quasiparticle population [Eq. (2)], the probability of quasiparticle tunneling across the barrier, and the area of the junction. The I–V characteristic becomes more linear (ohmic) as the temperature is raised, which increases n_Q and lowers Δ, [Eq. (1)].

A nonequilibrium flux of phonons with $h\nu > 2\Delta$ changes the I–V characteristic of a tunnel junction, as shown schematically in Figure 10(b). If the tunnel junction is connected in series with a voltage source V_0 and load resistance R_0, the change in current and voltage across the junction due to the phonons is indicated by the heavy arrow. (Note that the phonon-induced change in voltage across the junction is opposite in sign to that of a bolometer.) The impedance of a tunnel junction with a square-millimeter cross section is typically only a fraction of an ohm, requiring special impedance-matching techniques to derive the full signal out of the junction. Smaller junctions have the advantage of higher impedance but are generally more difficult to fabricate and easier to destroy with an inadvertent stray current pulse.

a) FABRICATION OF TUNNEL JUNCTIONS FOR PHONON IMAGING To obtain high angular resolution, it is desirable to fabricate tunnel junctions with 10–30 μm dimensions. Photolithographic techniques can easily achieve this and smaller scales; however, the chemistry necessary for microfabrication of small junctions has only been developed for a few systems. To maintain a flexibility in the choice of superconductors, the simple procedure of depositing the metal layers through masks that are cut with the desired patterns has proven to be quite effective. Generally the complete device can be made with only one or two pump-down cycles of an evaporation

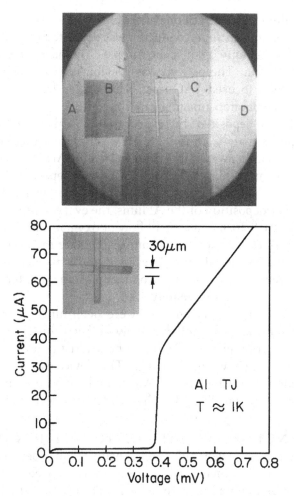

Figure 11 (above) Microscope photograph of an Al/oxide/Al tunnel junction fabricated by a Kinder/Dietsche mask changer fitted with Northrop
masks. Electrical leads are attached by In pads (not shown) pressed onto
the Cu films A and D. The sensitive junction area, shown in the expanded
view (below), is 30 × 30 μm². Plotted (below) is the I–V characteristic of
this junction, which was used in the experiments of Ref. 21.

chamber, whereas the photolithography process usually requires several cycles of photoresist application, metal evaporation, and chemical liftoff. The
critical step in all cases is the growth of a uniform, thin insulating layer that
acts as the tunneling barrier between the two superconducting films.

An example of an Al tunnel-junction detector used in a phonon imaging
experiment[21] is shown in Figure 11. This device required the use of a mask
changer with four different masks mounted on paddles that were sequentially
positioned against the sample (a Ge crystal) by magnetic solenoids. Film A is

2000 Å of Cu film used to attach one of the electrical leads. Film B, 1000 Å of Al that forms the bottom superconductor, was deposited through a stainless-steel mask pressed snugly against the sample. The narrowest linewidths in this mask are 30 μm. The pattern was previously cut into the chemically thinned metal by a focused YAG laser beam driven by the same scanning system used in the phonon-imaging experiments.

After film B was deposited, the evaporation chamber was filled with air at 1 atm for approximately a minute to grow the aluminum oxide insulating layer. Next, the chamber was pumped out and the top Al film (C) was deposited through a similar mask. Finally, a Cu film (D) was deposited for attachment of electrical leads. (This procedure avoided an oxide layer between the Cu and Al films.) For deposition of the Al films, the evaporator was pumped to a pressure of 2×10^{-6} Torr, and a small flow of oxygen was directed at the Al evaporation boats, causing a rise in chamber pressure to 4×10^{-6} Torr. The purpose of adding oxygen during Al evaporation is to increase granularity of the film, thereby raising the superconducting transition temperature to a convenient point (here approximately 1.2 K).

The detailed conditions given here were determined empirically for a particular apparatus and are provided to give a flavor for the important variables in the fabrication process. The I–V characteristics for a junction fabricated in this way are shown in the figure. This device was used in the experiments in Section A of Chapter 12. An example of a photolithographically produced tunnel junction (PbIn) for phonon detection is found in Ref. 22.

b) **FREQUENCY SELECTIVITY OF TUNNEL JUNCTIONS** Tunnel junction detectors are generally more sensitive than superconducting bolometers; however, their principal utility arises from their frequency selectivity.* They are sensitive to phonons with $h\nu > 2\Delta$ (1 meV = 242 GHz). This onset, or threshold, frequency can be used to study the frequency dependence of ballistic phonon propagation and scattering. Frequently, the crystal itself acts as a low-pass filter because the scattering rate of phonons from defects depends strongly on frequency ($\tau^{-1} \propto \nu^4$; see Chapters 7–9). Combined with the detector's onset frequency, this effect can produce a quasi-monochromatic distribution of detected phonons, as noted by Dietsche et al.[23] in studying the dispersive propagation of phonons in Ge.

A systematic tuning of the onset frequency can be achieved by selecting metal alloys of different concentrations for the tunnel-junction detectors. Figure 12(a) shows a series of I–V characteristics for junctions fabricated with PbTl and PbBi alloys of various concentrations. The derivatives of these curves, plotted in Figure 12(b), were used to pinpoint the superconducting gap, and

---

* In addition to their utility as frequency-sensitive phonon *detectors*, tunnel junctions have been developed as quasi-monochromatic phonon *generators*, as described by Kinder.[19]

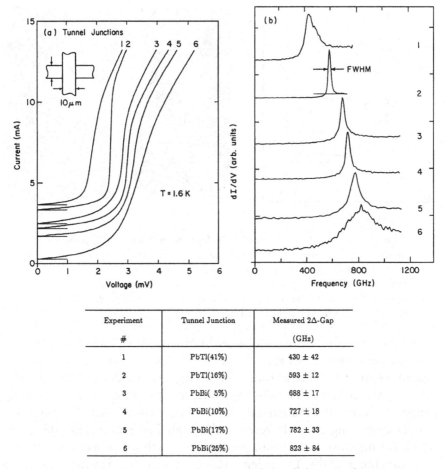

Experiment	Tunnel Junction	Measured 2Δ-Gap
#		(GHz)
1	PbTl(41%)	430 ± 42
2	PbTl(16%)	593 ± 12
3	PbBi(5%)	688 ± 17
4	PbBi(10%)	727 ± 18
5	PbBi(17%)	782 ± 33
6	PbBi(25%)	823 ± 84

Figure 12 (a) Characteristics of a series of PbTl and PbBi tunnel junctions used in the study of dispersive phonons in InSb. The table gives the percentage of alloying metal added to the Pb and the resulting onset frequency of the junction. The onset frequencies and their uncertainties were obtained from the derivative curves shown in (b). Experiments by Hebboul and Wolfe,[24] are described in Chapter 7.

thus the onset frequency, for a particular device. The accompanying table shows the results.*

c) **SCANNABLE TUNNEL-JUNCTION DETECTOR** So far we have only considered the use of tunnel junctions as fixed single-element detectors. Phonon images are produced by scanning the heat source. One can imagine instances where it would be useful to have a spatially scannable *detector* – for

---

* These detectors were produced with masks similar to those used for Figure 11, except the width of the junction lines were only 10 μm. This series of tunnel junctions was employed by Hebboul and Wolfe[24] to study the effects of frequency dispersion on phonon focusing in InSb, as described in Chapter 7.

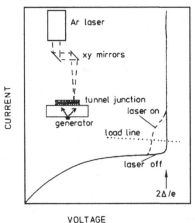
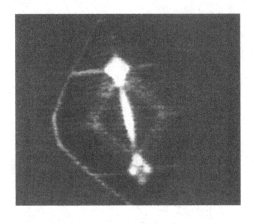

Figure 13 *Idea behind the spatially resolving tunnel-junction detector introduced by Schreyer et al.*[25] *A small portion of the large-area junction is heated by a focused cw laser beam, causing a precursor in the I–V characteristic. Along the dashed portion of the characteristic, the tunneling current flows mainly in the photoexcited region. The image obtained with this device is a focusing pattern of Ge.*

example, to study the phonon emission from a fixed phonon source, such as a semiconductor device. Schreyer et al.[25] have demonstrated a novel scheme for gaining spatial resolution from a single large Al tunnel junction. The idea is to scan a focused laser beam across the surface of the junction, locally heating the superconductor and changing the I–V characteristic. Figure 13 shows that the superconducting gap is lowered in the heated region, making this point on the junction more sensitive to phonon flux, with the proper current bias. The phonon image of the (110) oriented Ge crystal, obtained by raster scanning the laser beam across the tunnel junction, is shown. This scheme has been used to study the anisotropic flux emitted by the two-dimensional electron gas in a field-effect transistor on Si.[26]

Before closing this section, I mention several related facts. Extremely high sensitivities have been achieved by cooling Al tunnel junctions to milli-Kelvin temperatures. These devices are capable of detecting single particles or photons with energies as little as 100 eV (or even less) and are being developed for x-ray and particle detectors (see Chapter 15). Also, superconducting tunnel junctions are used as tunable phonon *generators* to achieve a form of phonon spectroscopy – a technique reviewed by Eisenmenger,[18] Kinder,[19] Bron,[27] and Wybourne and Wigmore.[28]

C. Phonon generation

There are myriad ways to generate nonequilibrium phonons in crystals. Any electromagnetic excitation that is absorbed by the crystal may directly or indirectly generate phonons. Far infrared photons, for example, can be directly converted into phonons via absorption in the crystal's vibrational bands. The absorption of visible, UV, and x-ray photons, however, is usually associated with *electronic* excitations (sometimes with the simultaneous creation of phonons). The relaxation of the excited electrons back to their ground states occurs by phonon emission (which tends to raise the crystal temperature) and/or by photon emission (such as luminescence).

Beams of electrons or other particles are also used to generate phonons in a crystal. Focused electron beams were used to produce the phonon image in Figure 6 on page 19 and in a number of other experiments.[8,9,22,29] One possible advantage of the electron-beam technique is that electrons are absorbed near the surface of any crystal, whereas transparent crystals must be covered with an optically absorbing film when optical excitation is used. Another electron-injection method of producing high-frequency phonons utilizes metallic point contacts.[30]

High-energy particles that are absorbed or scattered in a crystal also produce electronic and vibrational excitations. Particles of sufficient energy are able to dislocate atoms in the lattice, creating copious amounts of vibrational energy. It has been suggested that the missing dark matter in the universe – perhaps in the form of weakly interacting massive particles (WIMPs) – may preferentially generate vibrational, rather than electronic, excitations.[32] For this reason, some physicists say, conventional electronic detectors will not be effective in detecting such particles. Phonon-based detectors[31] may be the answer (see also Chapter 15).

In all these excitation processes a particularly important issue arises: What is the frequency distribution of the nonequilibrium phonons that are generated? Because phonon generation depends on the particular electron–phonon interactions, and because high-frequency phonons can anharmonically decay into lower frequency phonons, there is no generic answer to this question. The answer certainly depends on the duration and type of excitation, the time and position of observation, and, of course, the type of material being excited. In Chapter 10 we attempt to answer this question for photoexcitation of a semiconductor. A number of complex processes are involved, including frequency downconversion, elastic scattering, quasidiffusion, interaction of phonons with photoexcited carriers, and potentially the formation of a hot spot.

i) Phonon emission by a heated metal overlayer

The most common means of injecting nonequilibrium phonons into a non-metallic crystal is to heat a metal film that has been deposited onto its surface. The metal film may be ohmically heated with an electrical current or, in the case of a phonon-imaging experiment, optically heated with a laser beam. The excited electrons relax by phonon emission, and, if the characteristic electron–phonon interaction time is shorter than the time for a typical phonon to escape from the film, the electrons and phonons approach a common effective temperature. According to the acoustic-mismatch model, this steady-state heater temperature depends on the balance between the excitation power and the transmission of phonons through the interfaces.[33,34]

The potential problem with this simple model is that the electron–phonon coupling, the mean free path of a phonon in the metal film, and the transmission probability for crossing the interface all depend upon the phonon frequency. Maris[35] considered this problem and attempted to quantify the range of validity of the Planck-radiator model in terms of physical parameters of the film and the interface. The results depend on the metal in question, the film thickness, impurities, and the probability for transmission through the interface.

Perrin and Budd[36] pointed out that low-frequency phonons have long scattering times and are unlikely to attain good thermal contact with the heated electronic gas. Using phonon spectroscopy, Berberich and Kinder[37] found experimental evidence for this effect in granular Al films at heater temperatures less than 2 K. For higher frequencies, Bron and Grill[38] discovered significant deviations from blackbody radiation in a SrF$_2$ crystal when a constantan film was ohmically heated above approximately 10 K. A subsequent spectroscopic study of constantan/CaF$_2$ and SrF$_2$ systems by Eisenfeld and Renk,[39] however, showed reasonable agreement with the blackbody theory over their entire measurement range of heater temperatures, 12–21 K.

Discrepancies in such results may be due to the somewhat different measurement techniques, or perhaps differences in the interfacial conditions. Contaminants at the interface or subsurface damage in the crystal, such as that produced by polishing, may cause scattering and frequency conversion that modify the distribution emitted by the heated film.[40,41] (See the discussion of diffusive scattering at an interface in Chapters 11 and 12.) Using subnanosecond pulses, Wybourne et al.[42,43] have noted anomalies in the thermal relaxation time of metal-alloy films on sapphire, which they suggest is related to anharmonic phonon decay in the metal film. Experiments and theoretical simulations described below suggest that for heater temperatures above 10 K elastic scattering and anharmonic decay of phonons *in the crystal itself* can greatly modify the frequency spectrum at a distant point.

A series of quantitative studies by Weis[44] strongly supports the model of a Planck emitter based on the acoustic-mismatch model for a wide range of films and substrates. In particular, Müller and Weis[45] have found extremely good quantitative agreement between the power radiated by heated constantan films on sapphire and Si and a theoretical description that includes elastic anisotropies. While a detailed understanding of the physical relaxation processes in the metal is generally incomplete, there is no question that modeling the heated metal film as a Planck radiator is a useful approximation for many systems.

ii) Example of an optically excited metal-film radiator

The idea of a Planckian emitter of phonons was quantitatively described by Weis.[34] Below we consider an application of this theory to a typical case: a Cu film evaporated onto a Si crystal. The results depend on whether the sample is placed in a vacuum or a liquid-helium bath. In the latter case, phonons from the heated portion of the film are emitted into both the crystal and the bath. Isotropic media are assumed. The numerical examples and accompanying experiments are based on a paper by Shields et al.[46]

As discussed briefly in Chapter 2, the vibrational energy per unit frequency emitted from a Planck radiator goes as $\omega^3 \bar{n}$, where $\bar{n}$ is the thermal occupation number given there by Eq. (1). Integrating this expression over frequency gives the Stefan–Boltzmann Law for the total power radiated per area for a Planckian emitter at temperature T. Assuming a transmission medium at 0 K, i.e., no backflow of heat from crystal to generator, we find

$$P/A = \sigma T^4. \tag{3}$$

For phonons emitted from Cu to Si, the constant is given by,

$$\sigma_{Cu/Si} = (\pi^5 k_B^4 / 15h^3)(e_L/c_L^2 + e_{FT}/c_{FT}^2 + e_{ST}/c_{ST}^2), \tag{4}$$

where the c's are the average phonon velocities for the three modes in the crystal and the e's are the spectral emissivities that take into account reflection, refraction, and mode conversion at the interface.* Because sound velocities are approximately 10^5 times smaller than that of light, thermal radiation is approximately 10^{10} times more effective than electromagnetic radiation at the same temperature.

For a steady-state absorbed laser power, P, deposited in an excitation area, A, conservation of energy flux requires that

$$P/A = \sigma_{Cu/Si} T^4 + \sigma_{Cu/He} T^4 = \sigma' T^4, \tag{5}$$

where $\sigma_{Cu/He}$ represents any contributions from phonons emitted from Cu to the He bath. Therefore, the steady-state source temperature is given by

$$T = (P/\sigma' A)^{1/4}. \tag{6}$$

It is useful to note that if the laser power is instantaneously shut off, the relaxation of energy density, E, in the metal film occurs at a rate of

$$dE/dt = -\sigma' AT^4. \tag{7}$$

For a film of density ρ and thickness ℓ, specific-heat constant α, and excited volume $V = A\ell$, the energy density is given by $E = \rho \alpha V T^4/4$, so

$$dT^4/dt = -(4\sigma'/\rho\alpha\ell)T^4 = -T^4/\tau, \tag{8}$$

which gives the decay rate of the energy density.

---

* In the acoustic-mismatch model (Chapter 11), these emissivities are frequency independent.

Using emissivities calculated by Rösch and Weis[47] ($e_L = 0.23$, $e_{ST} = e_{FT} = 0.152$), the known physical constants of the media, an excitation radius of 20 μm, a film thickness of 2000 Å, and an excitation power of 1 W, one finds a film temperature of approximately 70 K if the sample is in vacuum ($\sigma_{Cu/He} = 0$). Alternatively, assuming an emissivity $e = 0.4$ for the copper/liquid-helium interface,[48] a temperature of approximately 50 K is obtained for a sample immersed in liquid helium. The calculated cooling times are roughly $\tau = 2$ ns and 0.5 ns, respectively.[46] The precise values of these numbers should not be taken too seriously, considering the assumptions and approximations used.

With optical excitation one can conveniently vary the excitation density (thus the heater temperature) without changing the total energy emitted into the sample: by defocusing the laser beam, the excitation area, A, can be varied at a constant input power. Figure 14(a) shows the Planck distributions ($\omega^3 \bar{n}$) corresponding to several heater temperatures, assuming constant P and variable A. Due to scattering from naturally occurring isotopes in the crystal, phonons in the terahertz frequency range may not travel ballistically across a macroscopic crystal. The calculated transmission factor for a 5.5-mm crystal of Si is shown as the dotted line in Figure 14(a), indicating that phonons above 1 THz have little chance of traversing the crystal without scattering. Therefore, the transmitted *ballistic* flux is predicted to be greatly reduced as the laser beam is focused to a tighter and tighter spot.

This effect is shown in Figure 14(b), which plots the product of the Planck spectra with the transmission factor, assuming the sample is in a vacuum. The total ballistic flux, calculated as a function of film temperature, is plotted as the triangles in Figure 14(c). Corresponding experimental data[46] taken with an Al bolometer detector are shown as the solid dots in this figure and display a much weaker dependence of the ballistic signal on the calculated heater temperature in the experiment. This discrepancy is largely removed by considering the anharmonic downconversion of phonons in the bulk of the Si. Some of the phonons emitted with frequencies above 1 THz are downconverted to subterahertz phonons very close to the heat source, and these daughter phonons will propagate ballistically to the detector as though they had been directly emitted by the heater. The ballistic signal is therefore augmented by these quickly downconverted phonons. The result of a Monte Carlo calculation that uses the theoretical anharmonic decay rate for Si is shown as the squares in Figure 14(c). We see that reasonable agreement between the data and the simple theory outlined above is obtained when anharmonic decay of phonons is included.

Another interesting implication of Figure 14(b) is that, contrary to common perceptions,* the average frequency of ballistic phonons is not easily altered by changing the power density. Due to isotope scattering, the average frequency

---

* It is sometimes assumed that a Planck radiator provides a frequency-tunable source of phonons with $\omega_{char} \propto k_B T_{heater}$. Clearly, that is not a good assumption in the present case.

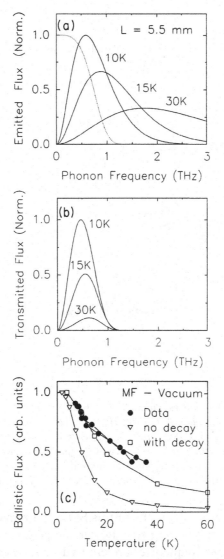

Figure 14 (a) Computed Planck distributions superimposed onto the transmission factor for ballistic flux (dotted line) in a 5.5-mm crystal of Si. The
integrals of the Planck spectra are equal in order to simulate constant input power and varying excitation area. (b) Flux transmitted ballistically
through the crystal, calculated by taking the product of the transmission
factor and the Planck spectra in (a). The integrals of these and similar
curves predict the heat flux arriving at the ballistic time of flight and are
plotted as the triangles in (c). Solid dots in (c) show the experimentally
measured peak intensities that occur at near-ballistic times of flight. These
data were obtained by varying the laser focal area while keeping the total power constant on the Cu film with a vacuum interface. A Monte
Carlo calculation of the peak intensities that includes anharmonic decay
of high frequency phonon in Si yields the squares, which agree well with
the experimental data. (From Shields et al.[46])

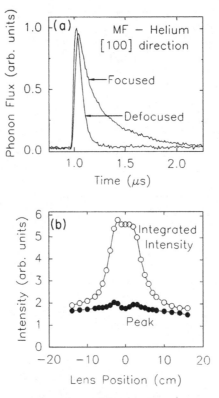

Figure 15 (a) Heat pulses obtained by optically exciting a copper metal film (MF) on Si immersed in a superfluid He bath. (Note the expanded time scale.) Focused $\simeq$90 W/mm^2, defocused $\simeq$3 W/mm^2. (b) Plot of the peak intensity and integrated intensity as the focal spot size is changed by moving the lens. 0 cm corresponds to optimal focus ($\simeq$20-μm spot size), and 15 cm corresponds to $\simeq$500-μm spot size. (From Shields et al.[46])

of ballistically transmitted phonons in this Si crystal remains below approximately 0.7 THz, corresponding to a heater temperature of approximately 12 K. Of course, higher frequencies are emitted by the heater, but they may scatter a number of times before reaching the detector and contribute to the diffusive tail of the heat pulse.

iii) Bubbles in the bath

The helium bath can lead to some interesting complications at high excitation densities, which we mention here because most heat-pulse experiments use samples immersed in superfluid liquid helium for effective cooling and temperature control. Figure 15(a) shows that as the laser beam is tightly focused, a broad tail with $\tau_{1/e} \simeq$ 250 ns appears in the heat pulse. Remarkably, the integrated intensity increases by about a factor of three [Figure 15(b)] despite the constant input power. What is happening here?

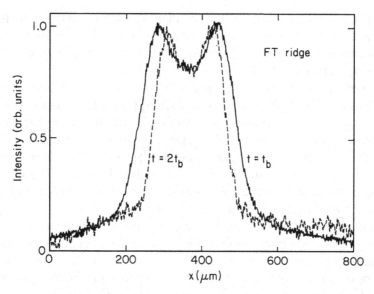

Figure 16 Spatial scans across the FT ridge in Si for focused excitation of the metal film in liquid He, showing a narrowing of this phonon-focusing structure with time after the pulse. This caustic shift is used to gauge the cooling of the metal film as the He gas bubble dissipates. The decay time of the heated metal film was enhanced by a confining geometry of the liquid He, as explained in Ref. 46.

Close inspection of the excitation surface through a side window of the cryostat reveals that when the laser is focused a small gas bubble is formed at the excitation point, reducing the contact of the film with the liquid-helium bath. The onset of the bubble is accompanied by a large increase in the scattered laser light at the excitation point. The experiment of Figure 15(b) suggests that when the bubble is not present, about twice as much heat is lost to the bath as is emitted into the crystal.

What is the source of the extended tail in the focused case of Figure 15(a)? One might initially guess that it is due to high-frequency phonons that scatter several times in the bulk of the crystal. A time-resolved imaging experiment, however, shows that sharp caustics are observed even when the boxcar gate is set at twice the ballistic time of flight, $t_b = 1$ μs. Figure 16 shows scans across the FT ridge [analogous to Figures 2(b) and 4, pages 14, 16] at two delay times. If the tail were composed of scattered phonons, the spatial scan at $t = 2t_b$ would show a broad diffusive distribution. Instead, a sharp structure remains, characteristic of ballistic phonons originating from a highly localized source.

Indeed, the FT ridge at $t = 2t_b$ is *sharper and narrower* than that sampled at $t = t_b$. As will be shown in Chapter 7, this shift in the angular positions of the phonon-focusing caustics is due to frequency dispersion in the lattice.[49] Shields et al.[46] conclude that when the bubble is present the film temperature slowly falls leading to a slow shift of the Planck spectrum to lower frequency. Comparison with a lattice-dynamics model indicates that the frequencies in this experiment range from approximately 0.7 to 0.3 THz, corresponding to Planck temperatures of 12 K and 5 K, respectively.*

The extended source lifetime [Figure 15(a)] is not observed when the sample is situated in vacuum. Why is the film cooling much more slowly than predicted by Eq. (8) for the vacuum (or liquid-helium) interface? Apparently, a fair fraction of the initial deposited energy goes into vaporizing the He liquid, and the backflow of this energy into the film keeps its temperature elevated for several microseconds (on the order of the bubble lifetime). The amount of energy required to sustain the metal film at the observed 5–12 K temperatures is consistent with that deposited by the laser pulse.

The complicated dynamics of this system – involving the superfluid properties of the He – is a fascinating problem that has not been solved. From the heat-pulse perspective, it should be realized that the lifetime of the gas bubble, and hence the duration of the extended tail, depends on the local pressure of the liquid, which, in turn depends on the height of the liquid column above the sample. Empirical studies of the bubble dynamics which elucidate these aspects have been reported by Greenstein and Wolfe[50,51] and others.[52,53]

The practical message that one must remember is that the shapes and intensities of the crystalline heat pulses can depend significantly upon the environment of the sample.

References

1. G. A. Northrop and J. P. Wolfe, Phonon imaging: theory and applications, in *Nonequilibrium Phonon Dynamics*, W. E. Bron, ed. (Plenum, New York, 1985), p. 165.
2. Data taken by J. A. Shields, University of Illinois at Urbana-Champaign.
3. G. A. Northrop and J. P. Wolfe, Ballistic phonon imaging in Ge, *Phys. Rev. B* **22**, 6196 (1980).
4. W. Eisenmenger, K. Lassmann, H. J. Trump, and R. Krauss, Intrinsic and experimental quasiparticle recombination times in superconducting films, *Appl. Phys.* **12**, 163 (1977).
5. J. P. Maneval, J. Desailly, and B. Pannetier, Subnanosecond bolometry using niobium films, in *Phonon Scattering in Condensed Matter*, W. Eisenmenger, K. Lassmann, and S. Dottinger, eds. (Springer-Verlag, Berlin, 1984), p. 43.
6. Data from M. Msall, University of Illinois at Urbana-Champaign.
7. R. J. von Gutfeld and A. H. Nethercot, Heat pulses in quartz and sapphire at low temperature, *Phys. Rev. Lett.* **12**, 641 (1964).
8. W. Metzger and R. P. Huebener, Phonon focusing in [001] Ge, *Z. Phys. B* **73**, 33 (1988). The deduced source size may be smaller than reported due to dispersive shifts in the ST caustics. See also Ref. 29, Figure 16, and Chapters 6 and 7 for a discussion of dispersive effects.

---

* It is notable that while the shift of the Planck spectrum to low frequencies is observable for the broadband detector (Al bolometer) used here, a Pb tunnel junction would detect only a broader FT ridge (Figure 16) corresponding to the subset of phonons near the frequency onset of the detector.

9. R. P. Huebener, Applications of low-temperature scanning electron microscopy, *Rep. Prog. Phys.* **47**, 175 (1984).

10. J. A. Shields and J. P. Wolfe, Measurement of a phonon hot spot in photoexcited Si, *Z. Phys. B* **75**, 11 (1989). A subsequent interpretation of these experiments is contained in Ref. 46.

11. K. Weiser, U. Strom, S. A. Wolf, and D. U. Gubser, Use of granular NbN as a superconducting bolometer, *J. Appl. Phys.* **52**, 4888 (1981).

12. M. Obry, J. Tate, P. Berberich, and H. Kinder, Fabrication of thin-film high-T superconductors: application as a phonon detector, in *Phonons 89*, S. Hunklinger, W. Ludwig, and G. Weiss, eds. (World Scientific, Singapore, 1990), p. 328.

13. P. Berberich, Phonon transport in high-T_c superconductors, *ibid.*, p. 227.

14. M. Obry, J. Tate, P. Berberich, and H. Kinder, Study of phonon pulse propagation in silicon and the effect of N-processes, in *Phonon Scattering in Condensed Matter VII*, M. Meissner and R. O. Pohl, eds. (Springer-Verlag, Berlin, 1993).

15. R. J. von Gutfeld, Heat pulse transmission, in *Physical Acoustics, Vol. 5*, W. P. Mason, ed. (Academic, New York, 1968).

16. See, for example, M. Tinkham, *Introduction to Superconductivity* (McGraw-Hill, New York, 1975).

17. B. Mitrovic, H. G. Zarate, and J. P. Carbotte, The ratio $2\Delta/kT$ within Eliashberg theory, *Phys. Rev. B* **29**, 184 (1984).

18. W. Eisenmenger, Superconducting tunnel junctions as phonon generators and detectors, in *Physical Acoustics, Vol. 12*, W. P. Mason, ed. (Academic, New York, 1976), p. 79.

19. H. Kinder, Monochromatic phonon generation by superconducting tunnel junctions, in *Nonequilibrium Phonon Dynamics*, W. E. Bron, ed. (Plenum, New York, 1985), p. 129.

20. W. Eisenmenger and A. H. Dayem, Quantum generation and detection of incoherent phonons in superconductors, *Phys. Rev. Lett.* **18**, 125 (1967).

21. C. Höss, J. P. Wolfe, and H. Kinder, Total internal reflection of high frequency phonons – a test of specular refraction, *Phys. Rev. Lett.* **64**, 1134 (1990).

22. H. Kittel, E. Held, W. Klein, and R. P. Huebener, Dispersive anisotropic ballistic phonon propagation in [100] GaAs, *Z. Phys. B* **77**, 79 (1989).

23. W. Dietsche, G. A. Northrop, and J. P. Wolfe, Phonon focusing of large-k acoustic phonons in Ge, *Phys. Rev. Lett. B* **26**, 780 (1982).

24. S. E. Hebboul and J. P. Wolfe, Lattice dynamics of InSb from phonon imaging, *Z. Phys. B* **73**, 437 (1989).

25. H. Schreyer, W. Dietsche, and H. Kinder, Laser induced nonequilibrium superconductivity – a spatially resolving phonon detector, in *Proceedings of LT-17*, U. Eckern, A. Schmid, W. Weber, and H. Wühl, eds. (North-Holland, Amsterdam, 1984), p. 665.

26. H. Schreyer, W. Dietsche, and H. Kinder, Phonon emission spectroscopy of a two-dimensional electron gas, in *Phonon Scattering in Condensed Matter V*, A. C. Anderson and J. P. Wolfe, eds. (Springer-Verlag, Berlin, 1986), p. 88.

27. W. E. Bron, Spectroscopy of high frequency phonons, *Rep. Prog. Phys.* **43**, 20 (1980).

28. M. N. Wybourne and J. K. Wigmore, Phonon Spectroscopy, *Rep. Prog. Phys.* **51**, 923 (1988).

29. R. Eichele, R. P. Huebener, and H. Seifert, Phonon focusing in quartz and sapphire imaged by electron beam scanning, *Z. Phys. B* **48**, 89 (1982).

30. M. J. van Dort, K. Z. Troost, J. I. Dijkhuis, and H. W. de Wijn, The phonon spectrum injected by a metallic point contact: an FLN study, *J. Phys: Condensed Matter* **2**, 6721 (1990).

31. B. Sadoulet, B. Cabrera, H. J. Maris, and J. P. Wolfe, Phonon-mediated detection of particles, in *Phonons 89*, S. Hunklinger, W. Ludwig, and G. Weiss (World Scientific, Singapore, 1990), p. 1383.

32. B. Cabrera, Phonon mediated detection of elementary particles, *ibid.*, p. 1373.

33. W. A. Little, The transport of heat between dissimilar solids at low temperatures, *Can. J. Phys.* **37**, 334 (1959).

34. O. Weis, Thermal phonon radiation, *Z. Angew. Phys.* **26**, 325 (1969).

35. H. J. Maris, Radiation of phonons by metallic films, *J. Phys.* **C4**, 3 (1972).

36. N. Perrin and H. Budd, Phonon generation by joule heating in metal films, *Phys. Rev. Lett.* **28**, 1701 (1972).

37. P. Berberich and H. Kinder, Electron–phonon interaction near the metal-insulator transition of granular Al, in *Phonon Scattering in Condensed Matter V*, A. C. Anderson and J. P. Wolfe, eds. (Springer-Verlag, Berlin, 1986), p. 106.

38. W. E. Bron and W. Grill, Phonon spectroscopy. I. Spectral distribution of a phonon pulse, *Phys. Rev. B* **16**, 5303 (1977).

39. W. Eisenfeld and K. F. Renk, Tunable optical detection and generation of terahertz phonons in CaF_2 and SrF_2, *Appl. Phys. Lett.* **34**, 481 (1979).

40. J. M. Worlock, Thermal conductivity in sodium chloride crystals containing silver colloids, *Phys. Rev.* **147**, 636 (1966).

41. H. Kinder, Kapitza conductance of localized surface excitations, *Physica* **107B**, 549 (1981).

42. M. N. Wybourne, C. G. Eddison, and J. K. Wigmore, Phonon emission from a metallic film excited by nanosecond electrical pulses, *Sol. State. Commun.* **56**, 755 (1985).

43. M. N. Wybourne, J. K. Wigmore, and N. Perrin, Phonon frequency down-conversion observed in thin metallic films, *J. Phys: Condensed Matter* **1**, 5347 (1989).

44. See the review by O. Weis, Phonon radiation across solid/solid interfaces within the acoustic mismatch model, in *Nonequilibrium Phonons in Nonmetallic Crystals*, W. Eisenmenger and A. A. Kaplyanskii, eds. (North-Holland, Amsterdam, 1986).

45. G. Müller and O. Weis, Quantitative investigation of thermal phonon pulses in sapphire, silicon, and quartz, *Z. Phys. B* **80**, 15 (1990).

46. J. A. Shields, M. A. Msall, M. S. Carroll, and J. P. Wolfe, Propagation of optically generated acoustic phonons in Si, *Phys. Rev. B* **47**, 12510 (1993).

47. F. Rösch and O. Weis, Phonon transmission from incoherent radiators into quartz, sapphire, diamond, silicon, and germanium within anisotropic continuum acoustics, *Z. Phys. B* **27**, 33 (1977).

48. P. Berberich, private communication.

49. S. Tamura, J. A. Shields, and J. P. Wolfe, Lattice dynamics and elastic phonon scattering in silicon, *Phys. Rev. B* **44**, 3001 (1991).

50. M. Greenstein and J. P. Wolfe, Modulation of ballistic phonon fluxes in Ge by a He gas film at a solid/superfluid interface, *J. Phys.* **C6**, 274 (1981).

51. M. Greenstein, Anisotropy and time evolution of the electron-hole droplet cloud in germanium, Ph.D. Thesis, University of Illinois at Urbana-Champaign (1981), Chapter 4.

52. M. Scheuermann, I. Iguchi, and C. C. Chi, Oscillation of a bubble in superfluid helium, in *Proceedings of LT-17*, U. Eckern, A. Schmid, W. Weber, and H. Wuhl, eds. (Elsevier, 1984).

53. V. A. Tsvetkov, A. S. Alekseev, M. M. Bonch-Osmolovskii, T. I. Galkina, N. V. Zamkovets, and N. N. Silbel'din, Transfer of electronic-exitation energy of germanium to liquid helium, *JETP Lett.* **42**, 335 (1985).

54. J. P. Wolfe and M. R. Hauser, Acoustic wave front imaging, *Ann. Physik* **4**, 99 (1995).

4

Focusing in cubic crystals

A. Introduction – Eigenvalues and eigenvectors for general symmetry

In Chapter 2 the formal solution to the elastic wave equation was presented. In the next two chapters we will examine the solutions for specific crystal symmetries. First we derive phase velocities for high symmetry directions and planes. Then we treat the elastic constants as variables and systematically study phonon focusing in cubic crystals. The basic topological structures of the slowness and wave surfaces that we will encounter in the cubic system are also common to lower symmetry systems. This approach is quite phenomenological, but the systematic sampling of the elastic parameter space of cubic crystals will lead us to useful insights into the topology of acoustic-wave surfaces that relate directly to caustic patterns.

Recall that the eigenvalue equation describing an elastic wave with wavevector direction (n_x, n_y, n_z) is

$$(D_{il} - v^2 \delta_{il}) e_l = 0, \tag{1}$$

with the Christoffel matrix,

$$D_{il} = C_{ijlm} n_j n_m / \rho. \tag{2}$$

The characteristic equation is

$$\begin{vmatrix} D_{xx} - v^2 & D_{xy} & D_{xz} \\ D_{yx} & D_{yy} - v^2 & D_{yz} \\ D_{zx} & D_{zy} & D_{zz} - v^2 \end{vmatrix} = 0, \tag{3}$$

a third-degree polynomial equation in v^2, which has roots v_α^2 for $\alpha = $ L, FT, and ST. Explicitly writing the elements of Eq. (2) using the abbreviated subscripts defined by Eq. (9) on page 41,

$$\rho D_{xx} = C_{11} n_x^2 + C_{66} n_y^2 + C_{55} n_z^2 + 2 C_{16} n_x n_y + 2 C_{15} n_x n_z + 2 C_{56} n_y n_z. \tag{4}$$

The other D_{il} can be written similarly. (For a complete list, see Auld,[1] p. 211.) All terms are needed for the lowest symmetry (triclinic) crystals. For high symmetry crystals, many terms are zero.

The polarization vectors $\mathbf{e} = (e_x, e_y, e_z)$ for the roots v_α^2 are found from Eq. (1), which may be written in matrix form:

$$\begin{bmatrix} D_{xx} - v^2 & D_{xy} & D_{xz} \\ D_{xy} & D_{yy} - v^2 & D_{yz} \\ D_{xz} & D_{yz} & D_{zz} - v^2 \end{bmatrix} \begin{bmatrix} e_x \\ e_y \\ e_z \end{bmatrix} = 0. \tag{5}$$

Explicitly, we have for the first two equations,

$$(D_{xx} - v^2)e_x + D_{xy}e_y + D_{xz}e_z = 0, \tag{6}$$

$$D_{xy}e_x + (D_{yy} - v^2)e_y + D_{yz}e_z = 0. \tag{7}$$

Eliminating e_z, we find

$$e_y/e_x = [D_{yz}(D_{xx} - v^2) - D_{xz}D_{xy}]/[D_{xz}(D_{yy} - v^2) - D_{yz}D_{xy}], \tag{8}$$

and, using the third equation,

$$e_z/e_x = [D_{yz}(D_{xx} - v^2) - D_{xz}D_{xy}]/[D_{xy}(D_{zz} - v^2) - D_{xz}D_{yz}]. \tag{9}$$

With the condition $e_x^2 + e_y^2 + e_z^2 = 1$, all three components of $\mathbf{e}$ for a given eigenvalue $v^2 = v_\alpha^2$, can be found. The eigenvectors are denoted $\mathbf{e}(\alpha)$.*

B. Application to cubic crystals

For cubic crystals the independent elastic constants reduce to three. When the coordinate axes are chosen along the cube axes, the elastic tensor is given by

$$C_{IJ} = \begin{bmatrix} C_{11} & C_{12} & C_{12} & 0 & 0 & 0 \\ C_{12} & C_{11} & C_{12} & 0 & 0 & 0 \\ C_{12} & C_{12} & C_{11} & 0 & 0 & 0 \\ 0 & 0 & 0 & C_{44} & 0 & 0 \\ 0 & 0 & 0 & 0 & C_{44} & 0 \\ 0 & 0 & 0 & 0 & 0 & C_{44} \end{bmatrix}. \tag{10}$$

Thus, Eq. (4) simplifies to

$$\rho D_{xx} = C_{11}n_x^2 + C_{44}(n_y^2 + n_z^2)$$

or

$$\rho D_{xx} = C_{11}n_x^2 + C_{44}(1 - n_x^2), \tag{11}$$

---

* In this chapter, the dependence on $\mathbf{n}$, or (θ_k, ϕ_k), is suppressed.

with $n_x^2 + n_y^2 + n_z^2 = 1$. In general, for $i, j = x, y, z$, we have,

$$\rho D_{ii} = C_{11} n_i^2 + C_{44}(1 - n_i^2)$$

and $\hspace{10cm}$ (12)

$$\rho D_{ij} = (C_{12} + C_{44}) n_i n_j \qquad (i \neq j).$$

The simplest application of these equations is for a wave with **k** along the fourfold axis [100]. In this case, $n_x = 1$ and $n_y = n_z = 0$; so $\rho D_{xx} = C_{11}$, $\rho D_{yy} = \rho D_{zz} = C_{44}$, and off-diagonal elements are zero. Equation (3) yields

$$(C_{11} - \rho v^2)(C_{44} - \rho v^2)(C_{44} - \rho v^2) = 0, \tag{13}$$

which has the roots

$$v_{1,2}[100] = (C_{44}/\rho)^{1/2}, \tag{14}$$

$$v_3[100] = (C_{11}/\rho)^{1/2}, \tag{15}$$

and polarization vectors [from Eqs. (8) and (9)]

$$\mathbf{e}(1, 2) = \hat{y}, \hat{z}, \tag{16}$$

$$\mathbf{e}(3) = \hat{x}. \tag{17}$$

Because the wavevector **k** is parallel to $\hat{x}$, we see that the mode with velocity v_3 has longitudinal polarization and the modes with $v_{1,2}$ are transverse. In nearly all crystals C_{11} is greater than C_{44}, so the longitudinal mode has a higher velocity than the transverse mode. The physical basis for this is that a longitudinal, or compression, wave involves a local change of volume that usually produces a greater restoring force than the shear distortions of transverse waves. Table 1 lists the lattice constants for many of the cubic crystals cited in this book. Also given are the mass densities and some useful parameters defined below.

A direction in the crystal for which the two transverse velocities are equal is called an *acoustic axis*. In a cubic crystal, [100] and [111] are acoustic axes. The degeneracy of the two transverse modes along [100] implies that any unit vector in the y–z plane is also an eigenvector.

Now consider the more general case of wavevectors in the x–z plane of a cubic crystal, i.e., $n_y = 0$, $n_x = \cos \phi$, and $n_z = \sin \phi$. In this case, $D_{xy} = D_{yz} = 0$ from Eqs. (12), and,

$$\rho D_{xx} = C_{11} \cos^2 \phi + C_{44} \sin^2 \phi$$

$$\rho D_{yy} = C_{44}$$

$$\rho D_{zz} = C_{11} \sin^2 \phi + C_{44} \cos^2 \phi \tag{18}$$

$$\rho D_{xz} = (C_{12} + C_{44}) \cos \phi \sin \phi.$$

Table 1. *Elastic constants and parameters* $a = C_{11}/C_{44}$, $b = C_{12}/C_{44}$, *and* $\Delta = a - b - 2$ *for various crystals*

Crystal	Temperature	C_{11}	C_{12}	C_{44}	ρ	a	b	Δ	Ref.
CaF$_2$	4 K	1.7400	0.5600	0.3593	3.21	4.84	1.56	1.28	a
GaAs	4 K	1.1260	0.5710	0.6000	5.34	1.88	0.95	−1.07	b
InSb	4 K	0.6660	0.3350	0.3140	5.79	2.12	1.07	−0.95	a
InAs	4 K	0.8980	0.5025	0.3924	5.67	2.29	1.28	−0.99	a
LiF	4 K	1.2445	0.4264	0.6471	2.64	1.92	0.66	−0.74	a
Si	4 K	1.6772	0.6498	0.8036	2.33	2.09	0.81	−0.72	a
NaCl	4 K	0.5834	0.1192	0.1337	2.22	4.36	0.89	1.47	a
NaF	4 K	1.0850	0.2290	0.2899	2.85	3.74	0.79	0.95	a
SrF$_2$	4 K	1.2870	0.4748	0.3308	4.32	3.89	1.44	0.45	a
Diamond	300 K	10.764	1.2520	5.7740	3.51	1.86	0.22	−0.36	c
GaP	300 K	1.4120	0.6253	0.7047	4.13	2.00	0.89	−0.89	a
Ge	73 K	1.3150	0.4948	0.6840	5.34	1.92	0.72	−0.80	a
InP	300 K	1.0220	0.5760	0.4600	4.78	2.22	1.25	−1.03	a
KBr	300 K	0.3421	0.0436	0.0513	2.76	6.67	0.85	3.82	a

[a] G. Simmons and H. Wang, *Single Crystal Elastic Constants and Calculated Aggregate Properties: A Handbook* (MIT Press, Cambridge, MA, 1971).
[b] J. S. Blakemore, *J. Phys. Chem. Solids* **49**, 627 (1988).
[c] M. H. Grimsditch and A. K. Ramdas, Brillouin scattering in diamond, *Phys. Rev. B* **11**, 3139 (1975).
Note: The C_{ij}'s are in units of 10^{12} dynes/cm^2, and ρ is in units of g/cm^3. A more extensive list is given by A. G. Every and A. K. McCurdy, *Landolt-Börnstein Numerical Data and Functional Relationships in Science and Technology,* Vol. 29a, D. F. Nelson, ed. (Springer-Verlag, Berlin, 1992).

The determinant equation is

$$(D_{xx} - v^2)(D_{yy} - v^2)(D_{zz} - v^2) - D_{xz}^2(D_{yy} - v^2) = 0, \qquad (19)$$

which contains the common factor

$$(D_{yy} - v^2) = 0, \qquad (20)$$

corresponding to a transverse mode with a velocity that is isotropic in the (100) plane,

$$v_1 = (C_{44}/\rho)^{1/2}. \qquad (21)$$

The remaining factor has the two roots,

$$v_{2,3}^2 = (1/2\rho)\{D_{xx} + D_{zz} \mp [(D_{xx} + D_{zz})^2 - 4(D_{xx}D_{zz} - D_{xz}^2)]^{1/2}\}$$
$$= (1/2\rho)\{C_{11} + C_{44} \mp [(C_{11} - C_{44})^2\cos^2 2\phi + (C_{12} + C_{44})^2\sin^2 2\phi]^{1/2}\}. \qquad (22)$$

The slowness curves $s_\alpha = 1/v_\alpha$ for these three modes are shown in Figure 1, with the elastic constants of GaAs (see Table 1). Using Eq. (22) and similar

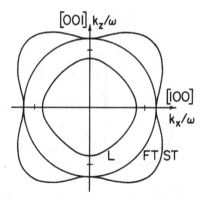

Figure 1 Slowness curves for GaAs in the (010) plane. The marks on the axes represent a slowness of 2.5×10^{-4} s/m. (From Auld.[1])

expressions, the velocities for **k** along the principal symmetry axes are found to be

$$v_1^2[110] = C_{44}/\rho, \tag{23}$$

$$v_2^2[110] = (C_{11} - C_{12})/2\rho, \tag{24}$$

$$v_3^2[110] = (C_{11} + C_{12} + 2C_{44})/2\rho, \tag{25}$$

$$v_{1,2}^2[111] = (C_{11} - C_{12} + C_{44})/3\rho, \tag{26}$$

$$v_3^2[111] = (C_{11} + 2C_{12} + 4C_{44})/3\rho. \tag{27}$$

The subscripts 1, 2, and 3 on the velocities correspond to FT, ST, and L for elastic constants similar to GaAs. For crystals like CaF$_2$ with $(C_{11} - C_{12}) > 2C_{44}$, the subscripts 1 and 2 correspond to ST and FT, according to our convention that $\alpha =$ L, FT, ST are the inner, middle, and outer slowness sheets.

The polarization vectors of these modes are readily determined by Eqs. (8) and (9). The fast transverse mode with isotropic velocity $v_1 = (C_{44}/\rho)^{1/2}$ in Figure 1 has a constant polarization $\mathbf{e}(1) = \hat{y}$, perpendicular to the (010) plane of propagation. In a general wavevector direction $(n_x, 0, n_z)$ the L and ST modes are quasi-longitudinal and quasi-transverse waves. That is, while $\mathbf{e}(1)$, $\mathbf{e}(2)$, and $\mathbf{e}(3)$ are always orthogonal for a general **k** direction, $\mathbf{e}(3)$ is not exactly along **k**, nor is $\mathbf{e}(2)$ exactly perpendicular to **k**.

There is a condition among the elastic constants that corresponds to perfect isotropy. Indeed, there is no angular variation of v_2 and v_3 in Eq. (22) if

$$C_{11} - C_{44} = C_{12} + C_{44},$$

i.e.,

$$\tag{28}$$

$$C_{11} - C_{12} - 2C_{44} = 0,$$

which can also be obtained by setting $v_1[110] = v_2[110]$ above. This condition

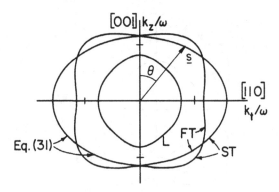

Figure 2 Slowness curves for GaAs in the (1̄10) plane. The marks on the axes represent a slowness of 2.5×10^{-4} s/m. (From Auld.[1])

yields two spherical slowness surfaces, corresponding to a longitudinal mode with velocity $v_3 = (C_{11}/\rho)^{1/2}$ and two degenerate transverse modes with velocity $v_{1,2} = (C_{44}/\rho)^{1/2}$. It is useful to define a dimensionless anisotropy factor that tells the degree of elastic anisotropy that a particular crystal possesses. The factors

$$\Delta = (C_{11} - C_{12} - 2C_{44})/C_{44} \tag{29}$$

or

$$A = 2C_{44}/(C_{11} - C_{12}) \tag{30}$$

are commonly used. For perfect isotropy, $\Delta = 0$ and $A = 1$. These two factors contain the same information because they are related by $\Delta = 2/A - 2$. In this text, we choose to use Δ. For a given anisotropy factor, $\Delta =$ constant, the *ratios* of the [100], [110], and [111] velocities for a particular transverse mode are determined. This does not mean, however, that the *shapes* of the transverse slowness surfaces depend only on the anisotropy factor. The shapes of these surfaces depend on *two* elasticity parameters, for example, $a = C_{11}/C_{44}$ and $b = C_{12}/C_{44}$. (See also the discussion in Appendix I.)

Figure 2 shows intersections of the slowness surface of GaAs with the (1̄10) plane, which contains the [001], [110], and [111] symmetry axes. Analytic expressions for these surfaces are derived by Auld.[1] For example, the eigenvalue of the characteristic equation that has a pure transverse polarization [$e \perp (\bar{1}10)$] has the functional form

$$\rho v_2^2 = [(C_{11} - C_{12})/2] \cos^2 \theta + C_{44} \sin^2 \theta. \tag{31}$$

This is a smooth curve that actually runs from one sheet of the slowness surface (middle, or FT, sheet) to another (outer, or ST, sheet), so the single subscript, 2, is a bit misleading. Recall from Chapter 2 that these two transverse sheets touch at a conic point along the [111] direction and are separated for a cross section just outside the (1̄10) plane. In view of this complicated topology, it is remarkable that the slowness sheets can be expressed in a closed analytic form for arbitrary propagation directions, as discussed in Appendix I. Understandably, the general expressions[2] are much more complicated than Eq. (31).

Analytic expressions for cross sections of the group-velocity surfaces in a symmetry plane may also be determined, and the interested reader is referred to Auld's text. The simplest procedure is implicit differentiation of the Christoffel equation. One defines the function

$$\Omega(\omega, k_x, k_y, k_z) = |C_{ijlm}k_j k_m - \rho\omega^2\delta_{il}| = 0. \tag{32}$$

The total differential $d\Omega$ must also be zero. If we set $\delta k_y = \delta k_z = 0$, this condition yields

$$(\partial\Omega/\partial\omega)\delta\omega - (\partial\Omega/\partial k_x)\delta k_x = 0,$$

which implies that

$$V_x = (\partial\omega/\partial k_x)|_{k_y,k_z} = (\partial\Omega/\partial k_x)/(\partial\Omega/\partial\omega). \tag{33}$$

The derivatives on the right side of this equation can be calculated from Eq. (32). Cross sections of the ST wave surface in the (100) plane, obtained by this method,[3,4] are shown in Figure 8(b) on page 31.

In practice, it is not necessary to write out analytical expressions for the components of **V**. For a given wave normal and elastic tensor the polarization vectors and phase velocity are computed using the Christoffel equation, and the components of **V** are given by Equation (26) on page 48 (recalling $\mathbf{k}/\omega = \mathbf{n}/v$).* A cross section of the velocity surface such as Figure 8(b) on page 31 is then obtained by connecting discrete calculations of **V** as a function of angle. A three-dimensional plot [Figure 9(b) on page 33] requires a graphics program that projects cross-sectional curves in an oblique plane onto the plane of the paper.

C.   Phonon focusing in cubic crystals

In cubic crystals three elastic constants and the crystal density are required to calculate the phase velocities. However, the *shape* of the slowness surface is determined by only two parameters. For example, if the Christoffel matrix in Eq. (2) is divided by the scale factor C_{44}/ρ, the resulting matrix depends only on the parameters

$$a = C_{11}/C_{44} \quad \text{and} \quad b = C_{12}/C_{44}. \tag{34}$$

The eigenvalues of this matrix are the same as those of D_{il}, apart from an overall scale factor. Because the phonon focusing pattern depends only on the shape of the slowness surface, it is completely defined by the parameters a and b.

The beauty of this result is that all cubic crystals may be mapped into a two-dimensional elastic-parameter space, each point of which corresponds to a well-defined acoustic anisotropy. Such a graph is shown in Figure 3, which

---

* The components of **V** for cubic crystals are given in Appendix II, using abbreviated tensor notation.

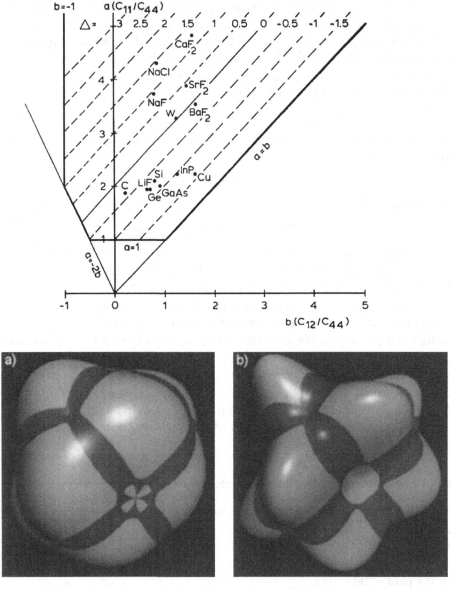

Figure 3 Top: Elastic-parameter space for cubic crystals. The axes are the ratios of elastic constants $a = C_{11}/C_{44}$ and $b = C_{12}/C_{44}$. The lines $a = b$ and $a = -2b$ bound the region of thermodynamic stability for a crystal. The spaces to the left of $b = -1$ and below $a = 1$ are the subject of Section D of this Chapter. The dashes show lines of constant anisotropy factor Δ, which is zero for perfect isotropy. (From Hurley and Wolfe.[7]) Shown below are ST slowness surfaces for (a) Si and (b) Cu. (Computed by Wolfe and Hauser.[14])

contains the positions of several different materials. The anisotropy factor [Eq. (29)],

$$\Delta = a - b - 2, \tag{35}$$

is extremely helpful in categorizing the anisotropy of various crystals. It is found that crystals with nearly the same value of Δ have similarly shaped slowness surfaces. The line defining $\Delta = 0$ corresponds to perfect isotropy. The elastic parameter space is thus divided into positive- and negative-Δ regions, which, as we have seen from the discussion of Figure 1, correspond to qualitatively different slowness surfaces. Dashed lines of constant Δ are plotted in Figure 3.

Also plotted in this figure are heavy lines, inside of which nearly all known crystals lie.[5-7] The lines $a = b$ and $a = -2b$ are the boundaries of thermodynamic stability, outside of which a crystal would possess negative shear or bulk modulus. Below $a = 1$ and to the left of $b = -1$ the transverse phase velocities exceed the longitudinal phase velocities near the $\langle 100 \rangle$ and $\langle 111 \rangle$ directions, respectively. (This occurs for only a few materials, as discussed in Section D, Origins of the Phonon Caustics.)

A good deal of insight about phonon focusing can be gained by treating a and b as variable parameters. Systematic theoretical studies of the focusing properties of cubic crystals, using this approach, have been conducted by Every[6] and by Hurley and Wolfe.[7] Following these studies, we now present a detailed look at the interesting topology that underlies phonon focusing in cubic crystals. With this information the reader will be able to predict quite accurately the phonon caustic pattern in just about any cubic crystal. However, no continuity is lost in going directly to Section D, which examines the origins of the caustic patterns and forms the basis for Chapter 5.

i) Negative-Δ regime

We begin by considering the region in elastic-parameter space that contains semiconductors noted for their technological importance: Si, Ge, and the III–V compounds, GaAs, InSb, InP, etc. From Table 1 and Figure 3 we see that these crystals have negative values of the anisotropy parameter Δ.

Figure 4 shows the evolution of the ST slowness surface for an increasing degree of anisotropy, corresponding to points in a–b space that are increasingly farther from the isotropy line. The chosen points are plotted in Figure 5. The solid curves labeled A, B, C, and D demarcate where qualitative changes in the focusing structure occur, as discussed below.

For perfect isotropy, the ST and FT surfaces are spherical and degenerate. As one moves slightly toward negative Δ, these two sheets move apart, but remain in contact along $\langle 111 \rangle$ and $\langle 100 \rangle$ axes. The ST sheet is the outer surface, so one can imagine the ST sheet being pulled inward to touch the FT sheet at these symmetry points. The effect is different for $\langle 111 \rangle$ and $\langle 100 \rangle$ because the

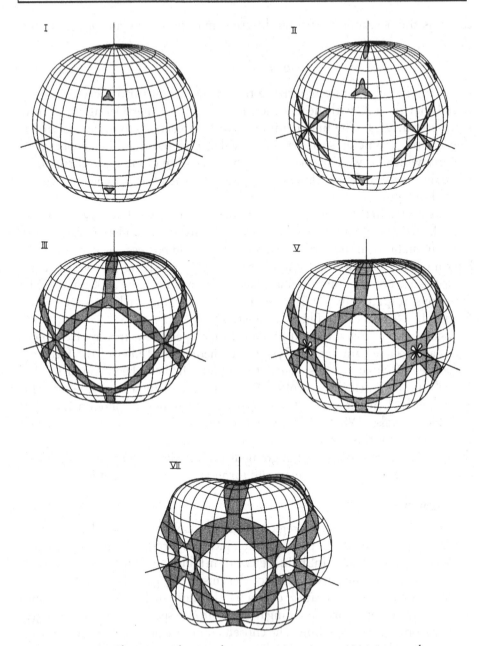

Figure 4 *Slowness surfaces in the negative-Δ regime, as |Δ| is increased. The corresponding values of (a, b), as plotted in Figure 5, are I (2.40, 0.60), II (2.35, 0.65), III (2.275, 0.725), V (2.20, 0.80), and VII (2.0, 1.0). The shaded areas are regions of saddle curvature which are bounded by parabolic lines of zero Gaussian curvature. (From Hurley and Wolfe.[7])*

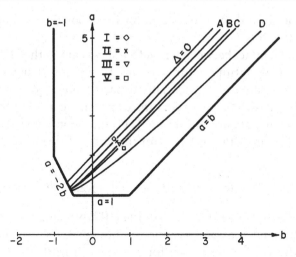

Figure 5 Critical lines in elastic-parameter space that mark the onset of new caustic structures in the phonon-focusing pattern. The symbols relate to the slowness surfaces in Figure 4. (From Hurley and Wolfe.[7])

contact is a conical one along the ⟨111⟩, which has threefold axial symmetry, and a tangential one along the ⟨100⟩, which has fourfold axial symmetry.

To see the initial distortion of the ST surface, one can imagine a balloon with a string attached to a point on the inside. Pulling radially inward causes a distortion from sphericity like that of the ST surface at a ⟨111⟩ conic point. Mathematically or topologically there is no continuous transition between the spherical surface and the one distorted by pulling on the string, no matter how gently. The conically distorted region has a negative Gaussian curvature. To see this, imagine a surface element containing the conic point that is small enough to look almost flat when there is no pull on the string. One can draw a series of concentric circles around the conic point. When the string is pulled, there is a region around the conic point where the curvature *tangential* to the circles is negative. The curvature perpendicular to the circles remains positive, so the Gaussian curvature (the product of the two principal curvatures) is negative. No matter how small the conic distortion, there is a region of saddle curvature around it.

This topological property has a profound influence on the phonon focusing. Because the majority of the ST slowness surface is convex (K positive), there must be parabolic ($K = 0$) lines in the neighborhood of the conic point that separate the convex and saddle regions, as shown by surface I in Figure 4. Thus, even an infinitesimal deviation from sphericity causes phonon-focusing caustics. We conclude that all cubic crystals possess phonon-focusing caustics.

If we continue to pull harder on the imaginary string, no new curvatures appear at the conic point. The negative-Δ ST surface always has a saddle curvature near the conic point. The curvature exactly at the conic point is infinite, or rather, undefined – which explains how the curvature near this region can undergo a discontinuous jump from positive to negative when anisotropy is introduced.

The tangential contact between ST and FT sheets along the $\langle 100 \rangle$ directions does not display a discontinuous jump in curvature as anisotropy is introduced, and no caustic lines appear as Δ initially departs from zero. However, as the distortion is increased, a negative curvature appears near $\langle 100 \rangle$ in the form of furrows that radiate outward toward the $\langle 111 \rangle$ directions, as shown by surface II in Figure 4. The analytic condition for the onset of this negative curvature region is

$$2(a - b)(b + 1) + (a - b - 2)(a + 2b + 1)(2a + b - 1) = 0, \qquad (36)$$

which is plotted as line A in Figure 5. For the $\langle 100 \rangle$ propagation directions, this is the condition that demarcates a change in the sign of the principal curvature transverse to the $\{110\}$ planes (see, for example, Musgrave[8-10] and Every[6]). As the anisotropy is increased beyond line A, the saddle regions originating along $\langle 111 \rangle$ and $\langle 100 \rangle$ meet, producing a continuous furrow between these symmetry axes.*

As the anisotropy increases beyond line B in Figure 5, the furrows broaden and surround the $\langle 100 \rangle$ axes, as shown by surface III in Figure 4. Beyond line C, new positive-curvature regions appear near the $\langle 100 \rangle$ axes, as indicated by surface V in Figure 4. These cloverleaf structures give rise to a Maltese-cross type of caustic pattern, shown previously in Figure 10(b) on page 34. The inside of the cloverleaf has a concave curvature.

It is useful at this point to recall the link between the slowness surface and the phonon caustic pattern: the group-velocity, or wave, surface. The wave surface is constructed from the normals of the slowness surface. Parabolic lines on the slowness surface produce folds in the wave surface and caustics in heat flux from a point source. A projection of the folds in the wave surface onto the detection surface produces a caustic map. Figure 6(a) shows a section of the ST slowness surface like that of surface V in Figure 4 (like Ge or Si). The section of group-velocity surface corresponding to Figure 6(a) is drawn in Figure 6(b), which is another view of the wave-surface topology illustrated in Figure 25 on page 57. By studying the details of these constructions, one can gain some insights into the multiple folded group-velocity surfaces and the caustic patterns in Figure 6(c) and in Figure 24 on page 56.

As the anisotropy is increased beyond line D in Figure 5, the cloverleaf caustic loses its central contact point (see surface VII in Figure 4). The parabolic and caustic lines for this structure are plotted in Figure 7. The double-headed arrows show directions of *zero* principal curvature. The cusps in the caustic pattern correspond to points where the vanishing principal curvature is

---

* This degree of anisotropy is seen in the experimental focusing pattern of diamond, which is shown in Chapter 12, page 309.

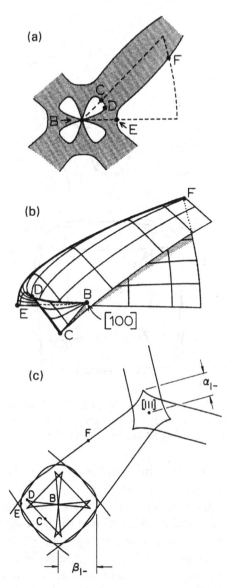

Figure 6 (a) Lines of zero Gaussian curvature on a small section of the slowness surface shown in Figure 4, Part V. (b) A section of the group-velocity surface constructed from the normals of the slowness surface section indicated by the dashed line in (a). Folds in the group-velocity surface give rise to the caustic pattern displayed in (c) and seen in the phonon image of Figure 24 on page 56.

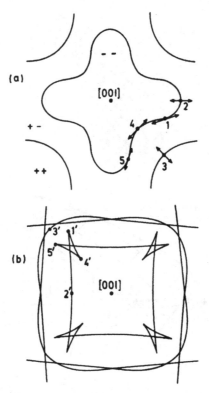

Figure 7 (a) Parabolic lines of zero Gaussian curvature on the slowness surface like that of Figure 4, Part VII, which occurs to the right of line D in Figure 5. (b) Corresponding caustic lines. (From Every.[6])

tangent to the parabolic line. The equations for lines B through D may be found in Every's paper.[6]

These are the topological features associated with the ST mode in the negative-Δ regime. A good approximation of the focusing pattern for a given crystal can be given by merely stating the angular sizes of the principal structures, for example, β_{1-} for the $\langle 100 \rangle$ structure and α_{1-} for the $\langle 111 \rangle$ cusp structure defined in Figure 6(c). Lines of constant α_{1-} and β_{1-} were empirically determined by Hurley and Wolfe[7] and are plotted in Figure 8. This type of graph provides a means for quickly determining the angular positions of the principal focusing caustics for an arbitrary set of elastic constants.

The FT mode in the negative-Δ region is much less complex than the ST mode just discussed. For small anisotropy, the surface is entirely convex. However, due to a local reduction in curvature, an enhanced flux appears between adjacent $\langle 100 \rangle$ directions. As $-\Delta$ is increased beyond curve H in Figure 9(c), furrows of saddle curvature appear that run between adjacent $\langle 100 \rangle$ directions, as displayed in Figure 9(a). These saddle regions produce the high-intensity ridges bounded by caustics in the wave surface, shown in

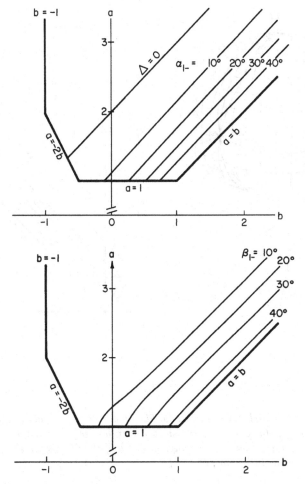

Figure 8 Curves in elastic-parameter space showing the angular dimensions α_{1-} and β_{1-} defined in the caustic map of Figure 6(c). (From Hurley and Wolfe.[7])

Figure 9(b). Figure 9(c) maps the maximum angular width of the ridges, which occurs in {110} planes.

Finally, Figure 10 contains Monte Carlo calculations showing the evolution of phonon-focusing structures for the ST mode in the negative-Δ region. A different perspective of the focusing patterns is given in Appendix III.

ii) Positive-Δ regime

Crystals with positive Δ, i.e., those lying to the left of the isotropy line in Figure 3, include the ionic insulators NaCl, NaF, CaF$_2$, and SrF$_2$. Notice that there is no one-to-one correspondence between crystal structure and position on the a–b map; indeed, LiF with face-centered-cubic (fcc) structure lies among

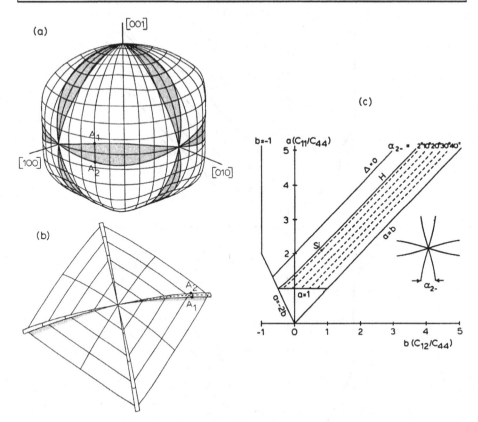

Figure 9 (a) FT slowness surface for Si. Shaded areas are saddle regions. (b) Section of the group-velocity surface near the [100] direction. (c) Lines of constant α_{2-}, which is the angular width of the FT ridge in the (110) plane. (From Hurley and Wolfe.[7])

the semiconductors with diamond structure in the negative-Δ regime. Few metals (W and Cu) are plotted because strong electron–phonon scattering precludes ballistic heat-pulse propagation over macroscopic distances.

An a–b map for this regime is shown in Figure 11. The critical lines E, F, and G, where new focusing caustics emerge from the symmetry axes, are shown. Figure 12 shows the evolution of the ST slowness surface as the anisotropy factor Δ is increased from zero. Also shown are the corresponding focusing patterns. As in the negative-Δ case, for nonzero Δ this outer transverse surface develops a region of negative (saddle) curvature around the $\langle 111 \rangle$ conic axes. The resulting caustics extend toward the $\langle 100 \rangle$ axes. When Δ is increased to line E on the a–b map, these caustics reach the $\langle 100 \rangle$ axes, as do the furrows of saddle curvature on the slowness surface. When Δ is increased sufficiently to cross critical curve F, three small concave islands appear around each $\langle 111 \rangle$ direction on the slowness surface.

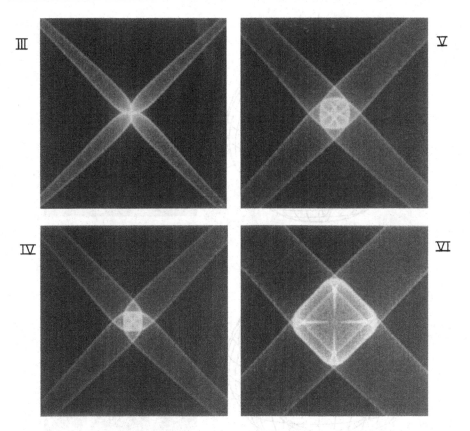

Figure 10 Calculated phonon-focusing pattern for the ST mode, negative-Δ regime, showing the evolution of the ST-box structure along [100] as $|\Delta|$ is increased. Values of (a, b) are from Figure 4 and IV (2.225, 0.775) and VI (2.10, 0.90). (From Hurley and Wolfe.[7])

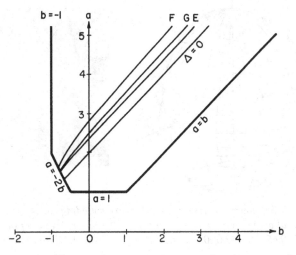

Figure 11 Critical lines in the positive-Δ regime that demarcate the onset of new caustics. (From Hurley and Wolfe.[7])

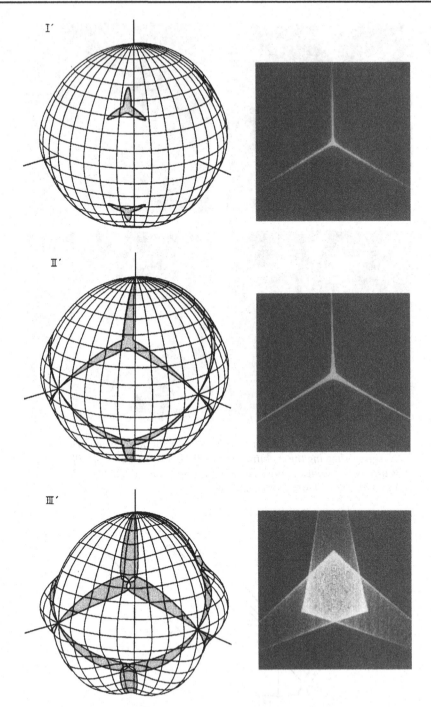

Figure 12 ST slowness surfaces and the corresponding focusing patterns for several values of a and b in the positive-Δ regime. Values of (a, b) are I′ (2.95, 0.65), II′ (3.00, 0.60), and III′ (3.64, −0.19). (From Hurley and Wolfe.[7])

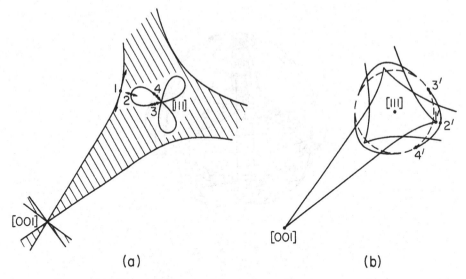

(a)   (b)

Figure 13 (a) Parabolic lines and (b) corresponding caustic map for the ST mode with slightly larger Δ than that in Figure 12-III'. (From Every.[6])

A detailed look at this interesting topology is provided in the schematic drawing of Figure 13. Figure 13(a) shows the parabolic lines on the ST slowness surface. The double-headed arrows indicate principal axes of vanishing curvature at selected points. At point 1 the vanishing principal curvature is tangential to the parabolic line, producing a cusp in the caustic pattern at point 1'. As points 3 and 4 on the parabolic line approach the [111] axis, the corresponding caustic line comes to a halt on the conic circle (dashed line)! Recall that the conic circle is where the ST and FT wave surfaces meet. We shall see that this caustic segment will continue smoothly in the FT caustic map. It appears that the parabolic line passes right through the conic point from one surface to the other.

Of course we must look at the FT topology to get the complete picture of how this little maneuver is accomplished, but we can start to get the picture by making a hidden-line drawing of the ST wave surface. Figure 14 shows the calculated slowness and wave surfaces for CaF_2. It is not easy to visualize this multifolded surface, but we do not need to know everything in detail. The circle in Figure 14(b) is where the ST wave surface terminates and joins the FT surface. At point F there is a fold in the surface just before it meets the conic circle. (Point E indicates a different region.) This fold continues on the FT surface, as the anxious reader may see in Figure 18 below. The effects of this fold on the ballistic heat flux may be seen in the Monte Carlo calculation of Figure 14(c) and the caustic map of Figure 13(b).

Let us now consider the FT mode. The evolution of the slowness surface with increasing Δ is shown in Figure 15. Monte Carlo simulations of heat

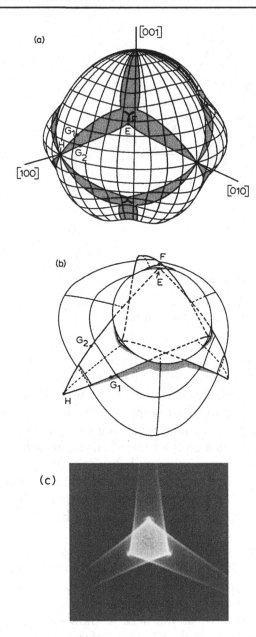

Figure 14 (a) ST slowness surface for CaF$_2$. (b) Section of the corresponding wave surface near the [111] direction. (c) Calculated phonon-focusing
pattern for a scan range of ±35° left to right. (From Hurley and Wolfe.[7])

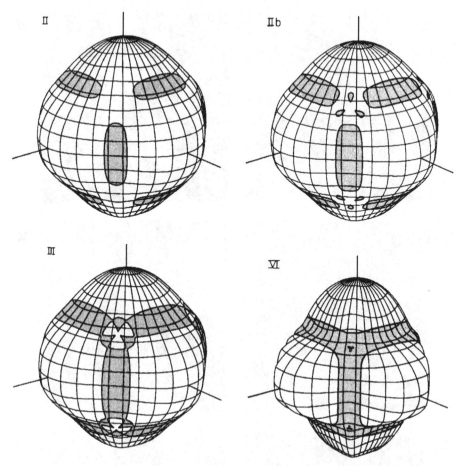

Figure 15 (a) FT slowness surfaces in the positive-Δ regime, showing topological changes with increasing Δ. The values of (a, b) are II (3.15, 0.47), IIb (3.18, 0.43), III (3.20, 0.40), and VI (3.75, −0.50). (From Hurley and Wolfe.[7])

flux are shown in Figure 16. For materials near the $\Delta = 0$ line, the surface is completely convex and no caustics appear, although the flux may be strongly concentrated near the $\langle 110 \rangle$ directions, giving rise to the precursors of caustics seen in Figure 16, Part I. These precursors can show rapid angular variation in flux without being mathematically singular. As the anisotropy is increased beyond line G in Figure 11, saddle regions appear around the $\langle 110 \rangle$ direction and corresponding caustics appear. The caustics are linked by a triangle of intense bands of flux that are precursors to caustics.

As the anisotropy is increased, three small droplets of saddle curvature appear around each $\langle 111 \rangle$ direction. At larger Δ these regions meet with the $\langle 110 \rangle$-centered lobes to leave a single island of convex curvature surrounding $\langle 111 \rangle$. At line F on the a–b map, this region separates into three lobes as

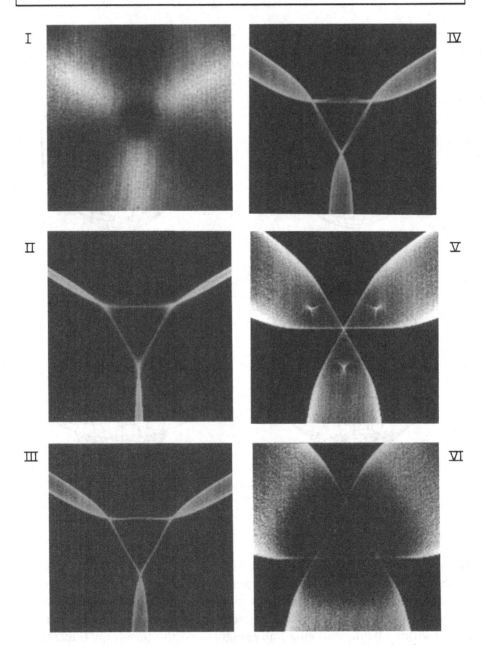

Figure 16 Calculated FT focusing patterns for positive-Δ values of (a, b) given in Figure 15 and I (2.95, 0.65), IV (3.25, 0.35), and V (3.75, −0.50). The images span ±35° left to right, with [111] at the center. (From Hurley and Wolfe.[7])

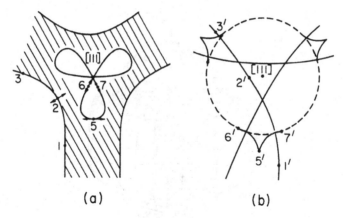

(a)   (b)

Figure 17 (a) Parabolic lines corresponding to image V in Figure 16, FT mode, positive Δ. (b) FT caustic map showing segments of the caustic lines that continue on the ST focusing pattern of Figure 13(b).

the parabolic boundary line touches the conic point. As previously shown, line F marks the emergence of parabolic lines from the conic point on the ST slowness surface. A three-leaf clover is formed on that surface that is rotated 60° from the one on the FT surface.

Close-ups of the parabolic and caustic lines near the conic point are shown in Figure 17. In contrast to the ST case, the FT cloverleaves are convex rather than concave. The corresponding caustic structures are cusped segments that end on the conic circle (dashed line) and exactly meet up with the aforementioned ST segments. The angular sizes of the caustic features of this positive-Δ regime have been characterized by Hurley and Wolfe.[7]

The slowness and wave surfaces for the FT mode in CaF_2 are given in Figure 18. One can see the conic circle where the ST and FT sheets meet, showing that these two sheets (Figure 18(b) and 14(b)) are really a part of a single continuous transverse wave surface. In this particular case the conic circle, defined by the locus of normals at the vertex of the slowness cone, cuts through a fold in the wave surface. The Monte Carlo simulation in Figure 18(c) clearly shows the caustic segments associated with the FT mode. Note also that there is no FT ballistic flux in the central triangular region.

What about reality? Figure 19 reproduces an experimental phonon image of CaF_2 from Ref. 7. A wide time gate was chosen to display both transverse modes simultaneously. This beautiful picture of the acoustic phonon flux exhibits all the theoretical features described above. Although experimental phonon images of negative-Δ crystals precipitated the detailed topological studies described in this chapter, caustic patterns for positive-Δ crystals were predicted by Every[6] before they were observed experimentally.

Experimental phonon images of several positive-Δ crystals with varying degree of anisotropy are shown in Figure 20. All these images are taken with

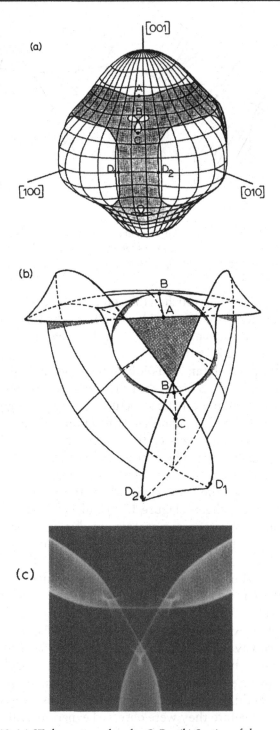

Figure 18 (a) ST slowness surface for CaF$_2$. (b) Section of the correspond-ing wave surface near the [111] direction. (c) Calculated phonon-focusing pattern for a scan range of ±35° left to right. (From Hurley and Wolfe.[7])

Figure 19 *Experimental phonon image of CaF₂. (From Hurley and Wolfe.[7])*

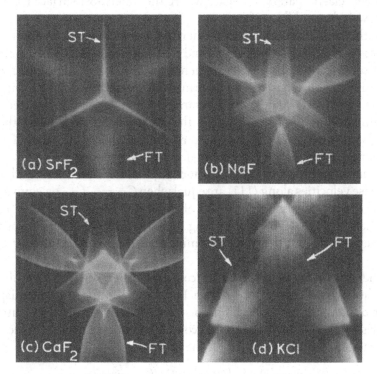

Figure 20 *Phonon images of several positive-Δ crystals. (From Hurley, Ramsbey and Wolfe.[11])*

wide time gates to show both transverse modes.[11] Various developmental stages of the focusing topology can be seen, including the highly anisotropic case of KCl, whose values of a and b are off the top of the scale in Figure 3.

Experimentally, there is no way to continuously change the elastic constants of a solid, as we have done in the gedanken studies above. At least, no one has yet figured out a practical way to do this.[12]

D. Origins of the phonon caustics

Valuable insights into the formation of phonon caustics can be gained by considering a special line in elastic parameter space – one where the Christoffel matrix is completely diagonal. This discussion involves also the inner sheet of the slowness surface, i.e., the L mode.

So far we have neglected the L mode, because in continuum elasticity theory the innermost slowness sheet possesses only convex curvature and cannot produce phonon-focusing caustics, even though its nonspherical shape can result in a highly anisotropic flux. In the cases we have considered, the innermost sheet of the slowness surface is topologically separated from the two outer sheets, which together produce a single continuous wave surface. Every and Stoddart[13] have pointed out that there are two distinct regions of the cubic elastic-parameter space where the innermost surface comes into contact with the middle surface.

Near the contact points of the inner two sheets, the polarization of the wave has highly mixed longitudinal and transverse character. Indeed, in these wavevector directions, the middle sheet has a mainly longitudinal polarization, and the inner sheet has a mainly transverse polarization. Thus, in these directions, the phase velocity of the transversely polarized wave exceeds that of the longitudinally polarized wave; the shear wave travels faster than the compression wave. (We still choose to retain the labels L, FT, and ST for the well-defined inner, middle, and outer slowness surfaces.)

One can readily see from Eqs. (26) and (27) that the transverse and longitudinal phase velocities along $\langle 111 \rangle$ are equal when $C_{12} = -C_{44}$, or, equivalently, $b = -1$. For crystals lying to the left of the $b = -1$ line in Figure 3, $v_3[111]$ is less than $v_{1,2}[111]$. Therefore, along this direction the outer slowness surface is separated from the inner and middle surfaces, which touch at a conic point.

To understand this topology better, consider the hypothetical case of a crystal that lies exactly on the $b = -1$ line. From Eq. (12) this condition diagonalizes the dynamical matrix for all propagation directions, and the solutions to the Christoffel equation are three uncoupled slowness surfaces. Specifically, Eq. (3) factors to give three roots of the form

$$v^2 = D_{ii} = [C_{11}n_i^2 + C_{44}(1 - n_i^2)]/\rho. \tag{37}$$

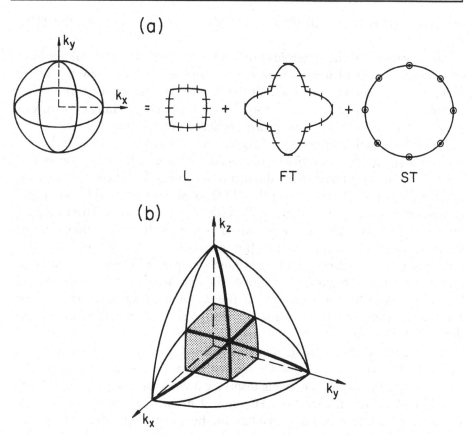

Figure 21 (a) (100) cross sections of the slowness surface when $b = -1$. Inner, middle, and outer sheets are separated on the right. Short lines indicate the polarizations; dots indicate polarization along z. (b) Three-dimensional representation of the slowness surface when $b = -1$. Three oblate spheroids intersect along the heavy lines. These line degeneracies reduce to point degeneracies for a slight modification of the elastic constants, and the slowness surface breaks naturally into inner (L), middle (FT), and outer (ST) sheets, as indicated in Figure 22. The inner sheet is shaded here.

Using the definition for the slowness, $s_i = n_i/v$, we have

$$\rho = C_{11}s_x^2 + C_{44}(s^2 - s_x^2), \tag{38}$$

which represents a spheroid that is rotationally symmetric about the x axis. The acoustic wave for this sheet has a constant polarization along the x axis. Because $C_{11} > C_{44}$ for this region of elastic-parameter space, the slowness sheets are pancake shaped (oblate). Intersections of the three sheets with the (001) plane are plotted in Figure 21(a). A three-dimensional representation is given in Figure 21(b), which shows the intersections of these slowness sheets as heavy lines. All three spheroids are degenerate along the ⟨111⟩ directions.

All curvatures of the spheroids are convex, so there are no parabolic lines separating regions of negative curvature. However, if we view the entire surface as composed of inner, middle, and outer sheets, we see three rather complex surfaces, joined along the solid lines and having infinite curvature at these intersections. The inner surface is shaded in the figure.

An infinitesimal displacement of elastic constants toward the left of the $b = -1$ boundary produces dramatic topological changes. The line degeneracies are immediately removed. Now, distinct inner and middle sheets contact each other only at conic points along the ⟨111⟩ axes. Distinct middle and outer sheets contact each other (tangentially) only along ⟨100⟩ axes. These *avoided intersections* of the three surfaces, caused by the off-diagonal coupling terms in the Christoffel matrix, produce (finite-sized) negative-curvature regions on the slowness surfaces. The consequence is parabolic lines and phonon-focusing caustics, no matter how small the displacement from the $b = -1$ boundary. Also, the polarizations of the waves with wavevectors near the avoided intersections are mixtures of the pure x, y, or z polarizations for the decoupled ellipsoids.

An example of this phenomenon is shown in Figure 22, which corresponds to a point between $b = -1$ and the thermodynamic-constraint line $a = -2b$. As expected, the two outer sheets display parabolic lines separating regions of different Gaussian curvature. Furrows of negative curvature correspond to the avoided intersections marked by the heavy lines in Figure 21(b). The directions of polarization are indicated by the short lines. The inner surface touches the middle surface at a conic point along ⟨111⟩. Near this direction, the inner surface displays significant transverse polarization and the middle surface is longitudinally polarized, as is the outer surface along ⟨111⟩. Along ⟨100⟩, the inner surface is longitudinally polarized and the two outer surfaces are transversely polarized. A Monte Carlo simulation of the heat flux for each mode is also shown in the figure.

At present, there have been no phonon-imaging experiments performed in this region of elastic-parameter space. (Examples of crystals falling into this category are the intermediate valence compounds, Tm-Se and Sm-Y-S.) The example given above, however, will prove to be useful in understanding the lower symmetry crystals discussed in Chapter 5. We shall see that this unusual topological behavior – in which the inner and middle slowness surfaces contact each other and trade polarization vectors – is observable in the tetragonal crystal, TeO_2.

The concept of avoided intersections gives insights into the formation of negative-curvature furrows in the slowness surfaces, which produce phonon-focusing caustics. For example, the positive-Δ surfaces in Figures 12 and 15 show the evolution of the avoided-intersection regions when the elastic

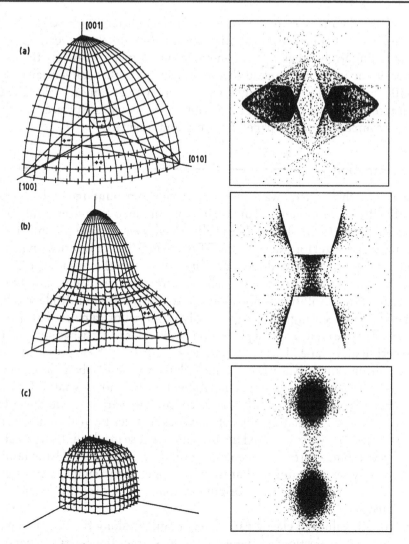

*Figure 22 Slowness sheets for b < −1 and the corresponding focusing patterns, (calculated by Every and Stoddart.[13] The heavy lines indicate zero Gaussian curvature, and + and − denote the signs of the two principal curvatures of the surface. The short lines indicate the directions of polarization. Near the conic point (along [111]) the inner two sheets exchange polarization, so the fastest wave with **k** near [111] has a largely transverse polarization. Calculated focusing patterns for each mode are shown at the right.*

parameters fall to the right of the $b = −1$ boundary. The two slowness sheets (ST and FT) repel each other but are intimately linked at ⟨111⟩ and ⟨100⟩ points.

The second region in the a–b plane where the inner and middle surfaces touch is below $a = 1$ (and, of course, between the thermodynamic-constraint lines, $a = -2b$ and $a = b$ in Figure 3). This condition, $C_{11} = C_{44}$, sets the longitudinal and transverse velocities along $\langle 100 \rangle$ equal. An example of a crystal falling in this region is the metallic alloy Mn–Ni–C. Below the $a = 1$ boundary, the inner and middle sheets contact each other tangentially along the $\langle 100 \rangle$ axis, and the middle and outer sheets contact conically along $\langle 111 \rangle$. The interested reader can find a discussion of the slowness surfaces for these types of compounds in the paper by Every and Stoddart.[13]

E. Summation

At first sight, the acoustic-wave topologies that we encountered in this chapter are difficult to comprehend. Fortunately, we are able to concentrate on the single-valued slowness sheets. The multivalued wave surfaces are much more difficult to visualize; for example, the ST mode in Si has 11 interleaving sheets along some group-velocity directions. This means that there would be 11 heat pulses arriving at the detector at slightly different times. Generally, the experimental time resolution is not sufficient to resolve this multitude of pulses.

The phonon-focusing pattern is usually observed or calculated as a time-integrated image. It provides a projection of the folds in the wave surface onto an experimental plane. Each fold produces a caustic in heat-pulse intensity. The resulting caustic pattern is simply related to the parabolic lines on the slowness surfaces. Simple, that is, for someone who knows what to look for.

By treating the elastic parameters as continuous variables, we are able to gain insight into the relationships of slowness surfaces for real cubic crystals. In the last section, we considered the informative case in which the inner slowness sheet becomes partially degenerate with the middle sheet. Using familiar concepts of wave mechanics – that is, matrix algebra and simple perturbation theory – we were able to see the origin of phonon caustics as the degeneracies were lifted.

For certain values of elastic parameters (namely, along the line defined by $C_{12}/C_{44} = -1$ in elastic-parameter space) the Christoffel matrix is diagonalized for all $\mathbf{k}$ directions. In this special case, the three modes are completely decoupled and the three slowness sheets *intersect* each other along curved lines [Figure 21(b)]. No phonon caustics are present; however, a slight displacement left or right in elastic parameter space introduces off-diagonal coupling terms in the Christoffel matrix and lifts the degeneracy along the intersection lines.

The lifted degeneracies, or avoided intersections, introduce radical curvatures in the slowness surfaces (Figure 22). Normals to the slowness surface are directions of energy flux. Hence, the changing curvatures produce concentrations of flux in certain directions, just as optical caustics are produced by the Sun's reflections off a wavy pool of water. On the slowness surface, the avoided intersections produce furrows of saddle curvature, which, in turn, produce folds in the group-velocity surface.

References

1. B. A. Auld, *Acoustic Waves and Fields in Solids, Vol. I* (Wiley, New York, 1973).

2. A. G. Every, General closed-form expressions for acoustic waves in elastically anisotropic solids, *Phys. Rev. B* **22**, 1746 (1980).

3. G. A. Northrop and J. P. Wolfe, Ballistic phonon imaging in germanium, *Phys. Rev. B* **22**, 6196 (1980).

4. G. A. Northrop, Acoustic phonon anisotropy: phonon focusing, *Comp. Phys. Comm.* **28**, 103 (1982).

5. H. J. Maris, Enhancement of heat pulses in crystals due to elastic anisotropy, *J. Acoust. Soc. Am.* **50**, 812 (1971).

6. A. G. Every, Ballistic phonons and the shape of the ray surface in cubic crystals, *Phys. Rev. B* **24**, 3456 (1981).

7. D. C. Hurley and J. P. Wolfe, Phonon focusing in cubic crystals, *Phys. Rev. B* **32**, 2568 (1985).

8. M. J. P. Musgrave, On whether elastic wave surfaces possess cuspidal edges, *Proc. Cambridge Philos Soc.* **53**, 897 (1957).

9. M. J. P. Musgrave, Further criteria for elastic waves in anisotropic media, *J. Elasticity* **9**, 105 (1979).

10. M. J. P. Musgrave, *Crystal Acoustics* (Holden-Day, San Francisco, 1970).

11. D. C. Hurley, M. T. Ramsbey, and J. P. Wolfe, Phonon focusing near a conic point, in *Phonon Scattering in Condensed Matter V*, A. C. Anderson and J. P. Wolfe, eds. (Springer-Verlag, Berlin, 1986).

12. One possibility that comes to mind is the application of stress to the crystal. However, the elastic constants are modified only by the third-order corrections, implying that immense pressures are required to change a phonon-focusing pattern. Another possibility suggested by Every is to grow a family of mixed crystals with continuously variable elastic constants. However, the nonperiodicity of a mixed crystal would produce strong phonon scattering.

13. A. G. Every and A. J. Stoddart, Phonon focusing in cubic crystals in which transverse phase velocities exceed the longitudinal phase velocity in some direction, *Phys. Rev. B* **32**, 1319 (1985).

14. J. P. Wolfe and M. R. Hauser, Acoustic wavefront imaging, *Ann. Physik* **4**, 99 (1995).

5

Acoustic symmetry and piezoelectricity

A. Degeneracy and the slowness surface

All the intricate focusing effects considered in Chapter 4 occur for crystals with very high (cubic) symmetry. One might rightly wonder what added complexity appears for crystals of lower symmetry. That is the subject of this chapter. We shall see that the phonon-focusing patterns of real crystals show many variations and yet have underlying similarities. We are aided by the fact that the acoustic symmetry – i.e., the symmetry displayed by the slowness surface – is generally higher than the point-group symmetry of the crystal. For example, all crystals possess centrosymmetric slowness surfaces (i.e., they are invariant to the inversion $\mathbf{k} \to -\mathbf{k}$), but not all crystal structures have a point of inversion.

Further good news is that we have already been introduced to all the basic topological features. At lower symmetries, there are still only three slowness sheets, which can contact each other conically, tangentially, or, in rare cases, along a line. Just how these degeneracies are distributed over the slowness surface, however, will vary from one crystal class to the next.

The accompanying bad news is that, with the exception of cubic and hexagonal crystals, it is virtually impossible to study systematically the multi-dimensional elastic parameter spaces. Figure 1 shows several crystal types and the elastic parameters required to describe each. While cubic crystals have only two independent ratios of elastic parameters ($a = C_{11}/C_{44}$ and $b = C_{12}/C_{44}$), hexagonal crystals have four ratios, implying a four-dimensional elastic-parameter space. There are five independent elastic ratios for tetragonal and trigonal.*

---

* The seven elastic constants shown in Figure 1 can be transformed to six elastic constants by rotation. I thank A. Every for this and other clarifications.

122

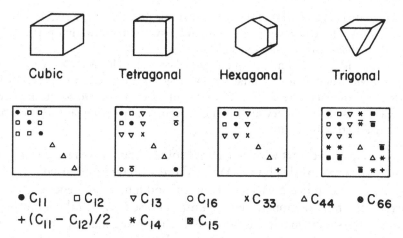

$\bullet\, C_{11}$ $\square\, C_{12}$ $\triangledown\, C_{13}$ $\circ\, C_{16}$ $\times\, C_{33}$ $\triangle\, C_{44}$ $\bullet\, C_{66}$

$+\, (C_{11} - C_{12})/2$ $*\, C_{14}$ $\boxtimes\, C_{15}$

Figure 1 Four crystal systems discussed in this chapter and the nonzero elements of the corresponding stiffness tensors. Each system has several classes of crystals, which may have more zero elements than shown (e.g., tetragonal and trigonal). A bar over a symbol implies a negative element. Triclinic, monoclinic, and orthorhombic systems are not shown.

Fortunately, the situation is not intractable. Despite the vast unexplored regions of the elastic-parameter spaces, known crystals span only a small part of these multidimensional spaces. Indeed, the focusing patterns of all the real crystals examined by phonon imaging to date can be traced to some unifying topology. Thus, we shall see that the caustic patterns observed for the trigonal crystals sapphire (Al_2O_3) and quartz (SiO_2) can be understood as evolutions from the higher hexagonal symmetry. There is a fascinating evolution of caustic structures as symmetry-lowering perturbations are introduced. Theoretical studies by Every[1-3] have graphically illustrated these ideas and shown the effectiveness of phonon imaging for displaying the acoustic symmetry of a crystal.

The concept that facilitates a global understanding of phonon-focusing patterns is *degeneracy*. We have seen that the lifting of degeneracies in the slowness surface produces phonon caustics. Recall that an infinitesimal displacement from the $\Delta = 0$ isotropy line (corresponding to completely degenerate transverse modes) instantly produces phonon caustics near the conic $\langle 111 \rangle$ directions in cubic crystals.

In addition to accidental degeneracies and those that occur for specific values of elastic constants (such as the triple degeneracy along $\langle 111 \rangle$ for $b = -1$ in cubic crystals), there are three types of degeneracies[1,4]:

(1) Conical degeneracies, which occur along all threefold axes and, as we shall see, frequently along nonsymmetric directions. Small changes in the elastic constants or even symmetry-lowering perturbations do not destroy a conical degeneracy. It is a structurally stable degeneracy.

(2) Tangential degeneracies of the two transverse branches along sixfold and four-fold axes. These degeneracies are lifted by symmetry-lowering perturbations (i.e., they are structurally unstable).

(3) Circular lines of degeneracy in hexagonal crystals. This is the only symmetry class with rotationally symmetric slowness surfaces; i.e., slices perpendicular to the sixfold axis are circles. Thus the intersections of two sheets are circles. Such lines of degeneracy are lifted by symmetry-lowering perturbations (i.e., they are structurally unstable).

In this chapter, we start with an example of phonon focusing in a tetragonal crystal, tellurium dioxide (TeO_2). While there is no representative crystal for this class, TeO_2 exhibits a variety of conical and tangential degeneracies and also experimentally demonstrates the unusual case in which the longitudinal and transverse sheets touch. Second, we discuss hexagonal crystals and show how the phonon-focusing patterns evolve when the symmetry is lowered to trigonal. Sapphire is the experimental example. Third, we examine how piezoelectricity affects ballistic heat flux. Finally, some general remarks are made about acoustic symmetry.

B. Tetragonal crystals

Extensive descriptions of phonon focusing in tetragonal crystals have been reported by Winternheimer and McCurdy[5] and by Every.[3] The former paper deals principally with the algebraic conditions for caustic formation along symmetry axes, and the latter work classifies the focusing patterns in the portion of elastic-parameter space that encompasses the currently known tetragonal crystals.

The Christoffel matrix for a tetragonal crystal is diagonalized for all k directions if $C_{13} + C_{44} = C_{12} + C_{66}$. Figure 2 shows the slowness sheets for a hypothetical tetragonal crystal that satisfies this condition. This condition is analogous to the condition $b = -1$ in cubic crystals. In this case, the slowness sheets are given by three ellipsoids,

$$\rho = C_{11}s_x^2 + C_{66}s_y^2 + C_{44}s_z^2 \qquad \text{(x polarized)}, \qquad (1)$$

$$\rho = C_{66}s_x^2 + C_{11}s_y^2 + C_{44}s_z^2 \qquad \text{(y polarized)}, \qquad (2)$$

$$\rho = C_{44}s_x^2 + C_{44}s_y^2 + C_{33}s_z^2 \qquad \text{(z polarized)}. \qquad (3)$$

These three sheets intersect each other along the heavy lines in Figure 2(a), analogous to the cubic case shown in Figure 21 on page 117. Here the velocity along z is not the same as that along y and x; hence, the tangential degeneracy found in the cubic case along these latter axes is removed.

Again we can picture the entire slowness surface as composed of inner (shaded), middle, and outer sheets. The focusing pattern associated with the two outer sheets is shown in Figure 2(b). The calculated phonon flux is proportional to the density of black dots. The square-shaped regions of

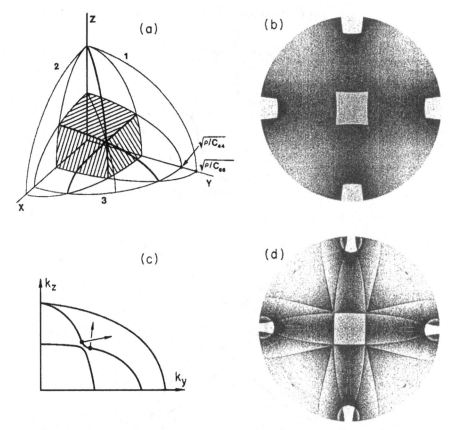

Figure 2 (a) Slowness surfaces for a hypothetical tetragonal medium with $C_{11} = 4$, $C_{33} = 3$, $C_{66} = 0.5$, and $C_{13} = -1$, all in units of C_{44}. Similar to Figure 21 of Chapter 4, the slowness surface consists of three ellipsoids that intersect along the heavy lines. (b) Combined phonon intensity pattern calculated for the middle and outer sheets. Contribution from the inner surface [shaded in (a)] would smoothly fill in the white squares. (c) and (d) For a small change in elastic constants, the line degeneracies are lifted and caustics are formed. Point degeneracies (e.g., in y–z plane) may still remain. (From Every.[3])

missing flux would be smoothly filled in if the flux from the inner sheet were included. There are no caustics for the ellipsoidal slowness surfaces of Eqs. (1)–(3). The surfaces are entirely convex.

If we relax the condition that exactly diagonalizes the Christoffel matrix (e.g., change one of the elastic constants slightly), the line degeneracies are removed. There are then distinct inner, middle, and outer sheets that are radically curved near the previous line intersections, as shown in Figure 2(c). The phonon flux regroups as in Figure 2(d). The sharp lines are caustics, with angular positions defined approximately by the normals to the unperturbed ellipsoids near the intersections. For small decoupling of the slowness sheets,

(a) (100) plane

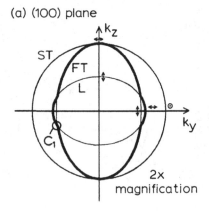

(b) (001) plane

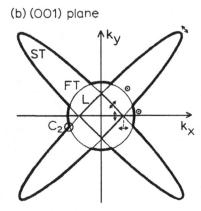

*Figure 3   Cross sections of the slowness surface of TeO$_2$ (paratellurite) in
(a) the y–z plane and (b) the x–y plane. Different line thicknesses identify
the three individual sheets. Polarizations at various locations are indicated
as double-headed arrows. Points labeled C$_1$ and C$_2$ are conic points where
two sheets touch. (From Hurley et al.[6])*

the positions of the caustics do not change much – but the intensity of the
caustics changes.

This gedanken experiment is a graphic example of how a small perturbation
can produce an immediate qualitative change in the flux pattern with only a
gradual change in phonon flux. As the perturbation increases from zero,
a faint caustic pattern is superimposed on the gradually modified focusing
pattern of the ellipsoids.

Unfortunately, the elastic constants of a crystal cannot be changed so easily
in an experiment. What topologies are observed in real tetragonal crystals?
The calculated intersections of the slowness surfaces with the (100) and (001)
planes are plotted for TeO$_2$ in Figure 3. The ⟨001⟩ axis is obviously the fourfold
axis. At first sight one sees intersecting ellipses, but the darker lines emphasize

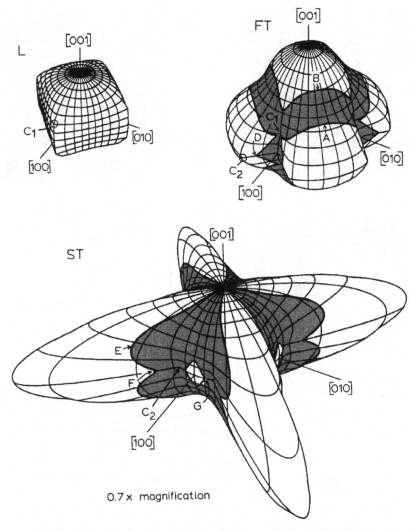

Figure 4 The three sheets of the slowness surface of the tetragonal crystal TeO$_2$. The shaded regions indicate saddle curvature, bounded by lines of zero Gaussian curvature. Conic points where the sheets touch are labeled C$_1$ and C$_2$. The ST sheet must be enlarged by 40% to match the scale of the other two sheets. (From Hurley et al.[6])

that the actual slowness sheets only touch at certain points. They do not pass through each other. This fact is clear from the three-dimensional plots of the three slowness sheets shown in Figure 4. These three surfaces nest inside each other, similar to the inner, middle, and outer surfaces of Figure 2(a).

For TeO$_2$, the inner surface (L) touches the middle surface (FT) at eight conic points, one of which is labeled C$_1$. The polarization of the waves near

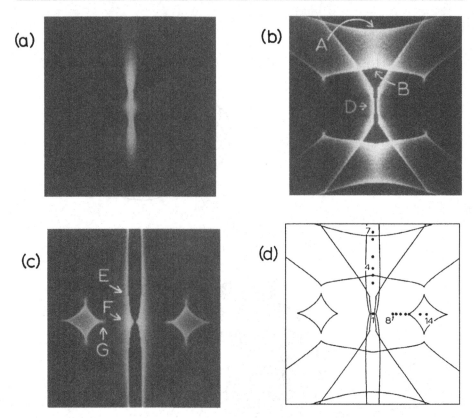

Figure 5 Monte Carlo calculations of phonon flux for TeO₂. (a) L mode, (b) FT mode, and (c) ST mode. (110) plane, ±65° left to right. The letters are correlated with features on the slowness surface in Figure 4. (d) Caustic map showing singular flux lines for all three modes. The dots labeled 1 through 14 are angular positions where the heat pulses shown in Figure 6 were recorded. (From Hurley et al.[6])

these contacts becomes highly mixed, with the FT mode taking on a largely longitudinal character and the L mode becoming largely transverse. The wave polarizations for propagation along the symmetry axes are indicated in Figure 3. The saddle regions on the FT sheet (shaded in Figure 4) are due to the avoided intersections with the L sheet. The FT sheet, in turn, conically meets the ST sheet at eight points in the (001) plane, one of which is labeled C_2, and tangentially meets the ST sheet along the [001] axis. The ST sheet is most spectacular, displaying pronounced lobes along the ⟨110⟩ directions. A microscopic reason for this extreme anisotropy will be given below.

Figure 5 contains the Monte Carlo simulations of phonon flux in the (110) plane for each mode of this crystal. Certain points on the caustics are labeled for comparison with the corresponding points on the parabolic lines of Figure 4. Notice that the L mode, Figure 5(a), has a strongly focused flux due

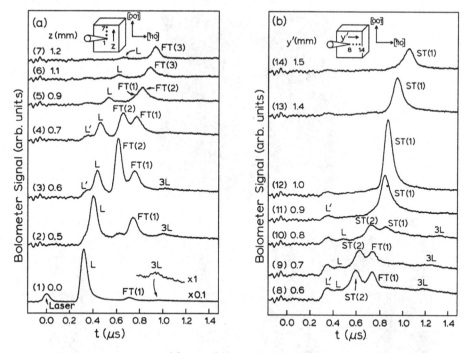

Figure 6 Experimental heat pulses (time traces) for various positions of the laser spot (Figure 5). The pulses are labeled according to the mode, as identified from the calculated times of flight. The numbers in parentheses identify phonons originating from different k's of the same mode. [001] is the fourfold axis of this tetragonal crystal. The label 3L identifies L pulses reflected from the surfaces. (From Hurley et al.[6])

to its flat surfaces, but it produces no caustics. A caustic map, including all modes, is given in Figure 5(d).

Experimental heat pulses in a TeO_2 crystal are shown in Figure 6. As the laser beam is translated in the ($\bar{1}10$) and (001) planes, the heat pulses undergo a variety of changes due to the focusing, and the pulses are identified and labeled with the help of velocity calculations in these planes. The laser positions for these time traces are shown as the dots in the caustic map of Figure 5(d). Notice that there are often two well-resolved heat pulses associated with one mode, arising from widely separated segments of the same wave surface. Phonon images for several selected time delays are shown in Figure 7. The principal features that were predicted in Figure 5 can be seen in the experimental images.

Finally, we address the question of why the ST mode of TeO_2 is so anisotropic. A simple representation of the atomic structure is given in Figure 8(a). Think of the structure in this plane as a simple square lattice with axes rotated by 45°, as drawn in Figure 8(b). A square lattice like this with principal

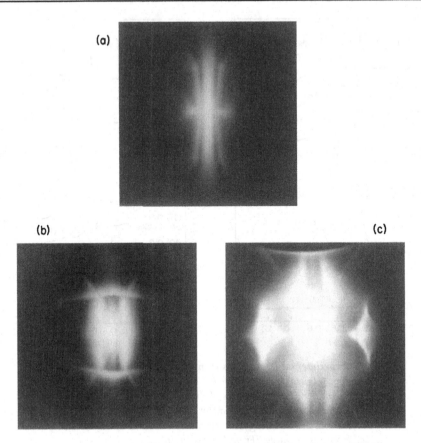

Figure 7 Experimental phonon images of TeO₂, for selected times of flights: (a) 390 ns, (b) 610 ns, and (c) 1000 ns. Boxcar gate = 200 ns. (110) surface, ±40° scan range for (a) and ±65° for (b) and (c), left to right. (From Hurley et al.[6])

bonding between near neighbors has very small resistance to a simple shear, resulting in a very slow transversely polarized wave with **k** along ⟨110⟩. This explains the long lobes of the ST sheet along these directions. The ST velocity in a ⟨110⟩ direction is less than one-seventh of the L velocity in that direction. This is one of the few examples in which the shape of a slowness surface can be simply related to the microscopic structure of the crystal. In general, the shapes of the slowness surfaces have a complicated relationship to the atomic structure and forces, as discussed in Chapter 6.

C. Hexagonal and lower symmetries

Hexagonal crystals are unique in having an axis of sixfold rotational symmetry, which leads to rotationally invariant slowness surfaces about that axis.

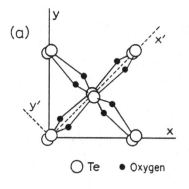

(a)

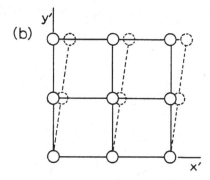

○ Te   ● Oxygen

(b)

Figure 8 (a) Microscopic explanation of the long lobes in the ST slow-ness surface of TeO$_2$ (Figure 4). By rotating the coordinate system to the x′–y′ axes along ⟨110⟩, the lattice can be viewed as a simple square lattice. (b) A square lattice with mainly nearest-neighbor bonding exhibits very little resistance to shear. Thus for $k_{ST} \parallel \langle 110 \rangle$, the wave velocity will be very slow. (From Hurley et al.[6])

This transverse isotropy is not apparent from simply looking at the elastic tensor (Figure 1), just as the condition $C_{11} - C_{12} = 2C_{44}$ does not obviously lead to complete isotropy in the cubic case. The transverse isotropy of hexagonal crystals makes a systematic classification of the slowness topologies possible,[1,5] despite there being four independent ratios of elastic constants.

The principal features of the slowness surfaces in this crystal class can be seen in Figure 9. These are cross sections of surfaces that are rotationally symmetric about the vertical axis. For hexagonal symmetry, it is useful to label the modes as quasi-longitudinal (QL), quasi-transverse (QT), and pure transverse (P). [Remember that in cubic crystals this designation was only possible for the (100) and (110) symmetry planes.] The two transverse surfaces are tangentially degenerate along the sixfold axis and, depending on the elastic constants, may also intersect each other, as shown in the figure. Such intersections are circular curves, such as those passing through points 4 and 6.

Parabolic lines of zero Gaussian curvature pass through points 1, 2, 3, 5, and 5′ in Figure 9. These circles give rise to circular caustic structures in

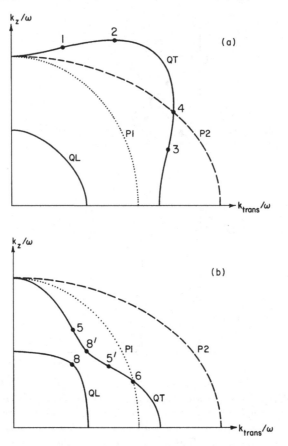

Figure 9 *Sections of the slowness surfaces of hexagonal media for several different values of elastic constants. These are surfaces of revolution about the z axis. For all curves, $C_{11} = C_{33} = 4$, in units of C_{44}. QL = quasi-longitudinal QT = quasi-transverse; P1, P2 = pure transverse for $C_{12} = 1.25$ and 3.0, respectively. (a) $C_{13} = 3$, (b) $C_{13} = 0$. (Adapted from Every.[1])*

the directions normal to the slowness surfaces, with one exception. All the normals to the surface along the circle passing through point 2 are along the same direction! This property results in a point caustic of very high intensity. (This effect has been observed in zinc by ultrasonic means.[7]) The phenomenon is also called external conical refraction, because waves propagating along this direction in real space would disperse into a cone of waves when incident on another medium. A single **V** along this direction in a hexagonal crystal corresponds (in part) to a cone of **k** vectors. Contrast this to internal conical refraction, where a single **k** corresponds to a cone of **V**'s, which is actually an anticaustic of zero intensity.

A slight lowering of the crystal symmetry acts to diffuse this point caustic into the usual line caustic. Every[1] has given an example in which a small value

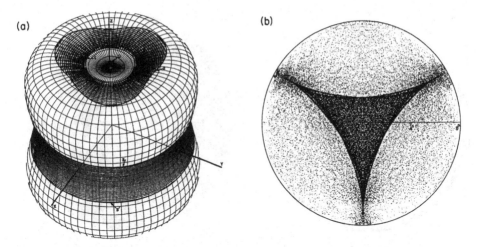

Figure 10 (a) Slow transverse surface for a nearly hexagonal crystal. [Formerly the QT sheet as in Figure 9(a), with P1 as the pure transverse mode.] A small nonzero value of C_{14} lowers the symmetry to trigonal and causes the parabolic line [labeled 2 in Figure 9(a)] to deviate from circular. (b) Calculated intensity pattern associated with the above-mentioned parabolic line. This six-cusped structure unfolds from a point at the center as C_{14} deviates from zero. (From Every.[1])

of C_{14} introduced into the elastic tensor makes the system slightly trigonal. The QT slowness sheet, with elastic parameters chosen so that the P sheet does not intersect it [e.g., like P1 in Figure 9(a)], is shown in Figure 10(a). The trigonal distortion deforms the parabolic line from circular and leads to the six-cusped caustic pattern calculated in Figure 10(b). To the author's knowledge, this intriguing phenomenon has not yet been observed experimentally.

If, for hexagonal symmetry, two slowness sheets are intersecting along a circular line, such as the case of QT and P1 in Figure 9(b), then a small trigonal distortion (nonzero C_{14}) lifts the line degeneracy. In this case, the designations QT and P1 become irrelevant as the surfaces separate into middle (FT) and outer (ST) sheets. These two sheets now contact each other at 12 conic points (six in each hemisphere), and the avoided intersections give rise to furrows on the ST sheet [Figure 11(a)], with a corresponding band of caustics [Figure 11(b)]. The polarizations indicated on the slowness surface of this figure graphically display the sections of the ST sheet that were previously QT and P2.

A small circuit about one of the conic points shows that the polarization may rotate either in a clockwise or counterclockwise sense, and adjacent conic points have opposite character. One may define a polarization index, n, where $2\pi n$ gives the total rotation of the polarization vector about a point on the slowness surface. Single conic points have an $n = \pm1/2$ character, corresponding to a

(a)   (b)

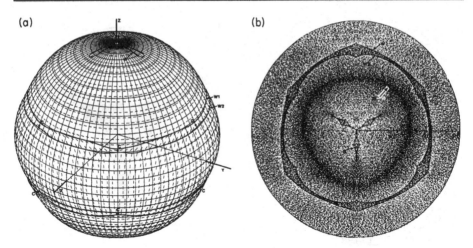

Figure 11 (a) ST slowness sheet of a trigonal medium, evolved from the P1 and QT sections shown in Figure 9(b). The polarizations, shown as the short line segments, retain the character of the original P1 and QT sheets, and the parabolic belt lines arise from the avoided intersection of these two sheets, similar to that in Figure 2(c). The original line degeneracy of P1 and QT is replaced by six conic points, labeled C, in each hemisphere. (b) Polar plot of the FT and ST phonon flux. (From Every.[1])

clockwise or counterclockwise rotation of the polarization as the conic point is circumnavigated in a clockwise sense. Of course, the polarization is undefined exactly at the conic point.

The polarization field near the sixfold axis has an index $n = +1$. This implies a singularity in the polarization field, which arises from the tangential degeneracy of the two transverse sheets along the axis. A small symmetry-lowering perturbation lifts the degeneracy of the two transverse sheets along this axis, but the long-range polarization field must still have the same character around this axis. To retain this singular field, some degeneracy near the z axis must remain. What happens is that the (structurally unstable) tangential degeneracy splits up into several (structurally stable) conic degeneracies, whose indices add up to $+1$. These conic points ensure the existence of caustics near this axis, which has changed from sixfold to threefold. Specifically, an $n = -1/2$ conical point on the axis and three equally spaced $n = 1/2$ conical points that are slightly off axis can give rise to the caustic pattern shown in Figure 12(a). As the separation between transverse sheets is increased, the caustics form a contiguous structure, shown in Figure 12(b), which resembles three neckties extending out from the symmetry axis.

Sapphire is a trigonal crystal that exhibits this caustic structure.[8] The transverse slowness and wave surfaces for sapphire are shown in Figure 13. We see that the various saddle regions associated with the avoided intersections and with the necktie structure have combined to form a continuous saddle region on the ST surface.

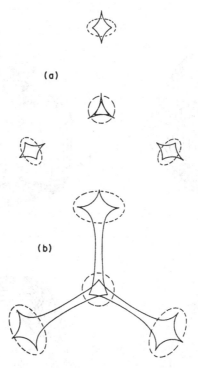

Figure 12 (a) Caustic structure that results from the lifting of the tangential degeneracy of QT and P sheets along the hexagonal axis, as the symmetry is lowered to trigonal. The four islands are caused by conic degeneracies between the ST and FT sheets (formerly QT and P). Dashed lines indicate ellipses of conical refraction. (b) Evolution of the above caustic structure as the separation of the two slowness sheets is increased. This pattern is associated with the polar structures (labeled B) in Figures 11(a) and 11(b). (From Every.[1])

Experimental phonon images of sapphire at several magnifications are shown in Figure 14. The necktie structure, as well as the caustics originating from the avoided intersections, is quite apparent. Also, for these time-integrated images, the caustics from the FT mode are observed. One can identify which structures belong to which modes by comparing the phonon images to the calculated wave surfaces of Figure 13. The caustics for this crystal are very sharp indeed, probably limited by the experimental resolution.

D. Piezoelectricity

Crystals that lack a center of inversion (with the exception of 432 symmetry) can exhibit piezoelectricity. That is, strain produces an electric polarization, and, conversely, an applied electric field produces a strain in the crystal. This coupling of electrical and mechanical fields is the basis for acoustic transducers, quartz clocks, radio-frequency filters based on surface acoustic waves,

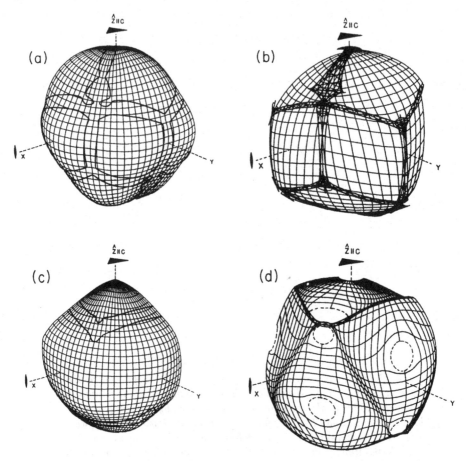

Figure 13 (a) ST and (b) FT slowness sheets for Al₂O₃, a trigonal crystal whose topology has evolved from that in Figure 11(a). (c) ST and (d) FT wave surfaces. The necktie structure described in Figure 12(b) can be seen in the ST wave surface. The dashed circles in (d) are due to the conic points in the slowness surface. (From Every et al.[8])

nonlinear optical devices, and many other useful devices. The addition of these piezoelectric forces modifies the simple Hooke's law relation between stress and strain. In short, the elastic constants of the crystal are stiffened by the piezoelectric coupling and depend on the phonon wavevector. The topic of this section is how this stiffening affects the ballistic heat flow – i.e., phonon focusing – in a crystal.

Because an applied stress produces both a strain field and an electric field, Hooke's law is generalized to[9-12]

$$\sigma_{ij} = C^E_{ijmn}\varepsilon_{mn} - e_{ijr}E_r, \tag{4}$$

where C^E_{ijlm} is the elastic tensor at constant electric field, ε_{mn} is the strain

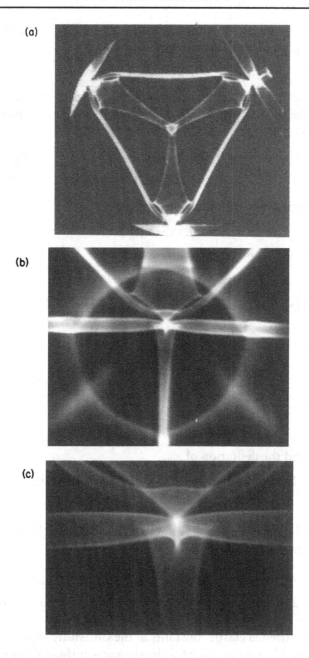

Figure 14 Experimental phonon images of Al_2O_3. (a) Wide-angle polar view showing the three neckties described in Figure 12(b). (b) Magnified view obtained with crystal having ($1\bar{1}02$) face. (c) Further magnified view, showing the sharpness of the caustics in sapphire. (From Every et al.[8])

tensor, e_{ijr} are the piezoelectric stress constants, and E_r is the electric field. The electric displacement field is given by

$$D_l = e_{lmn}\varepsilon_{mn} + e_{lr}^s E_r, \tag{5}$$

where e_{lr}^s is the dielectric constant at constant strain. Assuming that there are no free charges and that at sound velocities magnetic induction can be ignored, we have

$$\nabla \cdot \mathbf{D} = 0 = \partial D_l / \partial x_l \tag{6}$$

$$= e_{lmn} \partial \varepsilon_{mn} / \partial x_l + e_{pq}^s \partial E_q / \partial x_p. \tag{7}$$

In the quasi-static approximation,

$$\mathbf{E} = -\nabla\phi, \tag{8}$$

where ϕ is the electric potential. Thus,

$$e_{lmn}(\partial \varepsilon_{mn} / \partial x_l) = e_{pq}^s (\partial^2 \phi / \partial x_q \partial x_p). \tag{9}$$

Assuming a plane-wave solution,

$$\mathbf{u} = \mathbf{u}_o \exp[i(\mathbf{k} \cdot \mathbf{r} - \omega t)] \tag{10}$$

and

$$\phi = \phi_o \exp[i(\mathbf{k} \cdot \mathbf{r} - \omega t)], \tag{11}$$

Eq. (9) becomes

$$e_{lmn}(\partial \varepsilon_{mn} / \partial x_l) = e_{pq}^s k_q k_p \phi. \tag{12}$$

Using Eq. (8) and the definition of ε_{mn},

$$E_r = -i k_r \phi = -i k_r (e_{lmn}/e_{pq}^s k_p k_q)(\partial \varepsilon_{mn}/\partial x_l)$$
$$= -[(k_r e_{lmn} k_l)/e_{pq}^s k_p k_q]\varepsilon_{mn}, \tag{13}$$

and, with the definition of the wave normal $\mathbf{n} = \mathbf{k}/k$, we have the stiffened elastic constants C'_{ijlm} defined by

$$\sigma_{ij} = (C_{ijmn}^E + n_r e_{ijr} n_l e_{lmn}/e_{pq}^s n_p n_q)\varepsilon_{mn} = C'_{ijmn}\varepsilon_{mn}.$$

These stiffened elastic constants depend on the direction of the phonon wavevector.

Using these stiffened elastic constants in the Christoffel equation [Eqs. (18) and (19) on page 43], we can solve for the slowness surface of the material and compute phonon focusing patterns. Such calculations have been performed on a number of crystals, and the results may be found in the papers of Koos and Wolfe,[10,11] and Every and McCurdy.[12]

The effect of piezoelectricity on the phonon-focusing patterns of most noncentrosymmetric crystals is almost unobservable. The phase velocity is changed a few percent or less. For example, the piezoelectric constants of SiO_2 and of TeO_2 may be set to zero with negligible effect on the focusing pattern. In some crystals, however, piezoelectricity has a profound effect on the ballistic heat flow. This was demonstrated for lithium niobate ($LiNbO_3$) by Koos and Wolfe.[10,11] We shall now examine phonon focusing in this technologically important crystal. (For example, television receivers have surface-acoustic-wave radio-frequency filters using this material.)

LiNbO$_3$ is a trigonal crystal. To see the effect of piezoelectricity, we first plot the slowness surface by setting the piezoelectric constants to zero, i.e., using C_{ijlm}^E as the complete elastic tensor. Of course, this case cannot be realized in practice because it is not possible to set the electric field constant in an elastic wave,* especially one with a wavelength of a few hundred angstroms. The hypothetical ST and FT slowness sheets with no piezoelectricity are shown in Figures 15(a) and 15(b). The ST surface has the general features that we have seen for trigonal crystals, namely, two belts of parabolic lines (and conic points) that arise from avoided intersections with the FT surface. Also three furrows with parabolic lines connect the pole to the belt in each hemisphere. For these elastic constants, the FT sheet is completely convex; i.e., there are no parabolic lines or associated caustics, although the surface is almost flat near $\theta = 40°$, which produces the intense "x" seen in the Monte Carlo simulation of Figure 15(c).

Now for reality. By including the piezoelectric coupling constants measured by low-frequency techniques, we obtain the actual transverse slowness sheets of $LiNbO_3$ shown in Figures 15(d) and 15(e). The difference is striking. In particular, the continuous belts of saddle curvature are missing in the ST sheet, and the FT mode has acquired caustics [seen as the U-shaped belt at the bottom of Figure 15(f)], whereas it previously had none! The experimental phonon image (Figure 16), clearly shows this FT caustic structure brought about by the piezoelectric effect.

How do these topological changes come about? This question is largely answered by the gedanken experiment shown in Figure 17. These are cross sections of the slowness sheets in the y–z plane for a gradually increasing fraction f of the actual piezoelectric coupling constants. As f is increased from 0 (no piezoelectricity) to 1 (actual piezoelectricity), the conic contact, C_1, between FT and ST sheets moves from the southern to the northern hemisphere. Above $f = 0.9$, a saddle region is formed on the FT sheet, producing the parabolic lines running through points S_1 and S_2. Above $f = 0.95$, two new conic points, C_2 and C_3, are created in this plane. The conic points are created in pairs, with opposite polarization index n, in order to retain the global rotation index about a point on the surface.

---

* Unless one could add a high density of free carriers.

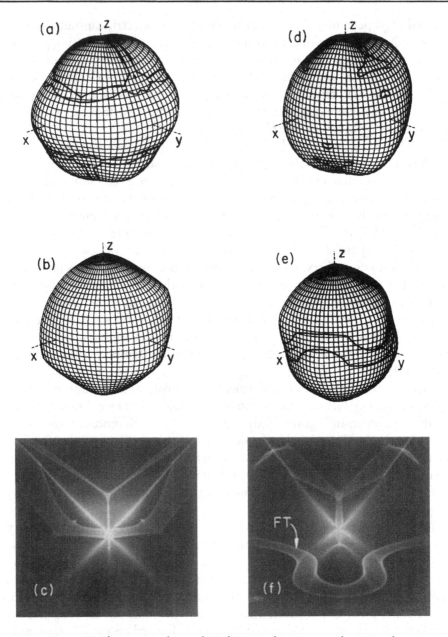

Figure 15 Slowness surfaces of LiNbO₃, neglecting piezoelectricity, for (a) ST and (b) FT modes. (c) Corresponding focusing pattern. (d) ST and (e) FT slowness sheets, including piezoelectric coupling. (f) Corresponding focusing pattern. (From Koos and Wolfe.[10])

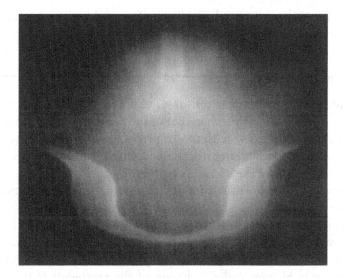

Figure 16 (a) Phonon image of LiNbO$_3$, showing the belt caustic that arises for the FT mode due to piezoelectricity. The image spans 50° left to right. (From Koos and Wolfe.[10])

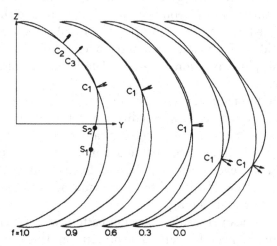

Figure 17 Cross sections of the LiNbO$_3$ slowness surface through the y–z plane as the piezoelectric stress constants are multiplied by the factor f. As f is increased, the conic degeneracy C$_1$ of the ST and FT sheets moves from the lower to the upper hemisphere. Also, new conic points, C$_2$ and C$_3$, appear, and singular points, S$_1$ and S$_2$, occur on the FT sheet. (From Koos and Wolfe.[11])

A systematic study of the topological effects of piezoelectricity on the slowness surface has been conducted by Every and McCurdy.[12] They find that for cubic crystals, the effect of piezoelectricity is greatest near the isotropy line, $\Delta = 0$, as defined in Eq. (29) on page 96. Frequently, the symmetry of the slowness surface is lowered by the addition of piezoelectricity – a fact that we shall now examine in more detail.

E. Acoustic symmetry

The symmetry of the elasticity tensor – i.e., its invariance to various rotations and reflections – is higher than that of the point-symmetry group of the crystal. This is often the case with tensor properties of crystals. For example, the scattering matrix describing diffraction of particles and waves from the lattice conforms to one of the 11 Laue symmetry groups, rather than to the 32 point-symmetry groups. Thus, an x-ray picture has higher symmetry than the crystal it describes.

The symmetry of the elastic tensor carries through to the properties derived from it: slowness and wave surfaces, phonon focusing patterns, etc. For long-wavelength phonons the acoustic symmetry is generally higher than that of the Laue group.[13] However, the inclusion of the piezoelectric tensor in the description of elastic waves [Eq. (14)] can lower the symmetry of phonon images to the Laue or even point group symmetry. Every[2] has examined this situation in detail, and here we give a brief description of his study.

Table 1 is a listing of the 32 point-symmetry groups and the corresponding Laue and acoustic-symmetry groups. The numbers $n = 2, 3, 4$ indicates twofold, threefold, and fourfold symmetry axes. A bar over the number indicates rotation plus inversion, $\mathbf{r} \to -\mathbf{r}$. The letter m indicates a mirror symmetry plane.

Let us take the tetragonal system as an example. There are seven crystal classes within this system, each with a different set of symmetry operations. From Table 1 we see that each of these crystal types belongs to one of two Laue groups, TI or TII. In fact, the stiffness matrix for TI has six independent elements, and TII has seven, including the additional component $C_{16} = -C_{26}$. But, remarkably, groups TI and TII have the same acoustic symmetry when there is no piezoelectric stiffening.

Specifically, this means that the focusing pattern of a TII crystal has two vertical reflection planes that are not present in the crystal structure. Figure 18(a) shows a phonon-focusing pattern for a TII crystal, calculated as described before. The vertical reflection planes are obvious, but they are rotated away from the principal crystal axes! This is in contrast to a TI crystal, such as TeO$_2$, whose phonon-focusing pattern has vertical mirror planes that contain the x and y crystal axes.

The angle of rotation between the x axis and the vertical reflection plane in the phonon image is given by

$$\tan 4\phi = 4C_{16}/(C_{11} - C_{12} - 2C_{66}). \qquad (14)$$

It can be shown that a rotation ϕ of the x and y axes about the fourfold z axis

Table 1. *The symmetry groups of phonon focusing patterns*

Crystal system	Crystal class	Laue group	Acoustic symmetry	
			No piezoelectric stiffening	With piezoelectric stiffening
Triclinic	1	$N = \bar{1}$	$\bar{1}$	$\bar{1}$
	$\bar{1}$	N	$\bar{1}$	
Monoclinic	2	$M = 2/m$	$2/m$	$2/m$
	m	M	$2/m$	$2/m$
	$2/m$	M	$2/m$	
Orthorhombic	222	$O = mmm$	mmm	mmm
	mm2	O	mmm	mmm
	mmm	O	mmm	
Tetragonal	4	$TII = 4/m$	$4/mmm$	$4/m$
	$\bar{4}$	TII	$4/mmm$	$4/m$
	$4/m$	TII	$4/mmm$	
	422	$TI = 4/mmm$	$4/mmm$	$4/mmm$
	4mm	TI	$4/mmm$	$4/mmm$
	$\bar{4}2m$	TI	$4/mmm$	$4/mmm$
	$4/mmm$	TI	$4/mmm$	
Trigonal	3	$RII = \bar{3}$	$\bar{3}m$	$\bar{3}$
	$\bar{3}$	RII	$\bar{3}m$	
	32	$RI = \bar{3}m$	$\bar{3}m$	$\bar{3}m$
	3m	RI	$\bar{3}m$	$\bar{3}m$
	$\bar{3}m$	RI	$\bar{3}m$	
Hexagonal	6	$HII = 6/m$	∞/mm	∞/mm
	$\bar{6}$	HII	∞/mm	$6/mmm$
	$6/m$	HII	∞/mm	
	622	$HI = 6/mmm$	∞/mm	∞/mm
	6mm	HI	∞/mm	∞/mm
	$\bar{6}m2$	HI	∞/mm	$6/mmm$
	$6/mmm$	HI	∞/mm	
Cubic	23	$CII = m3$	m3m	m3m
	m3	CII	m3m	
	432	$CI = m3m$	m3m	
	$\bar{4}3m$	CI	m3m	m3m
	m3m	CI	m3m	

transforms the C_{ij}'s among themselves in such a way that the element C_{16} is eliminated. Thus the stiffness matrices for TI and TII become identical in form, involving only six independent components.

As seen from Table 1, the addition of piezoelectricity reduces the acoustic symmetry of TII crystals to that of the Laue group. The vertical reflection planes are no longer present. A hypothetical phonon-focusing pattern for a TII crystal with piezoelectricity is plotted in Figure 18(b). Clearly the vertical reflection planes have vanished.

(a)   (b)

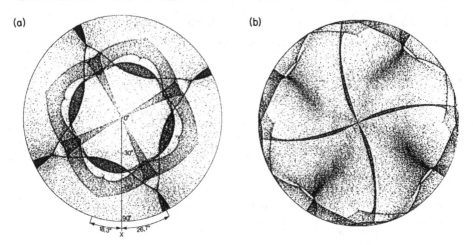

Figure 18 (a) Polar plot of the ST and FT phonon flux for AgClO$_3$. (b) Phonon flux for a hypothetical material, showing the symmetry lowering due to piezoelectric coupling. (From Every.[2])

It is hoped that this brief discussion gives an appreciation of how phonon-focusing patterns may be classified according to their symmetry. Indeed, phonon imaging is a vivid means of displaying the acoustic symmetry of a crystal.

References

1. A. G. Every, Formation of phonon-focusing caustics in crystals, *Phys. Rev. B* **34**, 2852 (1986).

2. A. G. Every, Acoustic symmetry in phonon imaging, *J. Phys. C: Solid State Phys.* **20**, 2973 (1987).

3. A. G. Every, Classification of the phonon-focusing patterns of tetragonal crystals, *Phys. Rev. B* **37**, 9964 (1988).

4. V. I. Al'shits and A. L. Shuvalov, Polarization fields of elastic waves near the acoustic axes, *Sov. Phys.-Crystallogr.* **29**, 373 (1984).

5. C. G. Winternheimer and A. K. McCurdy, Phonon focusing and phonon conduction in orthorhombic and tetragonal crystals in the boundary-scattering regime. II, *Phys. Rev. B* **18**, 6576 (1978).

6. D. C. Hurley, J. P. Wolfe, and K. A. McCarthy, Phonon focusing in tellurium dioxide, *Phys. Rev. B* **33**, 4189 (1986). See also A. G. Every and V. I. Neiman, Reflection of electroacoustic waves in piezoelectric solids: mode conversion into four bulk waves, *J. Appl. Phys.* **71**, 6018 (1992).

7. K. Y. Kim, W. Sachse, and A. G. Every, Focusing of acoustic energy at the conical point in zinc, *Phys. Rev. Lett.* **70**, 3443 (1993).

8. A. G. Every, G. L. Koos, and J. P. Wolfe, Ballistic phonons imaging in sapphire: bulk focusing and critical-cone channeling effects, *Phys. Rev. B* **29**, 2190 (1984); G. L. Koos, A. G. Every, G. A. Northrop, and J. P. Wolfe, Critical-cone channeling of thermal phonons at a sapphire-metal interface, *Phys. Rev. Lett.* **51**, 276 (1983).

9. B. A. Auld, *Acoustic Waves and Fields in Solids, Vol. I* (Wiley, New York, 1973).

10. G. L. Koos and J. P. Wolfe, Piezoelectricity and ballistic heat flow, *Phys. Rev. B* **29**, 6015 (1984).

11. G. L. Koos and J. P. Wolfe, Phonon focusing in piezoelectric crystals: quartz and lithium niobate, *Phys. Rev. B* **30**, 3470 (1984).

12. A. G. Every and A. K. McCurdy, Phonon focusing in piezoelectric crystals, *Phys. Rev. B* **36**, 1432 (1987).

13. F. I. Fedorov, *Theory of Elastic Waves in Crystals* (Plenum, New York, 1968).

6

Lattice dynamics

In the previous chapters we saw how continuum elasticity theory leads to a rich variety of phonon focusing patterns. The predicted phonon images are completely determined by the elastic constants, in some cases modified by piezoelectric stiffening. An alternative approach is to model the microscopic forces between atoms in the crystal lattice. Such lattice dynamics models are necessary to explain the propagation of phonons with wavelengths comparable to the lattice spacing. Of course, these microscopic models must also be compatible with the continuum theory in the limit of long wavelengths. The purpose of this chapter is threefold: (1) to introduce the mathematical underpinnings of lattice dynamics, (2) to examine the connection between continuum mechanics and the theories of lattice dynamics, and (3) to summarize the predictions of selected lattice dynamics models for phonon wavelengths approaching the atomic spacings.

A basic question that arises is how the intricate anisotropies in elastic-wave propagation relate to the crystal structure and the forces between atoms. In particular, what arrangements of atoms and types of interatomic forces lead to elastic stability? ... to positive or negative anisotropy factor, Δ? ... to an elastically isotropic medium? Can one construct a hypothetical lattice that has the same sound velocity in all directions? What kinds of slowness surfaces do the simple force models predict? How sophisticated must the microscopic models be in order to predict the measured elastic properties of real crystals?

The study of lattice dynamics is one of the principal areas of modern solid-state physics. In addition to describing the elastic, acoustic, and thermal properties of solids, it bears upon the many optical and electronic properties that involve the interaction of electromagnetic waves and electrons with phonons. The basis for the modern theory of lattice dynamics was forged by Born and coworkers. A selection of books and reviews on the subject are listed in the references.[1-9]

Equilibrium positions

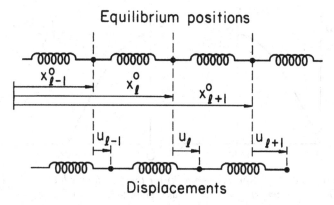

Displacements

Figure 1 Equilibrium positions and displacements for the linear chain of atoms, mass M, connected by springs with spring constant κ.

A.   Linear chain

The textbook example of lattice dynamics is the linear chain of atoms with equal masses and connecting springs,[10] as drawn in Figure 1. The lattice constant is given by a, so the equilibrium positions of the atoms are $x_\ell^o = \ell a$, where ℓ is an integer. The instantaneous positions of the atoms, x_ℓ, and their displacements, u_ℓ, are related by

$$x_\ell = x_\ell^o + u_\ell. \qquad (1)$$

With a radial spring constant, κ, the potential energy of the entire lattice is

$$\mathcal{U} = \mathcal{U}_o + \frac{1}{2}\kappa(u_{\ell+1} - u_\ell)^2 + \frac{1}{2}\kappa(u_\ell - u_{\ell-1})^2 + \mathcal{O}, \qquad (2)$$

where $\mathcal{U}_o$ is the potential energy of the undisplaced lattice and $\mathcal{O}$ represents all terms not involving the atom labeled ℓ. The force on atom ℓ is given by,

$$F_\ell = -d\mathcal{U}/dx_\ell$$
$$= -2\kappa u_\ell + \kappa(u_{\ell+1} + u_{\ell-1}). \qquad (3)$$

This equation has the general form

$$F_\ell = -\sum_{\ell'} \phi(\ell\ \ell') u_{\ell'}, \qquad (4)$$

where $\phi(\ell\ \ell')$ are the generalized force constants associated with atom ℓ and ℓ'. From Eq. (3), we see that the simplest case with spring coupling between nearest neighbors gives

$$\phi(\ell\ \ell) \equiv d^2\mathcal{U}/dx_\ell^2 = 2\kappa \qquad (5)$$
$$\phi(\ell\ \ell\pm1) \equiv d^2\mathcal{U}/dx_{\ell\pm1}dx_\ell = -\kappa. \qquad (6)$$

The equation of motion for the ℓth atom (mass M) is

$$M\ddot{u}_\ell = -2\kappa u_\ell + \kappa(u_{\ell+1} + u_{\ell-1}). \qquad (7)$$

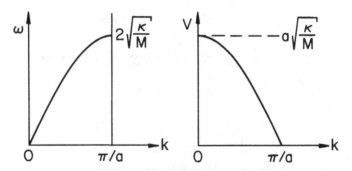

Figure 2 (a) Dispersion relation $\omega(k)$ for the monatomic linear chain of atoms. (b) Group velocity, $V = d\omega/dk$, versus wavevector.

Substituting the plane-wave solution,

$$u_\ell = u_0 \exp(i(kx_\ell^o - \omega t), \tag{8}$$

into Eq. (7) yields the condition

$$M\omega^2 = \kappa(2 - e^{ika} - e^{ika})$$
$$= 2\kappa(1 - \cos ka). \tag{9}$$

Noting the identity, $1 - \cos\theta = 2\sin^2(\theta/2)$, we have the dispersion relation,

$$\omega = 2(\kappa/M)^{1/2}|\sin(ka/2)|, \tag{10}$$

and the group velocity,

$$V = d\omega/dk = (\kappa/M)^{1/2}a\cos(ka/2). \tag{11}$$

These results are plotted in Figure 2. The maximum frequency and minimum group velocity (zero) of an elastic wave on a linear chain occur at the wavevector $k = \pi/a$, which represents a standing wave. For long wavelengths ($k \ll a$) the dispersion is linear, $\omega = (\kappa/M)^{1/2}ak = vk$.

The above calculation is for a longitudinal wave, having a polarization along the wavevector direction. It is possible to visualize a transverse wave of the linear chain, but the simple springs do not provide a restoring force to first order. If we imagine that the atoms are bound by directed electronic bonds, such as the covalent linear chain shown in Figure 3, a restoring force linear in the transverse displacement, u_ℓ, occurs, and we may write the relevant potential energy terms as Eq. (2) with κ replaced by a transverse coupling constant, κ_t, between neighboring atoms. The solution to this problem is identical to the previous one except that there are two degenerate transverse modes associated with the two transverse degrees of freedom. This simple model (Figure 3) is not very realistic, but it introduces the idea of noncentral forces between atoms.

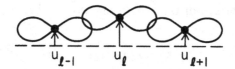

Figure 3 *Hypothetical linear chain with tangential restoring forces, which allow transverse acoustic waves.*

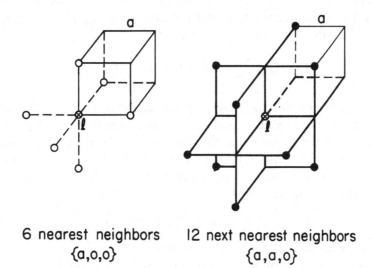

6 nearest neighbors
{a,o,o}

12 next nearest neighbors
{a,a,o}

Figure 4 *Atomic positions on the simple-cubic lattice, showing the nearest and next-nearest atoms to the atom labeled ℓ.*

B. Vibrations of Bravais lattices

Now consider the extension of the above ideas to a three-dimensional lattice, such as the simple-cubic lattice shown in Figure 4. Assume that there are only interactions between a given atom and its first and second nearest neighbors. The interactions are described by *central* forces, which depend solely upon the distance between atoms, as if there were springs connecting the atoms. The spring constant for nearest neighbors is κ, and that for next-nearest neighbors is κ'.

The instantaneous positions of the atoms of the lattice, $\mathbf{r}_\ell$, and their displacements, $\mathbf{u}_\ell$, are related by

$$\mathbf{r}_\ell = \mathbf{r}_\ell^0 + \mathbf{u}(\ell), \tag{12}$$

where $\mathbf{r}_\ell^0$ defines the equilibrium lattice positions.* A Bravais lattice is one

---

* The displacement $\mathbf{u}(\ell)$ is not written as the shorthand $\mathbf{u}_\ell$ because we must explicitly consider its cartesian components, $u_i(\ell)$.

in which the equilibrium distance between atoms labeled ℓ and ℓ' is defined by

$$\mathbf{r}_{\ell'}^0 - \mathbf{r}_{\ell}^0 = \ell_1 \mathbf{a} + \ell_2 \mathbf{b} + \ell_3 \mathbf{c}, \tag{13}$$

where ℓ_i are integers, and $\mathbf{a}$, $\mathbf{b}$, and $\mathbf{c}$ are basis vectors. There are 14 types of Bravais lattices in three dimensions, classified and defined by the relative orientations and magnitudes of the basis vectors.[10] For the simple-cubic lattice, the basis vectors are orthogonal and have equal length, a. The potential energy of the entire lattice is

$$\mathcal{U} = \mathcal{U}_o + \sum_{\ell'} \phi_{ij}(\ell\ \ell') u_i(\ell) u_j(\ell') + \cdots, \tag{14}$$

where i, j are cartesian coordinates x, y, and z, and $\phi_{ij}(\ell\ \ell')$ is a generalized force constant between atoms ℓ and ℓ'. (Sum over repeated coordinate indices, as usual.) Only terms involving the atom labeled ℓ are explicitly shown in Eq. (14). In addition, terms of higher order in the displacements are neglected (the harmonic approximation). The force-constant matrix, $\Phi(\ell\ \ell')$, between atoms ℓ and ℓ' has the elements

$$\phi_{ij}(\ell\ \ell') = \partial^2 \mathcal{U}/\partial u_i(\ell)\ \partial u_j(\ell'), \tag{15}$$

which are the Born–Von Kármán force constants of the lattice.

The components of force on atom ℓ are given by $F_i(\ell) = -\partial \mathcal{U}/\partial u_i(\ell)$, which, from Eq. (14), directly leads to the equation of motion for this atom,

$$M\ddot{u}_i(\ell) = -\sum_{\ell'} \phi_{ij}(\ell\ \ell') u_j(\ell'). \tag{16}$$

The right side of Eq. (16) is the ith component of the force on atom ℓ due to a displacement $\mathbf{u}(\ell')$ of atom ℓ', summed over all atoms ℓ' including ℓ. For example, the term containing the generalized force constant $\phi_{xy}(\ell\ \ell')$ determines the x component of the force on atom ℓ produced by a y displacement of atom ℓ'.

Assuming a plane wave with unit polarization $\mathbf{e}$ and wavevector $\mathbf{k}$,

$$\mathbf{u}(\ell) = u_0 \mathbf{e} \exp[i(\mathbf{k} \cdot \mathbf{r}_\ell^0 - \omega t)], \tag{17}$$

we obtain

$$M\omega^2 e_i = \sum_{\ell'} \phi_{ij}(\ell\ \ell') e_j \exp\{i[\mathbf{k} \cdot (\mathbf{r}_{\ell'}^0 - \mathbf{r}_{\ell}^0)]\}. \tag{18}$$

Defining the dynamical matrix for the atom ℓ,

$$\mathcal{D}_{ij} = M^{-1} \sum_{\ell'} \phi_{ij}(\ell\ \ell') \exp\{i[\mathbf{k} \cdot (\mathbf{r}_{\ell'}^0 - \mathbf{r}_{\ell}^0)]\}, \tag{19}$$

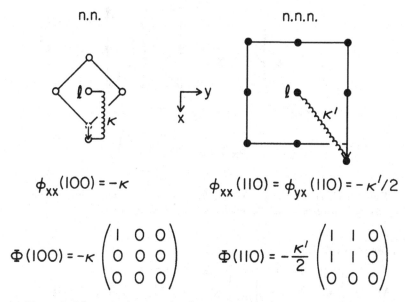

$$\Phi(100) = -\kappa \begin{pmatrix} 1 & 0 & 0 \\ 0 & 0 & 0 \\ 0 & 0 & 0 \end{pmatrix} \qquad \Phi(110) = -\frac{\kappa'}{2} \begin{pmatrix} 1 & 1 & 0 \\ 1 & 1 & 0 \\ 0 & 0 & 0 \end{pmatrix}$$

Figure 5 *Schematic diagram for determining the force-constant matrices between the atom labeled ℓ and neighboring atoms labeled $(\ell_1\ell_2\ell_3)$. This is a projection of the atoms in Figure 4 onto the z plane. Each atom pair has a 3×3 matrix, $\Phi(\ell_1\ell_2\ell_3)$, associated with it. Radial spring constants κ and κ' represent the central forces between nearest and next-nearest neighbors, respectively.*

reduces the wave equation to its simple eigenvalue form,

$$(\mathcal{D}_{ij} - \omega^2\delta_{ij})e_j = 0, \tag{20}$$

which can be solved for the three eigenvalues $\omega_\alpha(\mathbf{k})$ and eigenvectors $\mathbf{e}(\alpha, \mathbf{k})$, with $\alpha = $ L, FT, and ST as the mode index.

For a Bravais lattice, defined by Eq. (13), we denote $\Phi(\ell\ \ell') \equiv \Phi(\ell_1\ell_2\ell_3)$ to identify the force-constant matrix between atom ℓ at (000) and one of its neighbors ℓ' at $(\ell_1\ell_2\ell_3)$. The form of this matrix can be found from simple drawings such as those in Figures 5 and 6. General symmetry methods for determining the form of the force-constant matrices for various lattice types and force models are described in several textbooks.[2,6,7]

The dynamical matrix for a Bravais lattice is therefore written as

$$\mathcal{D} = M^{-1} \sum_{\ell_1\ell_2\ell_3} \Phi(\ell_1\ell_2\ell_3) \exp[i(\mathbf{k} \cdot \mathbf{a}\ell_1 + \mathbf{k} \cdot \mathbf{b}\ell_2 + \mathbf{k} \cdot \mathbf{c}\ell_3)]. \tag{21}$$

For a simple-cubic lattice with only nearest-neighbor forces, this sum contains seven terms, with the force-constant matrices given in Figure 5. As in the one-dimensional case [Eq. (9)], the terms combine pair-wise [except the single

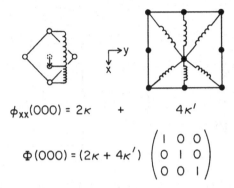

$$\phi_{xx}(000) = 2\kappa \qquad + \qquad 4\kappa'$$

$$\Phi(000) = (2\kappa + 4\kappa') \begin{pmatrix} 1 & 0 & 0 \\ 0 & 1 & 0 \\ 0 & 0 & 1 \end{pmatrix}$$

Figure 6 Determination of the zeroth-order force-constant matrix – the restoring force on atom ℓ caused by its own displacement.

zero-order term, $\Phi(000)$],

$$M\mathcal{D} = \begin{bmatrix} 2\kappa & 0 & 0 \\ 0 & 2\kappa & 0 \\ 0 & 0 & 2\kappa \end{bmatrix} + \begin{bmatrix} -\kappa & 0 & 0 \\ 0 & 0 & 0 \\ 0 & 0 & 0 \end{bmatrix} (e^{i\mathbf{k}\cdot\mathbf{a}} + e^{-i\mathbf{k}\cdot\mathbf{a}}) + \cdots \quad (22)$$

$$= 2\kappa \begin{bmatrix} (1 - \cos\mathbf{k}\cdot\mathbf{a}) & 0 & 0 \\ 0 & (1 - \cos\mathbf{k}\cdot\mathbf{b}) & 0 \\ 0 & 0 & (1 - \cos\mathbf{k}\cdot\mathbf{c}) \end{bmatrix} \quad (23)$$

From this simple example, we can see some general features of lattice dynamics in three dimensions. First, each atom in a Bravais lattice is at a center of inversion, so the force-constant matrices combine pair-wise to give elements in the dynamical matrix of the form $1 - \cos\mathbf{k}\cdot\mathbf{a} = 2\sin^2\mathbf{k}\cdot\mathbf{a}/2$. For wavevectors small compared to π/a, these terms approach zero, which, we shall see, leads to the familiar acoustic branches with $\omega = vk$. Second, if we consider the above case with orthogonal but unequal basis vectors, the maximum components of the phonon wavevectors are $\pi/a, \pi/b$, and π/c, respectively. These limits define the first Brillouin zone of the orthorhombic ($\mathbf{a} \perp \mathbf{b} \perp \mathbf{c}$) lattice.[10] Any wavevector outside this zone is equivalent to a wavevector inside the zone because displacement has no meaning between lattice sites. Therefore, only phonon wavevectors inside the first zone need to be considered. For electrons, whose wavefunctions are not only defined on the lattice sites but also between the nuclei, the higher zones *are* physically meaningful.

The eigenvalues of the dynamical matrix in the simple-cubic case, Eq. (23), are found by setting the determinant of Eq. (20) to zero, leading to,

$$(4\kappa \sin^2 k_x a/2 - M\omega^2)(4\kappa \sin^2 k_y a/2 - M\omega^2)(4\kappa \sin^2 k_z a/2 - M\omega^2) = 0. \quad (25)$$

The three solutions for ω correspond to the three acoustic branches. There are no off-diagonal terms in the dynamical matrix for this simple case; the

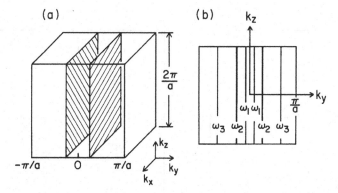

Figure 7 (a) The two shaded planes in the Brillouin zone of the simple-cubic lattice represent a constant-frequency surface for one of the three vibrational modes. This result assumes radial forces between nearest neighbors only. (b) Cross sections of the constant-frequency surfaces for several frequencies.

three modes are uncoupled. The constant-frequency surface for each mode is a pair of planes, corresponding to $\mp k_i$, as shown in Figure 7. For a closely spaced pair of y planes near $\omega = 0$, we may view this surface as a "pancake" with finite thickness along k_y and infinite lateral dimension along k_x and k_z, reminiscent of the oblate spheroids discussed in Chapter 4 (Section D). This mode has a diverging slowness (i.e., a phase velocity that approaches zero) for wavevectors with large k_x or k_z components. The polarization of this mode is independent of **k** and directed along the y axis. In other words, transverse waves – those with **e** nearly perpendicular to **k** – are extremely slow. Indeed, as can be seen from Figure 5, this simple-cubic lattice with only nearest-neighbor interactions has no resistance to shear – an unstable and unphysical situation.

Including central forces between next-nearest neighbors gives the simple-cubic lattice stability against shear forces. Now the dynamical matrix has both diagonal and non-diagonal terms, as shown in Figure 5. Using the shorthand notation, $S_{i\pm j} = \sin^2(k_i \pm k_j)a/2$, the elements of the dynamical matrix are,[7]

$$\mathcal{D}_{xx} = (4\kappa/M)\sin^2 k_x a/2 + (2\kappa'/M)(S_{x+y} - S_{x-y} + S_{x+z} + S_{x-z})$$
$$\mathcal{D}_{xy} = (2\kappa'/M)(S_{x+y} - S_{x-y}), \quad \text{etc.,} \tag{26}$$

where κ and κ' are the first- and second-neighbor force constants. Diagonalizing this matrix yields three well-behaved modes. Cross sections of the three constant-frequency surfaces in the (100) plane, assuming that $\kappa/\kappa' = 1.5$, are shown in Figure 8.

The similarities of this problem to the acoustic continuum problem are becoming apparent. The eigenvalue equation (Christoffel equation) for the

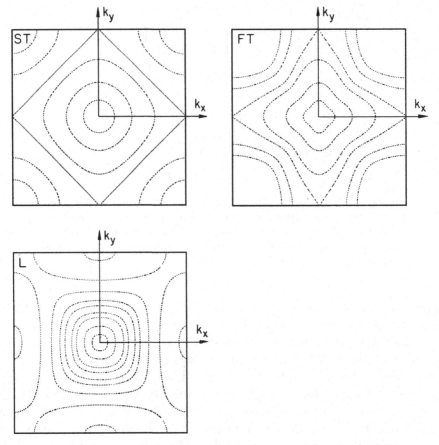

Figure 8 Cross sections of the constant-frequency surfaces in the (100) plane for the simple-cubic lattice with nonzero first- and second-neighbor radial forces. (Calculation by Markus Bauer.)

elastic continuum,

$$(C_{injm}k_n k_m/\rho - \omega^2\delta_{ij})e_j = 0 \qquad (27)$$

has the same form as the lattice-dynamics Eq. (20), but the Christoffel matrix involves only quadratic terms in the wavevector, $k_n k_m$, compared to the trigonometric functions in the dynamical matrix. The Christoffel matrix leads to a dispersionless solution, $\omega = vk$, for a given wavevector direction, which means there is a unique slowness surface, $s = 1/v = k/\omega_0$, having the same shape as any constant-frequency surface. As **k** approaches the Brillouin-zone boundary, however, Figure 8 shows us that the constant-frequency surfaces predicted by lattice-dynamics theory deform due to the proximity of the boundary. We shall now show that the low-frequency surfaces in this figure correspond directly to the continuum solution.

Comparing the Christoffel matrix to the dynamical matrix in the long-wavelength limit provides us with a connection between the microscopic force constants and the macroscopic elastic constants. Using the force-constant matrices in Figure 5, including first and second neighbors, and assuming the wavevector $\mathbf{k} = k\hat{\imath}$, two elements of the dynamical matrix are

$$M\mathcal{D}_{xx} = (2\kappa + 4\kappa')2 \sin^2 ka/2,$$
$$M\mathcal{D}_{yy} = 4\kappa' \sin^2 ka/2,$$

(28)

which, in the long-wavelength limit, $ka \ll 1$, become,

$$M\mathcal{D}_{xx} = (\kappa + 2\kappa')a^2k^2 \quad \text{and} \quad M\mathcal{D}_{yy} = \kappa'a^2k^2.$$

The corresponding elements of the Christoffel matrix are

$$C_{xxnm}k_nk_m = C_{xxxx}k^2 = C_{11}k^2$$
$$C_{yynm}k_nk_m = C_{yyxx}k^2 = C_{44}k^2.$$

(29)

Noticing that the density in Eq. (27) is $\rho = M/a^3$ and comparing Eqs. (20) and (27), we have $C_{11}k^2 = \rho\mathcal{D}_{xx} = M\mathcal{D}_{xx}/a^3$ and $C_{44}k^2 = M\mathcal{D}_{yy}/a^3$, leading to

$$C_{11} = (\kappa + 2\kappa')/a,$$
$$C_{44} = \kappa'/a,$$

(30)

for the simple-cubic lattice with central forces between first and second neighbors. As indicated previously, the component C_{44} associated with the shear-wave velocity $v = \sqrt{(C_{44}/\rho)}$ along $\langle 100 \rangle$ vanishes if there are no second-neighbor forces.

By comparing elements of the Christoffel and dynamical matrices for wave propagation along [110], one can show that $C_{12} = C_{44}$. This equality, known as the Cauchy relation, has a wider validity than the simple cubic case we have been examining; it is a property of the central (point-to-point) force law. There are additional relations for noncubic Bravais lattices.[6] The Cauchy relations are only valid for crystals in which the atoms are at a center of inversion, which includes all crystals with Bravais lattice structure, and only for central forces.

In terms of the a–b parameter space of cubic crystals, the Cauchy relation, $C_{12} = C_{44}$, means that all cubic crystals with central forces between the atoms should lie on the line, $b = 1$. Exactly where the crystal lies on this line depends on the ratios of force constants associated with the various neighbors – in the above case, κ'/κ.

Most of the cubic crystals we have been studying with heat-pulse methods have the rocksalt (NaCl) or diamond structures, which are not Bravais lattices. The rocksalt and diamond crystal structures are composed of a face-centered-cubic (fcc) Bravais lattice with a basis of more than one atom. It is intriguing that many of these crystals do lie close to the $b = 1$ line. For rocksalt, which has a center of inversion at each atom, this supports the argument for central forces. Atoms in a diamond structure do not sit at a center of inversion, yet $b \cong 1$ for Ge, Si, and most III–V compounds.

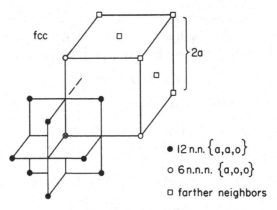

fcc

2a

- 12 n.n. $\{a,a,0\}$
- ○ 6 n.n.n. $\{a,0,0\}$
- □ farther neighbors

Figure 9 Atomic positions for the face-centered-cubic (fcc) lattice, showing the various neighbors of the atom labeled ℓ.

An fcc Bravais lattice is shown in Figure 9. Comparing Figures 4 and 9, we see that the positions of the 12 nearest-neighbor atoms in the fcc lattice are the same as the second neighbors in the simple-cubic lattice; they are given by $(\ell_1\ell_2\ell_3) = (110), (101), (\underline{1}10)$, etc. We have already solved this problem. The conventional cubic cell for fcc has a side of $2a$, as in Figure 9, and there are four atoms per cell. Therefore, taking $\rho = 4M/(2a)^3$ and neglecting the κ term due to (100) atoms in Eq. (30), the relationships between the elastic constants and the nearest-neighbor central-force constants are

$$C_{11} = \kappa'/a,$$
$$C_{44} = C_{12} = \kappa'/2a, \tag{31}$$

with an fcc nearest-neighbor spring constant κ'. Because only one force constant is chosen to describe the crystal, the position of this fcc lattice in a–b space is fixed. The coordinates are

$$a = 2, \qquad b = 1. \tag{32}$$

The force constant gives the overall velocity scale for the crystal but does not affect the shape of the slowness surface.

The next-nearest neighbors in the fcc lattice are six atoms along the $\langle 100 \rangle$ axes, so the form of the force-constant matrices are the same as for nearest neighbors in a simple-cubic lattice. The atomic coordinates of nearest neighbor and next-nearest neighbor are given by $(a, a, 0)$ and $(2a, 0, 0)$. The elements of the dynamical matrix for a wave traveling along the x axis are,

$$\mathcal{D}_{xx} = 4\kappa'' \sin^2 ka + 8\kappa' \sin^2(ka/2) \simeq (4\kappa'' + 2\kappa')k^2a^2,$$
$$\mathcal{D}_{yy} = 4\kappa' \sin^2(ka/2) \simeq \kappa'k^2a^2, \tag{33}$$

where κ'' is the radial force constant between second neighbors. The elastic constants in this case are

$$C_{11} = (2\kappa'' + \kappa')/a \qquad \text{and} \qquad C_{44} = C_{12} = \kappa'/2a.$$

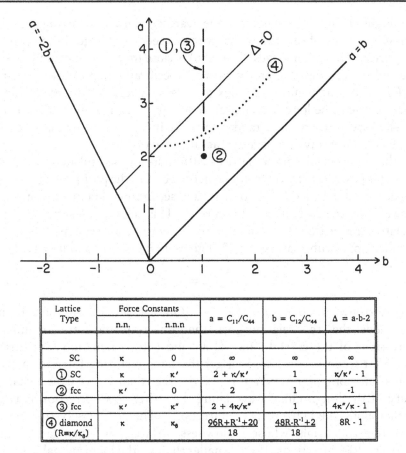

Lattice Type	Force Constants		$a = C_{11}/C_{44}$	$b = C_{12}/C_{44}$	$\Delta = a\text{-}b\text{-}2$
	n.n.	n.n.n			
SC	κ	0	∞	∞	∞
① SC	κ	κ'	$2 + \kappa/\kappa'$	1	$\kappa/\kappa' - 1$
② fcc	κ'	0	2	1	-1
③ fcc	κ'	κ''	$2 + 4\kappa/\kappa''$	1	$4\kappa''/\kappa - 1$
④ diamond $(R \equiv \kappa/\kappa_\theta)$	κ	κ_θ	$\dfrac{96R + R^{-1} + 20}{18}$	$\dfrac{48R - R^{-1} + 2}{18}$	$8R - 1$

Figure 10 Summary of the continuum-limit results for the simple rigid-ion models described in the text, showing the allowed coordinates of each model in elastic-parameter space. The main analytic results are given in tabular form. The model for the diamond lattice is a valence-force-field model, which has two parameters: a radial force constant κ between nearest neighbors and a torsional force constant κ_θ between adjacent covalent bonds. Variation of the ratio κ/κ_θ moves the system along the dotted line, $3a^2 - 7a + 3ab - 2b - 6b^2 = 0$, as shown by Musgrave.[6]

The results for this and the previous examples in terms of a, b, and Δ are summarized in the table accompanying Figure 10.

The allowed positions in a–b space of simple cubic and fcc lattices with central forces are plotted in Figure 10. From a purely theoretical point of view, it is fascinating that these simple ball-and-spring models yield all the topological complexity of real elastic surfaces in the long-wavelength limit. The most striking example is the fcc lattice with only nearest-neighbor central forces, represented as point 2 in Figure 10. The slowness surface of this simple lattice model is nearly identical to those of the common negative-Δ crystals, e.g., LiF, Ge, Si, GaAs, and InSb. This result is independent of the magnitude of the single adjustable parameter: the nearest-neighbor force constant. The ubiquitous pattern of caustics which we have seen (e.g., Figure 4 on page 16 and Figures 11 and 24 on pages 35 and 56) is close to that predicted by central forces and the *geometry* of an fcc lattice – specifically, that of the 12 nearest neighbors. Of course, more complex models with adjustable force constants can predict the same long-wavelength topology (i.e., the same a and b), and, as we shall see, more realistic modeling of the interatomic forces is necessary to understand the wave propagation at large wavevectors.

For the simple-cubic and fcc lattices with first- and second-neighbor inter- actions, it is possible to travel from the negative-Δ to the positive-Δ regions of a–b space by adjusting the ratio of first- and second-neighbor force constants, as indicated by the dashed line in Figure 10. This amounts to simply varying the relative interaction strength of (100)-type and (110)-type neighbors.

For the simple-cubic lattice, Eq. (30) gives $a = 2 + \kappa/\kappa'$ and $b = 1$, so,

$$\Delta \equiv a - b - 2 = \kappa/\kappa' - 1. \tag{34}$$

Perfect isotropy ($\Delta = 0$) results if $\kappa = \kappa'$. [One can verify from the ball-and- spring model that this relation gives the same restoring force for uniaxial compressions of the crystal along $\langle 100 \rangle$ and $\langle 110 \rangle$. Equation (34) further tells us that when $\kappa = \kappa'$, we will get the same restoring force for a given compression along *any* direction.] If κ is larger than κ', we expect that the velocity of the longitudinal wave along the stiffer (100) direction is larger than that along the (110) or (111) directions. This is precisely the case for the positive-Δ crystals, whose L slowness surfaces have their largest radii away from (100) directions. If κ is smaller than κ', the longitudinal velocities along (100) directions are relatively smaller than along (110) and (111), as observed for negative-Δ crystals. This interplay of the two sets of atoms gives us some idea how the slowness surfaces in crystals can assume their complicated shapes. The evolution of the transverse modes is more difficult to visualize.

C. Ionic and covalent crystals – A look at more realistic models

Few of the insulator and semiconductor crystals amenable to heat-pulse ex- periments are pure Bravais lattices. Alkali halides, for example, form in the rocksalt structure, which is an fcc lattice with a two-ion basis, such as Na^+ and F^-. These crystals are bound mainly by Coulombic forces between ions. Group IV crystals C, Ge, and Si and III–V crystals such as GaAs have the diamond or zinc-blende structure, which is an fcc lattice with a two-atom basis.[10] The diamond and zinc-blende crystals are characterized by tetrahedral covalent bonds, which implies highly directional interatomic forces.

For a lattice with a basis, the algebraic complexity of the lattice-dyna- mics theory escalates. The displacements and force-constant matrices are denoted as

$$\mathbf{u}\begin{pmatrix} \ell \\ s \end{pmatrix} \quad \text{and} \quad \Phi\begin{pmatrix} \ell & \ell' \\ s & s' \end{pmatrix},$$

where ℓ, ℓ' define the Bravais sites and s, s' label the basis atoms. So, for a two-atom basis, the dynamical matrix becomes a 6×6 matrix, producing 6 eigenvalues for a given wavevector $\mathbf{k}$. Generally there is no inversion symmetry about a particular basis atom. Nevertheless, the three acoustic modes associated with the Bravais lattice survive. The frequencies of the three additional modes do not approach zero as $\mathbf{k}$ approaches zero. These are known as the optical modes because they correspond to neighboring atoms vibrating out of phase and are thus optically active with respect to absorption or scattering of light. For a Bravais lattice with an N-atom basis, there are three acoustic modes and $3N - 3$ optical modes.

In addition to considering the particular lattice structure, realistic lattice-dynamics models must take into account the type of bonding between atoms. For insulators and semiconductors the two principal types of bonding are ionic (such as in NaF and CaF_2) and covalent (such as in Ge and Si), although the distinction becomes a bit blurred for intermediate compounds, such as II–VI and III–V compounds. We do not consider metallic or molecular crystals here, although extensive lattice-dynamics theories have been reported.

For ionic crystals, the first approach is simply to write the potential, $\mathcal{U}(\mathbf{r})$, in terms of Coulomb forces between positive and negative ions, plus a hard-core repulsion between nearest neighbors. One can then compute the Born force constants, $\partial^2 \mathcal{U}/\partial x_i \partial x_j$, and apply the methods outlined above. Sample calculations of these rigid-ion models for the common rocksalt structure are given, for example, in References 2, 6, and 9.

The general formalism can also be applied to covalent crystals,[4,5] but in this case the force constants that parameterize the theory have less physical meaning. An improvement is to consider the torsional character of the co-valent tetrahedral bonds associated with one atom.[6] Thus one can consider the radial force constant κ plus a torsional force constant associated with the bond angle, with potential $(1/2)r_0^2\kappa_\theta(\delta\theta)^2$, where r_0 is the bond length for a diamond lattice and $\delta\theta$ is the deviation of the bond angle from the equilibrium value. The long-wavelength results of this valence-force-field theory for a diamond lattice are shown in Figure 10. Because there are only two microscopic force constants, which involve both first and second neighbors, a constraining relation is generated between the three elastic constants, drawn as the dotted line in the figure. The accompanying table shows that isotropy occurs for $\kappa_\theta/\kappa = 1/8$. The constraint line comes reasonably close to (a, b) for Si and Ge, and C. However, this simple model does not agree well with the dispersion relations determined from neutron scattering.

The rigid-ion and valence-force-field models can be coerced into agreement with the dispersion curves by assuming a number of adjustable force constants

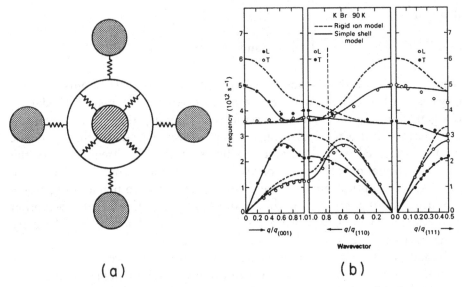

(a)   (b)

Figure 11 (a) Schematic diagram of the shell model (from Cochran[2]). The large circle represents a massless shell of valence electrons on the negative ion that interacts with the central core of the negative ion and the nonpolarizable positive ions. (b) Dispersion curves of KBr determined by inelastic neutron scattering (dots and circles). Dashed and solid lines are the fitted lattice-dynamics models. (From Woods et al.[13])

from further neighbors; however, it is generally agreed that these parameters have little physical significance. The shortcoming of such models is that they treat atoms as rigid units, whereas real atoms have polarizabilities which modify the interatomic forces. In an early attempt to remedy this, Tolpygo[11] presented a theory for incorporating atomic dipoles into the calculation (a dipole approximation). Dick and Overhauser[12] envisioned the ion as having an electronic shell that could be displaced relative to the atomic core. The shell model, first applied to ionic and covalent crystals by Cochran,[2] is formally equivalent to the dipole approximation, but gives a useful physical picture of the dynamics, as illustrated in Figure 11(a). In its simplest form for an alkali halide crystal, the polarizability of the positive ion is neglected and the negative ion is regarded as a rigid core surrounded by a massless shell representing the valence electrons.

The success of the shell model for an *ionic* crystal is illustrated in Figure 11(b).[13] The experimental dispersion for KBr is roughly predicted by a rigid-ion model (dashed curve) with only a single adjustable parameter, chosen to fit the longitudinal sound velocity. A much better fit is obtained with a simple shell model with three adjustable parameters, as indicated by the springs in Figure 11(a) plus core–core interactions. The excellent fit to the data indicates that the atomic polarizability contributes significantly to the lattice dynamics.

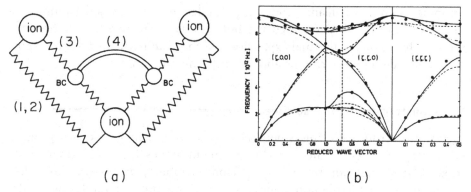

(a)   (b)

*Figure 12 (a) Schematic of the bond-charge model. The numbering cor-
responds to the description in the text. (b) Dispersion curves for Ge.
Dots are experimental data from Nilsson and Nelin[18]; dashed lines are a
5-parameter shell model by Cochran[15]; solid lines are a 4-parameter bond-
charge model by Weber.[16]*

The rigid-ion models require many adjustable force constants to explain
the dispersion curves of *covalent* crystals. For example, to obtain agreement
within experimental error to the dispersion curves in Ge, 15 independent
constants involving up to 5th nearest neighbors are required.[14] The shell
model yields satisfactory results with fewer parameters. The dashed lines
in Figure 12(b) show the results of a 5-parameter shell model by Cochran.[15]
Unfortunately, many more parameters in the shell model are required to
adequately fit the experimental dispersion of III–V compounds. This is not
too surprising because the shell model is not a physically realistic description
of the directed covalent bonds.

A significant improvement in the description of vibrations in covalent crystals was made by Weber.[16,17] He allowed for motion of "bond charges" in the
lattice dynamics, similar to the movable electronic shells, but localized on the
covalent bonds between the atoms and able to interact between themselves.
The interaction between two adjacent bonds helps to account for the torsional
strength of the tetrahedral bonds, as shown schematically in Figure 12(a).
With only four parameters – (1) repulsive core interaction between adjacent
atoms, (2) Coulomb interaction between point ions, (3) central potential between ions and point bond charges, and (4) potential between adjacent bond
charges – an excellent fit is obtained to the experimental dispersion of Ge
(Nilsson and Nelin[18]), as shown by the solid lines in Figure 12(b). Two
additional parameters are required for the III–V compounds, and equally
satisfactory fits are obtained. The bond-charge model is particularly successful in explaining the very flat regions of the transverse-acoustic dispersion
curves near the zone boundaries. Many more parameters in the rigid-ion and
shell models are required to explain these observations in the III–V crystals.

At the time of this writing, the most sophisticated modeling of a diamond-
type lattice has been conducted by Labrot et al.[19] For silicon, they have

combined both the shell and bond-charge models and included anharmonicity to fit not only the dispersion curves but also data sensitive to anharmonic forces: for example, thermal expansion and the Grünheisen constant. Their model includes eight harmonic and five anharmonic parameters.

D. Constant-frequency surfaces

Inelastic neutron scattering is generally used to determine $\omega(\mathbf{k})$ along high symmetry axes. As the theoretical modeling improves, it becomes important to know the dispersion relation away from symmetry directions. Also, information about the phonon eigenvectors (polarizations) is rarely obtained. Measurements of eigenvectors and off-axis $\omega(\mathbf{k})$ can test the physical validity of different lattice-dynamics models. Studies of this type have recently been conducted for the semiconductor GaAs, which we now consider.

The dispersion curves along symmetry directions in GaAs were first measured by Dolling and Waugh,[20] who found reasonably good agreement with a 14-parameter shell model. Good agreement is also found with a 6-parameter bond-charge model. Using these models, with parameters fit to the neutron-scattering data, Hebboul and Wolfe[21] plotted several constant-frequency curves in the symmetry planes. The results for the ST mode in the (100) plane are shown on the left side of Figure 13. Although, for a given frequency, the

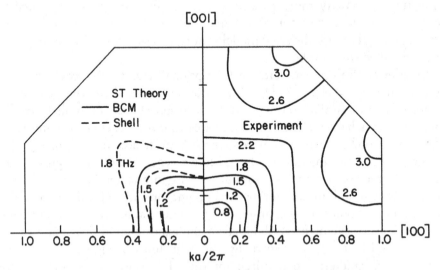

Figure 13 Constant-frequency curves in the (100) plane of the Brillouin zone for GaAs. On the left are contours plotted by Hebboul and Wolfe,[21] using a 6-parameter bond-charge model and a 14-parameter shell model fitted to the experimental dispersion curves along symmetry axes. On the right are the contours derived from off-axis neutron-scattering experiments of Dorner and Strauch.[22] The numbers are frequencies in terahertz (10^{12} Hz).

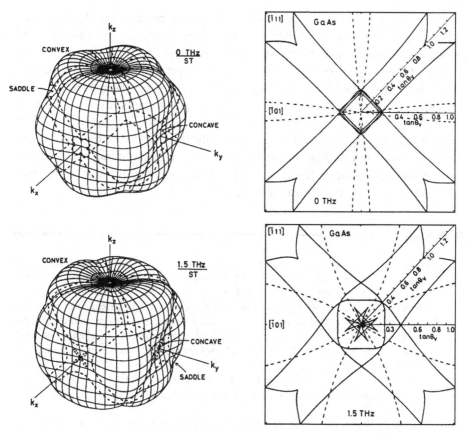

Figure 14 Slowness surfaces for GaAs at 0 and 1.5 THz, computed with a
23-parameter rigid-ion model of Tamura and Harada.[23] On the right are
the corresponding caustic maps in the (100) plane, showing the sensitivity
of phonon-focusing caustics to dispersion. Solid curves are for ST and
dashed curves are for FT.

shell and bond-charge models predict nearly the same wavevectors along the
[100] axis, the shapes of the corresponding curves are quite different. This
difference is related to the relatively flat dispersion curves along [110] at large
wavevectors for the bond-charge model.

Recently Dorner and Strauch[22] made detailed inelastic neutron scattering
measurements of the off-axis dispersion in GaAs. They measured $\omega(\mathbf{k})$ for 600
wavevectors in the (100) and (110) planes, respectively. Constant-frequency
contours in the (100) plane, obtained by least-squares fitting between discrete wavevectors, are shown on the right side of Figure 13. Comparison
to the theory curves on the left side of this figure shows that the bond-
charge model represents the data far better than the shell model. Dorner and
Strauch have used this data to calculate group velocities for wavevectors in the
symmetry planes. With this experimental information, they predict the times
of flight for this partial heat flux (**k**'s only in the symmetry plane), assuming

no phonon scattering. A theoretical study of the temporal and spatial dispersion of phonon flux in GaAs, including all wavevectors and the effects of phonon scattering, has been conducted by Hebboul and Wolfe.[21]

The latter study investigates the effects of dispersion on phonon focusing, which is the principal topic of Chapter 7. For wavevectors approaching the Brillouin-zone boundary, the constant-frequency surfaces distort, as shown in the calculation by Tamura and Harada[23] illustrated in Figure 14. The lines of zero Gaussian curvature are particularly sensitive to this distortion. Therefore, the predicted caustic patterns, shown in the right half of the figure, are greatly dependent on frequency in the tetrahertz regime, and, as we shall see in Chapter 7, on the choice of lattice-dynamics model.

References

1. In this chapter, we deal mainly with lattice-dynamics models based on empirically determined force-constants, as pioneered by M. Born and K. Huang, *Dynamical Theory of Crystal Lattices* (Oxford University Press, London, 1954). It is worth noting that *ab initio* calculations based on density functional theory are now possible and, thanks to high-speed computers and sophisticated program development, they provide quite accurate vibrational properties with no freely adjustable parameters. An excellent example of this approach is provided by P. Giannozzi, S. de Gironcoli, P. Pavone, and S. Baroni, Ab initio calculations of phonon dispersions in semiconductors, *Phys. Rev. B* **43**, 7231 (1991). See also, for example, A. Debernardi, S. Baroni, and E. Molinari, Anharmonic phonon lifetimes in semiconductors from density-functional perturbation theory, *Phys. Rev. Lett.* **75**, 1819 (1995).

2. W. Cochran, Lattice vibrations, *Rep. Prog. Phys.* **26**, 1 (1963); *The Dynamics of Atoms in Crystals* (Crane, Russak, New York, 1973).

3. R. F. Wallis, ed., *Lattice Dynamics* (Pergamon, Oxford, 1965).

4. R. W. H. Stevenson, *Phonons in Perfect Lattices and in Lattices with Point Imperfections* (Plenum, New York, 1966).

5. W. Cochran, Lattice vibrations, in *Lattice Dynamics*, A. A. Maradudin, I. M. Lifshitz, A. M. Kosevich, W. Cochran, and M. J. P. Musgrave, eds. (Benjamin, New York, 1969).

6. M. J. P. Musgrave, *Crystal Acoustics* (Holden-Day, San Francisco, 1970).

7. B. Donovan and J. F. Angress, *Lattice Vibrations* (Chapman and Hall, London, 1971).

8. W. Cochran, *The Dynamics of Atoms in Crystals* (Crane, Russak, New York, 1973).

9. P. Brüesch, *Phonons: Theory and Experiments I* (Springer-Verlag, Berlin, 1982).

10. C. Kittel, *Introduction to Solid State Physics*, 3rd ed. (Wiley, New York, 1966).

11. K. B. Tolpygo, Long-range forces and the dynamical equations for homopolar crystals of the diamond type, *Sov. Phys. Solid State* **3**, 685 (1961) [translated from Fizika Tverdogo Tela. **3**, 943 (1961)].

12. B. G. Dick and A. W. Overhauser, Theory of the dielectric constants of alkali halide crystals, *Phys. Rev.* **112**, 90 (1958).

13. A. D. Woods, W. Cochran, and B. N. Brockhouse, Lattice dynamics of alkali halide crystals, *Phys. Rev.* **119**, 980 (1960).

14. F. Herman, Lattice vibrational spectrum of germanium, *J. Phys. Chem. Solids* **8**, 405 (1959).

15. W. Cochran, Theory of the lattice vibrations of germanium, *Proc. Royal Soc. A* **253**, 260 (1959).

16. W. Weber, New bond-charge model for the lattice dynamics of diamond-type semiconductors, *Phys. Rev. Lett.* **33**, 371 (1974).

17. W. Weber, Adiabatic bond charge model for the phonons in diamond, Si, Ge, and α-Sn, *Phys. Rev. B* **15**, 4789 (1977).

18. G. Nilsson and G. Nelin, Phonon dispersion relations in Ge at 80 K, *Phys. Rev. B* **3**, 364 (1971).

19. M. T. Labrot, A. P. Mayer, and R. K. Wehner, in *Phonons 89*, S. Hunklinger, W. Ludwig, and G. Weiss, eds. (World Scientific, Singapore, 1990).

20. G. Dolling and J. L. T. Waugh, Normal vibrations in gallium arsenide, in *Lattice Dynamics*, R. F. Wallis, ed. (Pergamon, Oxford, 1965).

21. S. E. Hebboul and J. P. Wolfe, Focusing of dispersive phonons in GaAs, *Z. Phys. B* **74**, 35 (1989).

22. B. Dorner and D. Strauch, Phonon dispersion sheets and group velocities in GaAs, *J. Phys. Cond. Matter* **2**, 1475 (1990).

23. S. Tamura and T. Harada, Focusing of large-wave-vector phonons in GaAs, *Phys. Rev. B* **32**, 5245 (1985).

7

Imaging of dispersive phonons

The previous chapters have dealt mainly with phonon-focusing patterns that are independent of phonon frequency. This nondispersive behavior is expected when the wavelengths of the phonons are long compared to the spacing between atoms in the crystal. For phonons with wavelengths of only a few lattice spacings, large changes are expected in the slowness surface, group velocities, and focusing pattern. In practice, however, it is quite difficult to observe the ballistic propagation of such short wavelength (large-k) phonons. As their wavelength approaches twice the lattice constant (i.e., $k \equiv 2\pi/\lambda$ approaches π/a), the phonons become particularly sensitive to defects in the periodicity in the crystal: they scatter.

The crystalline defects can be impurity atoms or even atoms with different isotopic masses. The fractional difference in mass between two isotopes is usually small, but many elements have more than one isotope of large natural abundance. This implies a high density of weak-scattering centers, even in chemically pure crystals. For long-wavelength phonons, the rate of mass-defect scattering increases as the fourth power of the phonon frequency, similar to Raleigh scattering of light from particles in the atmosphere. For example, the mean free paths of 1-THz phonons in otherwise perfect Ge, InSb, and GaAs are limited to a fraction of a millimeter by isotope scattering.

In order to observe the ballistic propagation of large-wavevector phonons in such crystals, the experimenter must use thin samples of high chemical and structural purity. This fact puts severe constraints on a phonon-imaging experiment. One must not only devise a means of selecting phonon frequency but also fabricate extremely small detectors to preserve the necessary angular resolution for thin samples.

Once these difficulties are surmounted, phonon-imaging experiments can be used to observe the ballistic propagation of large-wavevector phonons and thereby gain information about the microscopic forces between atoms. As the wavelength approaches the lattice spacing, the velocity of the wave begins to reflect the detailed bonding and structure of the crystal. Lattice-dynamics models provide the necessary link between the microscopic force constants and the dispersion curves of the crystal.

As an alternative to the force-constant models discussed in Chapter 6, $\omega(\mathbf{k})$ can be predicted from ab initio calculations of the electronic wave functions. The accuracy of this newer approach, which may have no adjustable parameters (or parameters determined from other measurements), depends on how realistic the electronic wave functions are. At present, the empirical force-constant models provide more accurate characterizations of the phonon dispersion relation, which is used to derive the phonon density of modes and other vibrational properties of crystals. The experimentally measured dispersion relation of a crystal provides a stringent test of the lattice-dynamic theories.

The most powerful experiment for determining the phonon dispersion is inelastic neutron scattering. It measures both optical and acoustic branches of the dispersion over most of the Brillouin zone. Generally, the phonon frequencies are measured as a function of wavevector along the principal symmetry axes of the crystal. The transverse branch (degenerate ST and FT) of InSb along [100] and [111] axes is shown in Figure 1. The dots are the neutron scattering data, and the lines are different models that have been fitted

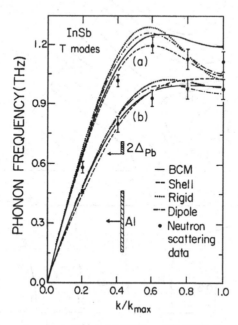

Figure 1 Dispersion curves along (a) [100] and (b) [111] directions for InSb. Dots are neutron scattering data from Price and Rowe.[1] The vertical bars indicate the effective detection frequencies of a Pb tunnel junction and an Al bolometer. (From Hebboul and Wolfe.[19])

simultaneously to these data and to those of other branches.[1] Considering also the fits along other symmetry axes, all the models show similar overall agreement with the data. The different models, however, do differ significantly in their prediction of polarizations, but this information is much more difficult to obtain experimentally.

With phonon-imaging methods, we approach the dispersion relation from quite a different viewpoint. The angular variations in heat flux are extremely sensitive to the *curvatures* of the constant-frequency surface. The various lattice-dynamics models predict significantly different shapes of the slowness surfaces, and, hence, different phonon caustic patterns. A measurement of the caustic positions as a function of frequency, therefore, can provide new information to compare with the theories.

The physical origin of this angular dispersion is displayed in Figure 2. These are constant-frequency contours of the FT phonons in the [110] plane

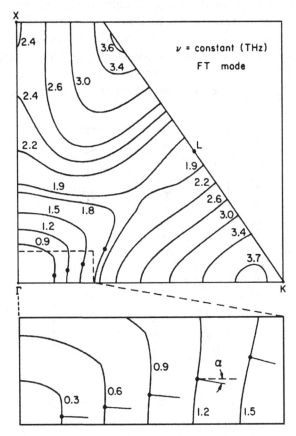

Figure 2 Constant-frequency contours of the FT mode in the (110) plane of Ge, using a rigid-ion model. The numbers give the frequency in tetrahertz. Enlargement of the low-frequency region shows how the group-velocity normal at the inflection point changes with frequency. (From Dietsche et al.[2])

of Ge.[2] The contours near the origin are the same as those previously predicted by the continuum theory. Inflection points are indicated by the black dots, and the group-velocity normals to the surfaces at these points indicate the directions of phonon caustics. As the frequency is increased, the shape of the surface changes due to the proximity of the Brillouin-zone boundary; the constant-frequency curves must meet the boundary normally. Even at the lower frequencies, these shape changes can lead to a significant deflection of the group velocity corresponding to the caustic, as shown in the magnified section of Figure 2. At higher frequencies, the predicted focusing pattern will be markedly different from the continuum pattern.

A. Early experiments

The first conclusive demonstration of the dispersive propagation of heat pulses in crystals was reported by Huet et al.[3] They compared the times of flight of heat pulses in InSb measured with detectors having different frequency responses: (1) an Al bolometer (broadband), (2) a Sn tunnel junction (onset frequency = $2\Delta/h$ = 280 GHz), and (3) a PbTl tunnel junction ($2\Delta/h$ = 545 GHz). The comparison between the bolometer and Sn junction data is shown in Figure 3. The slower velocity in the latter case is an indication of the decreasing slope, $V = d\omega/dk$, at higher frequency, as in Figure 1. For the

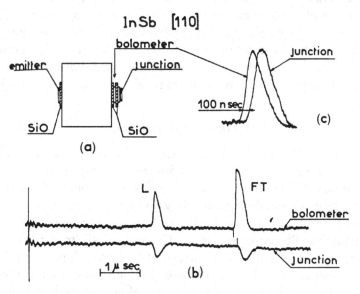

Figure 3 Heat-pulse experiment showing the increase in time of flight due to frequency dispersion in a 1.3-mm-thick crystal of InSb. (a) The sample arrangement, (b) heat pulses detected with Al bolometer and Sn tunnel junction, and (c) magnification of the FT pulses. The Sn junction selects against phonons with frequency below ≃280 GHz, whereas the bolometer is a broadband detector. (From Huet et al.[3])

Pb junction a 6% reduction in the FT velocity along [110], compared to the bolometer data, was observed.

The literature contains several reports of highly dispersive phonons traveling ballistically over millimeter distances – much further than isotope scattering theory permits. In an attempt to produce copious quantities of near-zone-boundary phonons, Ulbrich et al.[4] directly photoexcited a high-purity GaAs crystal with a movable laser beam and detected the resulting heat pulses with a Pb tunnel junction. Broad heat pulses were observed, and the times of flight of the *peaks* of the pulses scaled linearly with the source-to-detector distance, suggesting ballistic propagation.[5] Subsequent phonon-imaging experiments,[6–8] however, found only caustic structures of mildly dispersive phonons, in agreement with the isotope-scattering predictions. These latter experiments, and theoretical studies by Lax et al.[9,10] imply that the broadening of the pulses observed in GaAs and similar semiconductors[11,12] is mainly due to bulk scattering rather than highly dispersive ballistic transport. A review of these studies can be found in Ref. 13.

A unique time-of-flight study of phonon dispersion was conducted by Haavasoja et al.[14] They examined a crystal of solid ^{4}He, which is isotopically pure and has hexagonal close-packed structure. The low sound velocity of this crystal implies that dispersive phonons fall in the range of frequency-tunable tunnel-junction generators. With a Sn generator and Al tunnel-junction detector, Haavasoja et al. were able to observe up to 15% reduction in the times of flight of the leading edge of the heat pulses, as shown in Figure 4. The ^{4}He crystal was grown *in situ* at approximately 0.1 K in a helium dilution refrigerator, so the direction of propagation was not precisely known. The solid line in the figure is the derivative of the standard dispersion curve obtained by assuming Hooke's law between neighboring lattice planes (Figure 1 of Chapter 6):

$$\hbar\omega(k) = A\,\sin(ka_1/2), \tag{1}$$

using $A = 1.2\,\text{meV}$ and $a_1 = 2.6\,\text{Å}$, which is within 6% of the known distance between lattice planes for the assumed direction of propagation.

Experiments of Dietsche et al.[2] were the first to use phonon-imaging techniques to study phonon dispersion. They observed a difference in the caustic pattern of Ge obtained with (1) a (low-frequency) Al bolometer and (2) a (high-frequency) Pb tunnel junction. The angular shifts in the caustic positions are consistent with those calculated by lattice-dynamics models for Ge.[15,16] Similar caustic shifts are obtained for GaAs; the results of Northrop et al.[6] are shown in Figure 5. The principal effects at the higher frequency are a broadening of the FT ridge and a rounding of the ST box structure near ⟨100⟩. The ST caustics near [100] have been studied in Ge by Metzger and Huebener.[17] They observed that the caustic positions are measurably different from continuum theory predictions, even for bolometer detection with

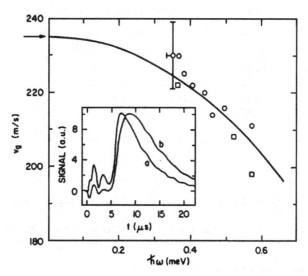

Figure 4 Group velocities in a crystal of ^{4}He, showing $\simeq 15\%$ reduction with frequency due to phonon dispersion. Circles and squares are velocities measured from the onset of the heat pulses. Typical pulses are with tunnel-junction detector gaps of (a) 0.46 meV and (b) 0.52 meV are shown in the inset. The arrow on the vertical axis shows the measured transverse acoustic velocity. The solid line is $d\omega/dk$, using Eq. (1). (From Haavasoja et al.[14])

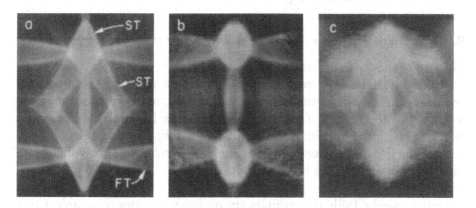

Figure 5 Phonon images of a (110)-oriented GaAs crystal, using photoexcitation with an Ar$^+$ laser ($\lambda = 514$ nm). (Similar data are obtained with metal-film excitation.) (a) Al-bolometer detector and broad time gate, (b) Pb tunnel-junction detector with $\nu_{onset} \simeq 0.7$ THz and narrow time gate to emphasize FT ridges and ST structure near $\langle 100 \rangle$ directions. Broadening of the FT ridges and rounding of the ST structures are due to frequency dispersion. (c) Same as (b) but sampling later times of flight to emphasize ST caustics. (From Northrop et al.[6])

Table 1. *Parameters for phonon scattering from isotopes*

Material	A_0 ($\times 10^{-42}$)	$\nu = 1$ THz, $V \parallel [100]$			Ref.
		τ (ns)	V (km/s)	ℓ (mm)	
InP	1.83	546.4	3.100	1.694	c
InAs	1.95	512.0	2.670	1.367	c
Si	2.43	411.5	5.870	2.416	a
GaAs	7.38	135.5	3.357	0.455	a, b
InSb	16.2	61.7	2.321	0.143	b
Ge	36.7	27.2	3.579	0.097	a

[a] S. Tamura, Spontaneous decay rates of LA phonons in quasi-isotropic solids, *Phys. Rev. B* **31**, 2574 (1985).
[b] S. Tamura, Isotope scattering of large-wavevector phonons in GaAs and InSb: deformation-dipole and overlap-shell models, *Phys. Rev. B* **30**, 849 (1984).
[c] S.Tamura, private communication.
Note: The scattering rate is given by $\tau^{-1} = A_0 \nu^4$ as derived in Chapter 8. The transverse group velocities, V, are the long-wavelength values. The mean free path is given by $\ell = V\tau$. (After S. E. Hebboul, Ph.D. Thesis, University of Illinois.)

2–3-mm-thick samples. Referring to Tamura's[15] dispersive calculations, they concluded that the caustic shifts were due to dispersion of phonons in the 300–400-GHz frequency range.

B. Dispersive phonon imaging in InSb

The semiconductor InSb has proven to be an excellent crystal for a systematic study of the effects of dispersion on phonon focusing. For a given detector frequency, InSb displays the largest dispersion of the III–V semiconductors, as shown in Figure 6(a). This is mainly due to the lower phonon velocities associated with the heavier atoms of InSb. The frequency range of Pb-alloy tunnel junctions is between approximately 400 and 850 GHz, which samples the InSb wavevectors between approximetely 18% and 40% of the zone boundary.

Unfortunately, InSb also exhibits strong isotope scattering. Table 1 lists the isotope scattering rates of several semiconductors. The rates are given by $\tau^{-1} = A_0 \nu^4$ (see Chapter 8). The intensity of ballistic flux travelling straight across a crystal of thickness d decreases with distance due to scattering, such that

$$I(\nu) = I_0(\nu) \exp[-d/V\tau(\nu)], \qquad (2)$$

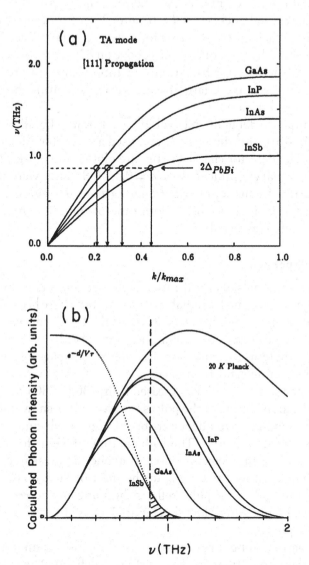

Figure 6 (a) Calculated dispersion curves for several compound semiconductors. The frequency onset (850 GHz) of a PbBi tunnel-junction detector is shown. (b) Transmitted ballistic flux for several crystals of thickness 0.5 mm, assuming an initial 20 K Planck distribution at the source. The transmission curves are the Planck distribution multiplied by a transmission factor (dotted line for InSb) given in Eq. (2). An 850-GHz detector onset frequency is shown for comparison. (From S. E. Hebboul, Ph.D. Thesis, University of Illinois, 1988.)

where V is the group velocity (approximated by the long-wavelength value). Assuming that the initial intensity distribution, $I_0(v)$, is a 20 K Planck distribution, and the crystal thickness is $d = 0.5$ mm, Figure 6(b) shows the intensity of ballistic flux reaching the detector for several crystals. The shaded region indicates the small fraction of phonons above an 820-GHz detector threshold that will reach the detector ballistically in InSb. Thus, in order to observe the effects of the relatively large frequency dispersion of this crystal, it is necessary to use thin samples.

For this purpose, Hebboul and Wolfe[18,19] fabricated Pb-alloy tunnel junctions with $10 \times 10 \ \mu m^2$ dimensions and used a laser spot of roughly $10 \ \mu m$ in diameter. The detector characteristics are plotted in Figure 12 on page 79. InSb crystals of varying thicknesses, $d = 0.43$–0.84 mm, were used in order to select a narrow frequency distribution for a given detector gap, as indicated by the shaded area in Figure 6(b). Both time-of-flight and phonon-imaging experiments were conducted.

i) Velocity dispersion

The rather weak signals obtained with a high-frequency detector on InSb are shown in Figure 7(a). The FT signal is relatively large here because the source-to-detector direction is along an FT caustic. If the laser is displaced slightly to a position off the caustic, the second trace is observed. By subtracting these two traces, one obtains the pure FT pulse. Figure 2 on page 14 shows a similar analysis at a lower detector frequency.

As the detection frequency is increased from 430 to 780 GHz, a noticeable increase in the arrival time of the pulses is observed, as shown in Figure 7(b). Also, the pulses at higher frequency are broader, primarily due to the increased fraction of scattered phonons. The ballistic time of flight for phonons with frequency equal to the detector onset frequency is measured at the onset of the heat pulse, shown as time t_o in the inset of Figure 7(c). The time t_p measured to the peak of the pulse reflects the scattering processes. The group velocity of the FT phonons, plotted in Figure 7(c), is observed to decrease with increasing frequency.

For propagation along a symmetry axis, the pulses cannot be isolated so cleanly, but their onset times are still fairly well defined. The measured group velocities for L and ST modes along the [111] axis are plotted in Figure 7(d). The ballistic pulse from ST phonons is even weaker than those shown in the FT case and must be extracted from a large scattered background. The measured velocities of these phonons at 730 GHz are reduced by 40% from their continuum values! These phonons have wavevectors of approximately 40% of the zone-boundary ST wavevector in this direction, as indicated in Figure 6(a).

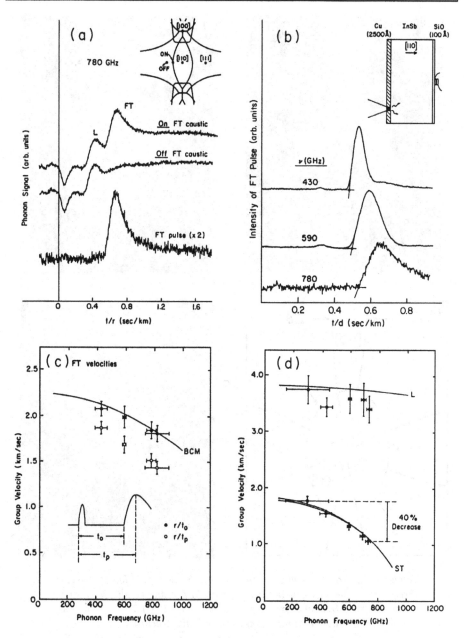

Figure 7 Heat-pulse signals in InSb for various detector frequencies, showing phonon dispersion. (a) Signals both on and off the FT caustic, as
shown in the inset. The difference signal shows the FT pulse alone, and
the scattered-phonon component (long tail) is much reduced. Early-time
signal is due to optical and electrical feedthrough at the time of the laser
pulse. (b) Similarly isolated FT pulses for several detector frequencies.
(c) Corresponding group velocities with comparison to the BCM. The pulse-
onset thresholds give the correct times of flight. Using the arrival time of
the peak of the pulse gives lower than ballistic velocities. (d) Group velocities for L and ST pulses along ⟨111⟩, showing a 40% dispersive effect for ST
phonons. Solid lines are BCM predictions. (From Hebboul and Wolfe.[19])

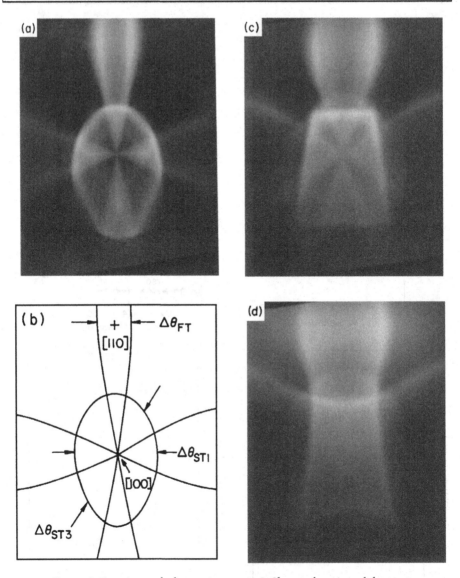

Figure 8 Experimental phonon images in InSb as a function of detector frequency. The shifts in the caustic pattern are due to phonon dispersion. Detector frequency onsets in GHz are (a) 430, (c) 590, and (d) 730; (b) defines the angular dimensions of the caustic structures.

ii)   Angular dispersion

Phonon images of InSb for increasing detector frequency are shown in Figure 8, with the laser beam scanned across the (110) face of the crystal. The detector position on the opposing face is marked with a cross near the top of the diagram in Figure 8(b), corresponding to [110] propagation. The

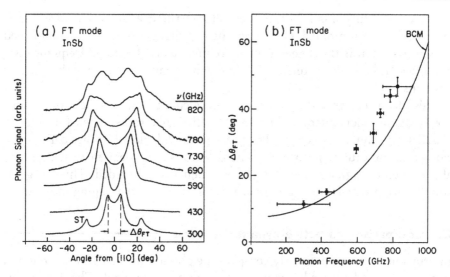

Figure 9 (a) Phonon intensity scans across the FT ridge (a horizontal line scan near the top of images in Figure 8) as a function of detector frequency. The dramatic widening of this feature is due to phonon dispersion. At 820 GHz some lower-frequency phonons are also detected, that give an additional interior structure. (b) Frequency dependence of the FT caustic separation, as defined in (a). The solid line is the BCM calculation. (From Hebboul and Wolfe.[19])

lowest-frequency image [Figure 8(a), 430 GHz] shows the FT ridges running horizontally and vertically along (100) planes and crossing along [100]. At 590 GHz [Figure 8(c)], the FT ridge has widened considerably and the ST structure becomes a square shape. These changes are even more pronounced at 730 GHz.

It is straightforward to quantify these striking changes by defining angular dimensions of caustic structures, as shown in Figure 8(b). Figure 9(a) shows line scans taken from a sequence of images. These are horizontal scans through the [110] direction near the top of the images. They give a profile of the FT ridge at its maximum width. The width of the ridge, $\Delta\theta_{FT}$, is defined as the separation of the two maxima, corresponding to the FT caustics. A Monte Carlo simulation[19] has shown that $\Delta\theta_{FT}$ is an accurate measure of the caustic separation. The caustic separation increases by about a factor of three for this frequency range, showing that the shape of the slowness and wave surfaces are indeed very sensitive to dispersion. Figure 9(b) is a plot of $\Delta\theta_{FT}$ versus frequency and the predictions of a bond-charge model (BCM) discussed below.

The images of Figure 8 are recorded for a constant velocity range, centered about the transverse velocity along [100]. At lower velocities, the ST ramps and [111] cusped structures are observable. The bright features in the images are quite sharp and indicate that the frequency range being sampled is very narrow. Only small shifts in these sharp structures are observed as different velocities are sampled, but the intensities change markedly. As indicated in Figure 6(b), it is the detector onset frequency and the rapid frequency dependence of the bulk scattering that define the frequency range of the ballistic phonons.

This idea is confirmed by Monte Carlo calculations that take into account the frequency-dependent reduction in ballistically transmitted flux due to elastic scattering, as in Eq. (2). The peaks in the calculated images correspond closely to the caustic positions at the detector onset frequency, as assumed above, even though a range of higher frequencies is included. This is a great simplification which enhances the utility of the phonon-imaging method.

C.   Comparison to lattice-dynamics models

In principle, one could directly compare the phonon flux $I(\theta, \phi)$ in a phonon image to that calculated by using a lattice-dynamics model. The differences between theoretical and experimental intensities could be minimized by adjusting the available force constants. However, each iteration of the fit requires a new calculation of $\omega(\mathbf{k})$ over the entire Brillouin zone (e.g., for a three-dimensional grid of 10^6 wavevectors). Moreover, the experimental flux depends on a number of poorly controlled parameters, such as detector and source size, as well as bulk scattering and phonon transmission at interfaces.

The practical approach, as indicated in Figure 9, is to use the angular positions of the phonon caustics. Thus, the phonon-imaging method is analogous to a magnetic resonance or optical spectroscopy experiment, where the *spectral position* of a resonance is at least as important as its intensity. With phonon imaging, it is the *angular positions* of caustic lines in two dimensions that are directly related to folds in the wave surface (and parabolic lines on the constant-frequency surfaces). The theoretical caustic positions may be determined from the positions of folds in a numerically generated wave surface (which is most practical for symmetry planes)[19] or measured directly from numerically generated phonon images.[6,15-21]

A dispersive phonon image is calculated with a particular lattice-dynamics model whose force constants have been (at least initially) determined by fitting to neutron-scattering data. From this model, a table of frequencies $\omega(\mathbf{k})$ and group velocities $V(\mathbf{k})$ is constructed for a fine grid of $\mathbf{k}$ vectors spanning the Brillouin zone. This look-up table is used as a basis for Monte Carlo calculations of the phonon flux as follows: (1) a random array of wavevectors is generated, (2) the group velocity for each $\mathbf{k}$ is determined by interpolating between values in the look-up table, and (3) the phonons are propagated to the experimental plane, where their points of intersection are recorded.

The reduction in ballistic flux due to scattering can be included in the model by defining a transmission probability for phonons of a given frequency,[19]

$\exp(-r/V\tau)$, where r is the path length and τ is the scattering time. For isotope scattering in the long-wavelength approximation, $\tau^{-1} = A_0 v^4$, with A_0 given in Table 1 [see Figure 6(b)]. A point is added to the image, with an intensity proportional to the transmission probability, where $V(k)$ intercepts the detection plane. In this case, the scattered phonons are ignored, so the computed images represent only the transmitted ballistic flux. Scattered phonons produce a background of flux, which usually has little effect in the determination of caustic positions.

Monte Carlo images calculated in this manner for InSb are shown in Figure 10. An initial Planck distribution of phonons with $T = 10$ K and a detector onset of 730 GHz are assumed (i.e., all phonons below this frequency are discarded). In all four cases, the force constants derived from fitting the neutron-scattering dispersion curves are assumed. The lattice-dynamics models used in each case are (letters correspond to the figure): (a) a 6-parameter BCM,[22,23] (b) an 11-parameter rigid-ion model,[24,25] (c) a 15-parameter deformation-dipole model,[24,25] and (d) a 10-parameter shell model incorporating valence force fields.[26,27] A description of the basic physical assumptions used in these models is included in Chapter 6, Section C. These are also the models used to plot the dispersion curves in Figure 1.

The striking differences in caustic patterns for the different lattice-dynamics models underscores the sensitivity of phonon focusing to the curvatures of a slowness surface. These results are made more quantitative by plotting the dimensions of various caustic structures as defined in Figure 11. A series of experiments are conducted with different detector frequencies and the results are plotted as the solid dots in Figure 11. Predictions of the various models are shown as the solid curves.

From a qualitative comparison of the experimental and theoretical images, Figure 8(d) and Figure 10, and the quantitative comparisons of Figure 11, we see that the BCM is clearly the best model for predicting the shapes of the dispersive wave surfaces. This is especially significant when one realizes that the BCM contains the fewest number of adjustable parameters. Figure 11(b) contains a particularly salient result: only the BCM predicts that the angle $\Delta\theta_{ST1}$ *decreases* with increasing frequency. At the present time there is no microscopic explanation of why Weber's BCM more correctly predicts this property, but a likely qualitative answer is that this model incorporates the torsional character of the covalent bonds more accurately.

The BCM is also used to predict the velocity dispersion, shown as the solid curves in Figures 7(c) and 7(d). Agreement with experiment is quite good. Attempts have also been made to refit the BCM, incorporating both the neutron scattering and phonon-imaging data.[19] It is found that adjustment of some of the parameters can bring the focusing predictions into better agreement with experiment, resulting in a small deterioration in the fit to the dispersion curves.

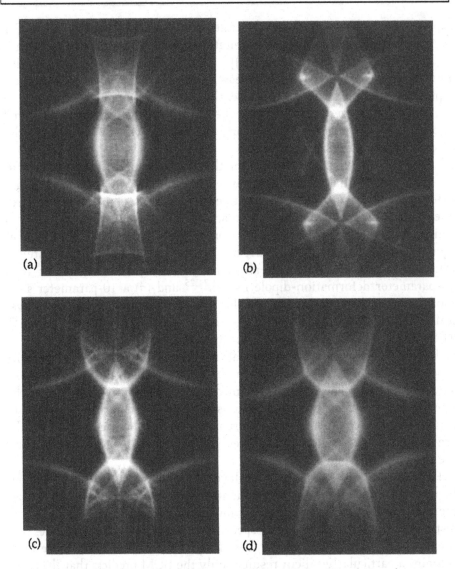

Figure 10 Calculated phonon-focusing pattern including FT and ST modes using (a) BCM, (b) rigid-ion model, (c) dipole model, and (d) shell model. The calculations assume a detector onset frequency of 730 GHz, a 10 K Planck source, and take into account the effect of isotope scattering on the transmission of ballistic flux. Scattered phonons are not included. (From Hebboul and Wolfe.[19])

Interesting experiments remain to be done. If ballistic propagation can be observed for phonons with wavevectors up to 50% or 60% of the zone boundary, some new caustic structures should be observed. This prediction was first made by Schreiber et al.,[28] who were attempting to explain an anomalous concentration of phonon flux along [100] in GaAs. The extra [100] flux was later

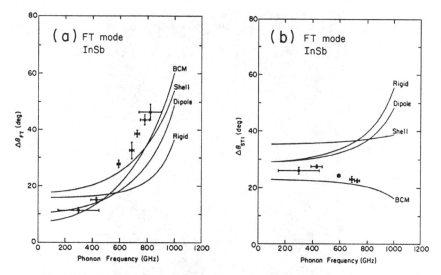

Figure 11 *The dimensions of selected caustic structures as a function of detector frequency. Dots show the data from images like those in Figure 8. Solid lines are the predictions of various lattice-dynamics models assuming the force constants fitted to neutron scattering data.*

Figure 12 *Phonon image of a (100)-oriented crystal of InP. Crystal thickness is 0.65 mm and the tunnel-junction detector is PbBi with a frequency onset of approximately 850 GHz. The broad FT ridges and large distortion of the ST structure (beyond circular) indicate significant frequency dispersion. (From S. E. Hebboul, Ph.D. Thesis.)*

explained by channeling of elastically scattered phonons,[29] but the focusing prediction remains valid. Hebboul and Wolfe[7] calculated the high-frequency focusing patterns for GaAs and discovered that, in addition to the [100] effect, there are other new caustic structures that are predicted by the BCM.[30]

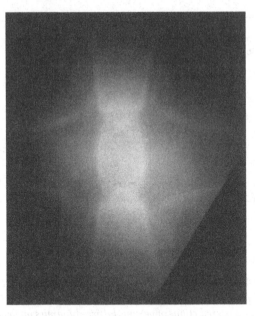

Figure 13 Phonon image of (110)-oriented InAs, taken with a 0.085-THz PbBi tunnel junction by Hebboul and Wolfe. As with InSb and InP, the caustic positions are significantly shifted by phonon dispersion. (See Ramsbey.[31])

Furthermore, the shell model makes qualitatively different predictions. It would be very difficult to observe these effects in InSb due to strong isotope scattering; clear images have been obtained only for frequencies below 800 GHz.

For probing deeper into the Brillouin zone, the semiconductors InP and InAs are good candidates. As seen in Figure 6 and Table 1, they are somewhat less dispersive than InSb for a given frequency, but their isotopic purity is far better. Phonon images for these two crystals at the highest frequency attainable with the PbBi-alloy detector (850 GHz) are shown in Figures 12 and 13. The caustic patterns are considerably shifted from their low-energy positions. It may prove fruitful to develop higher-frequency imaging detectors such as those based on NbN alloys. Another important possibility for large-k imaging is ionic crystals, some of which are highly dispersive in the frequency range of presently available detectors.

D.   First-order spatial dispersion

Because phonon imaging is sensitive to the curvatures and inflections in the slowness surfaces, it is a powerful method for observing the topological changes in these surfaces arising from phonon dispersion. A basic question one can ask is: Do the degeneracies that occur in the continuum limit

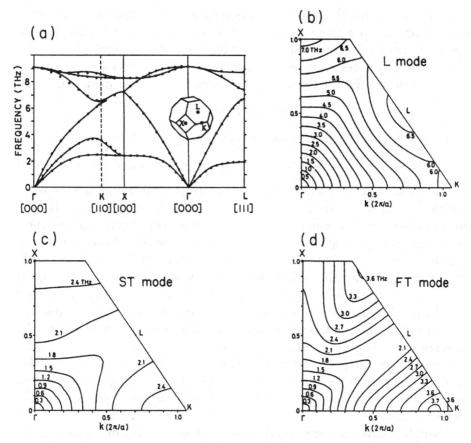

Figure 14 (a) Dispersion curves for Ge. Dots are neutron scattering data by Nilsson and Nelin. Curves are the results of a 31-parameter rigid-ion model. (b) – (d) Corresponding sections of the ST, FT, and L slowness surfaces in the (1̄10) plane. (From Tamura.²¹)

remain in the dispersive regime? Recall the definition of an acoustic axis as a wavevector direction along which two sheets of the slowness surface are degenerate in the continuum model. If dispersion causes a breaking of the degeneracy along an acoustic axis, interesting changes may be expected in the caustic structure.

To gain some insight into the problem, examine the dispersion relation, $\omega(\mathbf{k})$, for Ge shown in Figure 14. In these calculations, Tamura[15] uses a 31-parameter rigid-ion model to obtain accurate agreement with the inelastic neutron scattering data [Figure 14(a)]. The corresponding constant-frequency curves in (b) through (d) show that the two transverse modes remain degenerate along the ⟨100⟩ and ⟨111⟩ acoustic axes, although the relative curvatures of the two sheets change radically with increasing frequency (e.g., at $\nu = 2.4$ THz near ⟨100⟩). Also, there is an interesting surprise in Figure 14(b): The L mode above 3.5 THz develops a concave section, implying that there are inflection points that produce phonon caustics. The argument that rules out L mode caustics given in Chapter 2 is not in force here because it only applies to the continuum limit governed by Christoffel's equation.

To examine the degeneracies it is useful to approach dispersion as a perturbation on the continuum theory. We write the Christoffel equation as

$$(\Gamma_{il} - \rho\omega^2\delta_{il})e_l = 0, \tag{3}$$

with the Christoffel tensor including higher-order corrections,

$$\Gamma_{il} = C_{ijlm}k_jk_m + d_{ijlmn}k_jk_mk_n + f_{ijlmnr}k_jk_mk_nk_r + \cdots \tag{4}$$

The d tensor is known as the acoustic gyrotropic tensor and may be written in the contracted Voight notation, d_{IJN}. This tensor is identically zero in crystals that have a center of inversion, such as Ge (the center of inversion is halfway between adjacent atoms). In piezoelectric crystals, the gyrotropic tensor is nonzero, and the elastic tensor C_{ijlm} must be replaced by its stiffened form, given in Chapter 5. Nonzero d_{IJN} leads to a dispersive correction to the velocity that is linear in k along an axis where degeneracy is being lifted (elsewhere, the lowest-order correction is quadratic in k). In addition, d_{IJN} lifts the low-frequency degeneracy of the transverse surfaces along the acoustic axes. This latter effect gives rise to acoustical activity,[32,33] namely, the rotation of the polarization of a plane-polarized wave. The reason for this rotation (gyrotropic effect) is that the velocity of one circularly polarized wave is different from that of the other mode with circular polarization of the opposite sign.

That a first-order correction to the velocity does not occur in the centrosymmetric case can be illustrated by examining the result of the linear chain,

$$\omega = \omega_0 \sin(ka/2)$$
$$= \omega_0[(ka/2) - (ka/2)^3/3! + \cdots]$$
$$= v_0(1 - Ak^2 + \cdots)k, \tag{5}$$

with $v_0 = \omega_0 a/2$ and $A = a^2/24$. Considering further-neighbor interactions has the effect of adding higher-order odd functions, $\sin(nka/2)$ with $n =$ integer, which can only contribute quadratic or higher-order corrections to the velocity.[34]

Every[35] has pointed out that first-order spatial dispersion (i.e., a correction to the elastic constants that is linear in k) introduces radical changes in the shape of the slowness surfaces near an acoustic axis. As illustrated in Figure 15(a), if a point or line degeneracy is lifted, caustics are formed. For the line degeneracy occurring in hexagonal crystals, nonzero d_{IJN} produces a pair of circular caustics associated with the lifting of each line degeneracy. In the case of a conic degeneracy, first-order dispersion produces a caustic that replaces the conic circle.

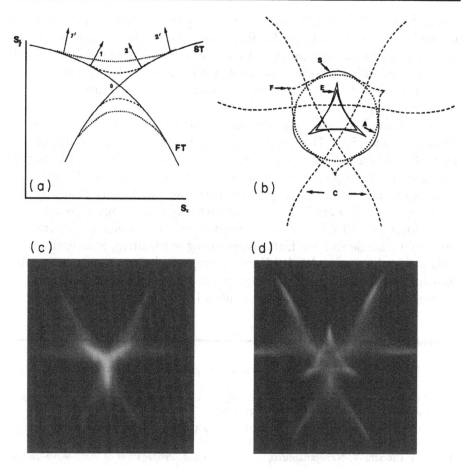

Figure 15 (a) Schematic diagram showing the effect of linear spatial dispersion on a conic degeneracy of ST and FT slowness sheets. At nonzero frequency, the degeneracy is lifted and new caustics are formed at the inflections (arrows). (b) Calculated caustic map for quartz near the trigonal axis, showing the FT (dashed) and ST (solid) caustics in the continuum limit. (From Every.[95]) (c) and (d) are experimental phonon images of quartz at 800-ns and 860-ns times of flight. Sample thickness is 3 mm, and the sampling gate is 100 ns wide. (From Koos and Wolfe.[37])

Quartz is an example in which the effect of first-order dispersion has been observed experimentally. Indeed, this mechanism was invoked[35] to explain unexpected discrepancies in the phonon images.[36-38] Normal quadratic corrections to the elastic constants due to dispersion are expected to be very small due to the large zone-boundary frequencies of the acoustic branches.

The continuum-limit caustic map along the c axis of quartz is shown schematically in Figure 15(b). ST and FT are represented by dashed and solid lines, respectively. The dotted line is the conic circle, or anticaustic, formed by normals to the ST surface exactly at the conic point, illustrated by the solid curve in Figure 15(b). Parabolic lines that run from one sheet to the other through the conic point give rise to the caustic structures labeled F and S in (b). This caustic pattern arises from a trigonal distortion of a hexagonal lattice, as discussed in Chapter 5, and is similar to the [111] caustic pattern of CaF_2 (page 115).

Time-resolved images of quartz along the trigonal axis are shown in Figure 15(c) and 15(d). There is no indication in either of these images of the caustic labeled F and S in Figure 15(b). Instead, at the early time, a Y-shaped structure that is not predicted in the continuum theory appears. Every[35] explains this feature as the dragging of the cusps F toward the axis, caused by the lifted degeneracy [broken lines in (a)], considering that a range of frequencies are produced by a 10 K Planck source and detected by a broadband Al detector. The bright triangle and line features appearing in the image at later time are explained by the caustics labeled C and E in (b), which correspond to parabolic lines that do not cross through the conic point. Monte Carlo calculations incorporating a nonzero value of d_{543} reproduce all of the observed features well.[35]

References

1. D. L. Price and J. M. Rowe, Lattice dynamics of grey tin and indium antimonide, *Phys. Rev. B* **3**. 1268 (1971).

2. W. Dietsche, G. A. Northrop, and J. P. Wolfe, Phonon focusing of large-k acoustic phonons in germanium, *Phys. Rev. Lett.* **47**, 660 (1981).

3. D. Huet, J. P. Maneval, and A. Zylbersztejn, Measurement of acoustic-wave dispersion in solids, *Phys. Rev. Lett.* **29**, 1092 (1972).

4. R. G. Ulbrich, V. Narayanamurti, and M. A. Chin, Propagation of large-wave-vector acoustic phonons in semiconductors, *Phys. Rev. Lett.* **45**, 1432 (1980).

5. R. G. Ulbrich, V. Narayanamurti, and M. A. Chin, Ballistic transport and decay of near zone-edge non-thermal phonons in semiconductors, in *International Conference on Phonon Physics*, W. E. Bron, ed. (J. Phys. (Paris) **42**, C6).

6. G. A. Northrop, S. E. Hebboul, and J. P. Wolfe, Lattice dynamics from phonon imaging, *Phys. Rev. Lett.* **35**, 95 (1985).

7. S. E. Hebboul and J. P. Wolfe, Focusing of dispersive phonons in GaAs, Z. *Phys. B* **74**, 35 (1989).

8. M. Fieseler, M. Wenderoth, and R. G. Ulbrich, The mean free path of 0.7 THz acoustic phonons in GaAs, *18th International Conference on the Physics of Semiconductors*, Warsaw (1988).

9. M. Lax, V. Narayanamurti, R. C. Fulton, and N. Holzwarth, Monte Carlo calculations of phonon transport, in *Phonon Scattering in Condensed Matter*, W. Eisenmenger, K. Lassmann, and S. Dottinger, eds. (Springer-Verlag, Berlin, 1984).

10. M. Lax, V. Narayanamurti, and R. C. Fulton, Classical diffusive photon transport in a slab, in *Laser Optics of Condensed Matter*, H. Z. Cummins and A. A. Kaplyanskii, eds. (Plenum, New York, 1988).

11. B. Stock, R. G. Ulbrich, and M. Fieseler, Direct observation of ballistic large-wavevector phonon propagation in gallium arsenide, in *Phonon Scattering in Condensed Matter*, W. Eisenmenger, K. Lassmann, and S. Dottinger, eds. (Springer-Verlag, Berlin, 1984).

12. B. Stock and R. G. Ulbrich, Measurements of dispersive phonon transport in GaP, *Physica* **117B–118B**, 540 (1983).

13. J. P. Wolfe, Propagation of large-wavevector acoustic phonons: new perspectives from phonon imaging, *Festkörperprobleme* **29**, 75 (1989).

14. T. Haavasoja, V. Narayanamurti, and M. A. Chin, Propagation of near-zone-boundary, acoustic phonons in solid ^{4}He with the use of superconducting tunnel junctions, *Phys. Rev. B* **27**, 2767 (1983).

15. S. Tamura, Focusing of high-frequency dispersive phonons, *Phys. Rev. B* **25**, 1415 (1982).

16. G. A. Northrop, Phonon focusing of dispersive phonons in Ge, *Phys. Rev. B* **26**, 903 (1982).

17. W. Metzger and R. P. Huebener, Phonon focusing in [001] germanium, *Z. Phys. B* **73**, 33 (1988); see also H. Kittel, E. Held, W. Klein, and R. P. Huebener, Dispersive anisotropic ballistic phonon propagation in [100] GaAs, *Z. Phys. B* **77**, 79 (1989).

18. S. E. Hebboul and J. P. Wolfe, Imaging of large-k phonons in InSb, *Phys. Rev. B* **34**, 3948 (1986).

19. S. E. Hebboul and J. P. Wolfe, Lattice dynamics of InSb from phonon imaging, *Z. Phys. B* **73**, 437 (1989).

20. G. A. Northrop and J. P. Wolfe, Phonon imaging: theory and applications, in *Nonequilibrium Phonon Dynamics*, W. E. Bron, ed. (Plenum, New York, 1985).

21. S. Tamura, Large-wavevector phonons in highly dispersive crystals: phonon-focusing effects, *Phys. Rev. B* **28**, 897 (1983).

22. K. C. Rustagi and W. Weber, Adiabatic bond charge model for the phonons in A^3B^5 semiconductors, *Sol. State Commun.* **18**, 673 (1976).

23. S. Yip and Y. C. Chang, Theory of phonon dispersion relations in semiconductor superlattices, *Phys. Rev. B* **30**, 7037 (1984).

24. K. Kunc, Nielsen, and O. Holm, Lattice dynamics of zincblende structure compounds using deformation-dipole model and rigid ion modes, *Comp. Phys. Comm.* **16**, 181 (1979).

25. K. Kunc, M. Balkanski, and M. A. Nusimovici, Lattice dynamics of several A^NB^{8-N} compounds having the zincblende structure, *Phys. Stat. Sol. B* **71**, 341 (1975); **72**, 229 (1975).

26. P. H. Borchards and K. Kunc, The lattice dynamics of indium ρ nictides, *J. Phys. C* **11**, 4145 (1978).

27. K. Kunc, Nielsen, and O. Holm, Lattice dynamics of zincblende structure compounds: II. Shell Model, *Comp. Phys. Comm.* **17**, 413 (1979).

28. M. Schreiber, M. Fieseler, A. Mazur, J. Pollman, B. Stock, and R. G. Ulbrich, Dispersive phonon focusing in gallium arsenide, in *Proceedings of 18th International Conference on the Physics of Semiconductors* O. Engstrom, ed. (World Scientific, Singapore, 1987), p. 1373.

29. M. T. Ramsbey, S. Tamura, and J. P. Wolfe, Phonon focusing of elastically scattered phonons in GaAs, *Z. Phys. B* **73**, 167 (1988).

30. In fact, a wide distribution of phonon frequencies is predicted for constant-velocity selection (see Ref. 7).

31. M. T. Ramsbey and J. P. Wolfe, Dispersive phonon imaging of InAs, *Z. Phys. B* **97**, 413 (1995).

32. A. S. Pine, Direct observation of acoustical activity in α-quartz, *Phys. Rev. B* **2**, 2049 (1970).

33. J. Joffrin and A. Levelut, Mise en evidence et mesure du pouvoir rotatoire acoustique naturel du quartz-α, *SSC* **8**, 1573 (1970).

34. C. Kittel, *Introduction to Solid State Physics*, 3rd ed. (Wiley, New York, 1966).

35. A. G. Every, Effects of first-order spatial dispersion on phonon focusing: application to quartz, *Phys. Rev. B* **36**, 1448 (1987).

36.   R. Eichele, R. P. Huebener, and H. Seifert, Phonon focusing in quartz and sapphire imaged by electron beam scanning, *Z. Phys. B* **48**, 89 (1982).

37.   G. L. Koos and J. P. Wolfe, Phonon focusing in piezoelectric crystals: quartz and lithium niobate, *Phys. Rev. B* **30**, 3470 (1984).

38.   A report of dispersive-like times of flight in quartz [Grill and Weis, *Phys. Rev. Lett.* **35**, 588 (1975)] has never been reproduced, despite attempts by Weis's group and others [W. E. Bron, M. Rossinelli, Y. H. Bai, and F. Keilmann, *Phys. Rev. B* **27**, 1370 (1983)].

8

Phonon dynamics

So far, our treatment of phonon propagation has relied on classical mechanics. We found that the low-frequency phonon-focusing patterns are well predicted by continuum elasticity theory. Even the high-frequency focusing patterns are explained by microscopic modeling of the lattice dynamics based on a classical equation of motion. Our interest now turns to the processes of phonon scattering, a subject that requires a quantum-mechanical description of the lattice vibrations. To calculate phonon scattering rates we must rigorously define what a phonon is and show how phonons are mathematically created and annihilated. The quantum theory of a harmonic crystal is found in many books, such as those listed in the references.[1-5] Here we present the essential ideas required to treat the scattering problem.

Just as with the lattice-dynamics problem, spring-like harmonic forces between atoms are assumed without delving into the quantum-mechanical origins of these forces. Viewing the atoms as distinguishable point-like particles and invoking the translational symmetry of the three-dimensional crystal allows us to reduce the N-body Hamiltonian of the crystal to a sum of $3N$ sub-Hamiltonians, each having the form of a simple harmonic oscillator.

A. The simple harmonic oscillator

The one-dimensional problem of a single mass connected to springs is illustrated in Figure 1(a). The energy of this system has the form

$$\mathcal{H} = (1/2)(p^2/M + \kappa x^2), \tag{1}$$

where the momentum is defined by $p = M\dot{x}$. Classically, we wish to solve for the position $x(t)$ and momentum $p(t)$. For this simple one-dimensional case, the equation of motion is obtained by setting the time variation of the

189

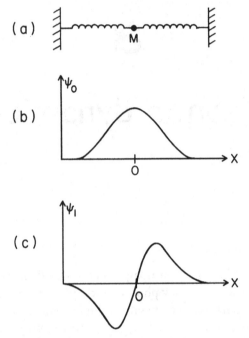

Figure 1 (a) Schematic of the simple harmonic oscillator. The effective spring constant κ is twice that of the individual springs. (b) Lowest-energy wave function with energy $\hbar\omega/2$. (c) First excited state with energy $3\,\hbar\omega/2$.

energy equal to zero,

$$d\mathcal{H}(t)/dt = (p/M)\,dp/dt + \kappa x\,dx/dt$$

$$= M\dot{x}\ddot{x} + \kappa\dot{x}x = 0. \tag{2}$$

The resulting equation, $M\ddot{x} + \kappa x = 0$, is Newton's law for the mass and springs. The solution is $x = x_0\sin(\omega t + \phi)$, with $\omega = \sqrt{(\kappa/M)}$. The amplitude of oscillation, x_0, and phase, ϕ, can take on any numerical values, depending on initial conditions.

The quantum-mechanical Hamiltonian looks the same as Eq. (1), but the wave nature of the particle leads us to solve for a wave function $\psi(x)$, where $\psi^*\psi$ is interpreted as the probability of observing the particle at position x. (ψ^* is the complex conjugate of ψ.) The wave function is postulated to be a solution to the eigenvalue equation,

$$\mathcal{H}\psi = E\psi, \tag{3}$$

where p and x are regarded as quantum-mechanical operators and E is the energy of the system, to be determined as an eigenvalue of this equation. The momentum operator is defined by the *ansatz*,

$$p = -i\hbar\nabla, \tag{4}$$

where $\nabla = \partial/\partial x$ in one dimension. The quantum-mechanical wave equation is now

$$(\hbar^2/2M)(-\nabla^2 + \beta^2 x^2)\psi = E\psi, \tag{5}$$

with $\beta^2 = (\kappa M/\hbar^2)$. One can attempt to "factor" this second-order differential equation with the intention of producing a simpler first-order equation. Assuming that ψ and E solve Eq. (5), try the combination

$$(\hbar^2/2M)(-\nabla + \beta x)(\nabla + \beta x)\psi = (\hbar^2/2M)(-\nabla^2 + \beta^2 x^2 + \beta x\nabla - \nabla\beta x)\psi$$

$$= (\hbar^2/2M)(-\nabla^2 + \beta^2 x^2 - \beta)\psi$$

$$= (E - \hbar^2\beta/2M)\psi$$

$$= (E - \hbar\omega/2)\psi, \tag{6}$$

where $\omega = \sqrt{\kappa/M}$ is the classical harmonic oscillator frequency. Comparison with Eq. (5) shows that $(-\nabla^2 + \beta^2 x^2)$ is not equal to $(-\nabla + \beta x)(\nabla + \beta x)$ because x and ∇ do not commute. Specifically, the second step in the above calculation uses $(x\nabla - \nabla x)\psi = x\nabla\psi - \psi - x\nabla\psi = -\psi$. This property of the two operators is compactly written as

$$[x, \nabla] \equiv x\nabla - \nabla x = -1$$

or, with Eq. (4),

$$[x, p] \equiv xp - px = i\hbar, \tag{7}$$

which is the general commutation relation for x and p.

For the special value of energy, $E = E_0 \equiv \hbar\omega/2$, the right side of Eq. (6) vanishes. The resulting homogeneous equation will certainly be solved if

$$(\nabla + \beta x)\psi = 0 \quad \text{or} \quad d\psi/dx = -\beta x\psi. \tag{8}$$

By integrating $d\psi/\psi = -\beta x dx$, we arrive at the solution

$$\psi_0 = A_0 \exp(-\beta x^2/2), \tag{9}$$

shown schematically in Figure 1(b). The constant $A_0 = \sqrt{\beta/2\pi}$ is determined from the normalization condition, $\int \psi_0^2 dx = 1$.

Equation (6) suggests a transformation of operators from $\{p, x\}$ to $\{a^+, a\}$, with the definitions

$$a^+ = \sqrt{(\hbar/2M\omega)}(-\nabla + \beta x) = \sqrt{(1/2M\hbar\omega)}(M\omega x - ip),$$

$$a = \sqrt{(\hbar/2M\omega)}(\nabla + \beta x) = \sqrt{(1/2M\hbar\omega)}(M\omega x + ip), \tag{10}$$

where the prefactors are chosen to simplify later equations. We may rewrite Eq. (6) in terms of these operators,

$$\hbar\omega a^+ a\psi = (E - \hbar\omega/2)\psi, \tag{11}$$

and, by interchanging the order of factors on the left side of Eq. (6),

$$\hbar\omega a a^+ \psi = (E + \hbar\omega/2)\psi. \tag{12}$$

Summing these last two equations, we obtain an expression for the Hamiltonian in the new representation,

$$\mathcal{H} = (\hbar\omega/2)(a^+ a + a a^+). \tag{13}$$

Subtracting Eq. (11) from Eq. (12) yields the commutation relation,

$$[a, a^+] = 1. \tag{14}$$

Using this relation, we can also write the Hamiltonian as

$$\mathcal{H} = \hbar\omega(a^+ a + 1/2). \tag{15}$$

One advantage of this new representation is that it provides a natural way of generating the complete set of eigenfunctions. Given that $\hbar\omega\, a^+ a\, \psi_0 = 0$ for $E = E_0 = \hbar\omega/2$, and using Eq. (12),

$$a a^+ \psi_0 = \psi_0. \tag{16}$$

Multiplying by a^+,

$$a^+ a (a^+ \psi_0) = (a^+ \psi_0), \tag{17}$$

which says that $\psi_1 = \text{(constant)} a^+ \psi_0$ is also an eigenfunction of $\mathcal{H}$. From Eq. (9) and the definition of a^+, $\psi_1 \propto (2x)\exp(-\beta x^2)$, as plotted in Figure 1(c). Using Eqs. (3) and (15), we see that ψ_1 has an energy $E_1 = (3/2)\hbar\omega$. Repeating this operation n times produces the eigenstate

$$\psi_n \propto (a^+)^n \psi_0, \tag{18a}$$

with

$$E_n = (n + 1/2)\hbar\omega. \tag{18b}$$

It can be shown that this procedure produces a complete set of orthogonal eigenfunctions of the form (Hermite polynomial) $\cdot \exp(-\beta x^2)$ tabulated in standard textbooks.

The energy of the harmonic oscillator is quantized into a ladder of levels separated by the quantum of energy, $\hbar\omega$. When the creation operator, a^+, operates on an eigenfunction, it adds $\hbar\omega$ to the energy of the system and produces the next-higher energy state in the ladder. Using the commutation relations, it is straightforward to show that the operator, a, acts as an annihilation operator, subtracting $\hbar\omega$ from the energy of the system and producing the next-lower energy state of the ladder. The lowest energy of the system is $E_0 = \hbar\omega/2$, because a operating on ψ_0 gives zero. E_0 is the zero-point energy of the system. The constant prefactors associated with a and a^+ are given by

$$a^+ \psi_n = (n + 1)^{1/2} \psi_{n+1},$$
$$a \psi_n = n^{1/2} \psi_{n-1}, \tag{19}$$

which satisfy Eqs. (3), (14), and (15) with $E_n = (n + 1/2)\hbar\omega$. The combination a^+a is the number operator, which, when operating on a particular eigenstate, gives the number, n, of energy quanta associated with that state.

B.   The linear chain – Creation and annihilation of phonons

The Hamiltonian for a one-dimensional chain of N masses coupled by springs is

$$\mathcal{H} = (1/2) \sum_{\ell} [p_{\ell}^2/M + \kappa(u_{\ell+1} - u_{\ell})^2], \tag{20}$$

where u_{ℓ} is the displacement of the mass at $x_{\ell}^0 = \ell a$, with $\ell = 1, \ldots N$. Periodic boundary conditions ($u_N = u_1$, $p_N = p_1$) are assumed. Due to the coupling between atoms, it does not make sense to describe the excitation of this system with annihilation and creation operators that operate on one mass of the system. We must first transform the single-particle operators p_{ℓ} and u_{ℓ} to operators that represent the normal modes of vibration. For a periodic system, the Fourier transformation,

$$u_{\ell} = N^{-1/2} \sum_{k} U_k e^{ik\ell a}, \tag{21a}$$

with the N coefficients,

$$U_k = N^{-1/2} \sum_{\ell} u_{\ell} e^{-ik\ell a}, \tag{21b}$$

describes the displacement of an individual atom as a superposition of N waves with wavevectors $k = \pm 2\pi n/Na$, where the integer n ranges from 1 to $N/2$. We can consider the coefficient U_k as the normal coordinate representing the amplitude of a wave with wavevector k, which is one of the N normal modes of the system. In quantum mechanics, the normal coordinate U_k is an operator. In a similar manner,

$$p_{\ell} = N^{-1/2} \sum_{k} P_k e^{ik\ell a}, \tag{22a}$$

with

$$P_k = N^{-1/2} \sum_{\ell} p_{\ell} e^{-ik\ell a} \tag{22b}$$

defining the normal momentum coordinate P_k of the system.

The operators for physical quantities, such as coordinates and momenta, must be Hermitian; i.e., their expectation values must be real. This implies that $u_{\ell}^+ = u_{\ell}$, where u_{ℓ}^+ is the Hermitian adjoint of u_{ℓ}, defined by $\int \phi^* u^+ \psi \equiv \int (u\psi)^* \phi$. From Eqs. (21) and (22), this condition requires that

$$U_k^+ = U_{-k} \quad \text{and} \quad P_k^+ = P_{-k}.$$

The squared terms in Eq. (20) represent the Hermitian products $p_\ell^2 = p_\ell^+ p_\ell$ and $(u_{\ell+1} - u_\ell)^2 = (u_{\ell+1} - u_\ell)^+ (u_{\ell+1} - u_\ell)$. [In Eq. (1), $p^2 = p^+ p$ is also implied.] The justification for all this formalism is that eventually it describes the real world.

Using Eq. (21), we find

$$(u_{\ell+1} - u_\ell) = N^{-1/2} \sum_k U_k e^{ik\ell a} (e^{ika} - 1),$$

$$\sum_\ell (u_{\ell+1} - u_\ell)^2 = N^{-1} \sum_{\ell k k'} U_k U_{k'}^+ e^{i(k-k')\ell a} (e^{ika} - 1)(e^{-ik'a} - 1)$$

$$= \sum_k U_k U_{-k} 2(1 - \cos ka), \qquad (23)$$

where the last step makes use of $U_k^+ = U_{-k}$ and the orthogonality relation,

$$\sum_\ell \exp[i(k - k')\ell a] = N\delta_{k,k'}, \qquad (24)$$

obtained by substituting Eq. (21a) into Eq. (21b). Similarly, the kinetic energy term in Eq. (20) is written in terms of normal momenta with Eq. (22a), and the Hamiltonian of the linear chain becomes

$$\mathcal{H} = (1/2) \sum_k [P_k P_{-k}/M + 2\kappa(1 - \cos ka)U_k U_{-k}]. \qquad (25)$$

This Hamiltonian is the sum of N uncoupled terms of the form

$$\mathcal{H}_k = (1/2)(P_k P_{-k}/M + 2\kappa(1 - \cos ka)U_k U_{-k}) \qquad (26)$$

$$= (1/2M)(P_k P_{-k} + M^2 \omega_k^2 U_k U_{-k}). \qquad (27)$$

The beneficial effect of defining the normal coordinates is to put the Hamiltonian into the same form as that of the simple harmonic oscillator, Eq. (1). Each normal mode of wavevector k is analogous to a simple harmonic oscillator with frequency given by $\omega_k^2 = 2(\kappa/M)(1 - \cos ka)$, which is the classical result we derived in Chapter 6 (page 148).* Amazing.

Physically, the simple form of Eq. (27) says that the harmonic-oscillator restoring forces act on an entire wave of wavevector k and amplitude U_k independently of all other waves in the system. This, of course, is the property of a normal mode, which, in the absence of nonharmonic or symmetry-breaking perturbations, vibrates independently of (i.e., is uncoupled to) the other normal modes of the system.

Now that we have reduced the Hamiltonian to a simple form by a transformation of coordinates, we must solve the eigenvalue Eq. (3). For the normal momentum P_k we cannot make a simple correspondence to the single-particle

---

* In this chapter we denote $\omega(k) = \omega_k$ to emphasize the discrete nature of k.

operator $-i\hbar\nabla$, but the U_k and P_k operators obey a similar commutation relation as x and p did in the single-particle case. That is,

$$[U_k, P_{k'}] = U_k P_{k'} - P_{k'} U_k = i\hbar\delta_{kk'}, \tag{28}$$

which follows directly from the definitions of U_k and P_k, Eqs. (21b) and (22b), and the fundamental commutation relations between the displacements and the momenta of the individual particles: $[u_\ell, p_{\ell'}] = i\hbar\delta_{\ell\ell'}$.

Once again, we can factor the Hamiltonian and [with the help of Eq. (28)] put it into the form

$$\mathcal{H} = \sum_k \hbar\omega_k (a_k^+ a_k + 1/2), \tag{29}$$

using the definition of the creation and annihilation operators,

$$a_k^+ = \sqrt{(1/2M\hbar\omega_k)}(M\omega_k U_{-k} - i P_k),$$
$$a_k = \sqrt{(1/2M\hbar\omega_k)}(M\omega_k U_k + i P_{-k}), \tag{30}$$

which, from Eq. (28), obey the simple commutation relation,

$$[a_k, a_{k'}^+] = \delta_{kk'}. \tag{31}$$

Eigenstates of the sub-Hamiltonian, $\mathcal{H}_k$, are denoted by $|n_k\rangle$. The eigenvalue equation is

$$\mathcal{H}_k|n_k\rangle = E_{n_k}|n_k\rangle,$$

with eigenvalues

$$E_{n_k} = (n_k + 1/2)\hbar\omega_k. \tag{32}$$

The integer n_k tells how many quanta of energy, $\hbar\omega_k$, are present in the normal mode with wavevector k, analogous to the case of the simple harmonic oscillator. A state $|n_k\rangle$ is said to contain n_k phonons, each of energy $\hbar\omega_k$, in addition to a zero-point motion with energy $\hbar\omega_k/2$. The phonon creation and anihilation operators have the properties

$$a_k^+|n_k\rangle = (n_k + 1)^{1/2}|n_k + 1\rangle$$
$$a_k|n_k\rangle = n_k^{1/2}|n_k - 1\rangle. \tag{33}$$

A wave containing n phonons and, therefore, having discrete energies and amplitudes is shown schematically in Figure 2.

The eigenstates of the full Hamiltonian given in Eq. (29) are written as $|\ldots n_k n_{k+1} \ldots\rangle$, which designates the number of phonons in each of the wavevector modes. The total energy of this state is the sum of the energies contained in each normal mode,

$$E(\ldots n_k n_{k+1} \ldots) = \sum_k (n_k + 1/2)\hbar\omega_k. \tag{34}$$

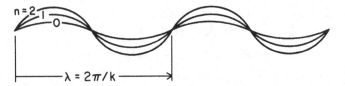

Figure 2 *Schematic picture of a normal mode, k, of the crystal containing n = 0, 1, and 2 phonons.*

C. Phonons in three dimensions

The potential energy of a three-dimensional Bravais lattice can be written in terms of the force-constant matrix $\phi_{ij}(\ell\ell')$ between pairs of atoms ℓ and ℓ', as given by Eq. (14) on page 150. The cartesian components for the displacement and momentum of the atom (mass M) from an equilibrium position $\mathbf{r}_\ell^0$ are $u_i(\ell)$ and $p_i(\ell)$. The Hamiltonian for the crystal is

$$\mathcal{H} = (1/2)\left[\sum_\ell \sum_i p_i^2(\ell)/M + \sum_{\ell\ell'}\sum_{ij}\phi_{ij}(\ell\ell')u_i(\ell)u_j(\ell')\right], \qquad (35)$$

where ℓ and ℓ' range over the N atomic sites in the crystal. Each atom has three degrees of freedom, so there are $3N$ normal modes of the system. Specifically, there are N distinct $\mathbf{k}$ vectors, each having three possible polarizations with index $\alpha = $ L, FT, and ST. (For a cubic crystal $\mathbf{k}$ has the cartesian components $k_i = \pm 2\pi n_i/L$, with the integers $n_i = 1, \dots, L/2a$, and $L = aN^{1/3}$.) The unit polarization vector of the wave labeled $\mathbf{k}\alpha$ is denoted $\mathbf{e}(\mathbf{k}\alpha)$, with cartesian components $e_i(\mathbf{k}\alpha)$. The normal coordinates $U_{\mathbf{k}\alpha}$ and normal momenta $P_{\mathbf{k}\alpha}$ are defined by the relations

$$u_i(\ell) = N^{-1/2}\sum_{\mathbf{k}\alpha}U_{\mathbf{k}\alpha}e_i(\mathbf{k}\alpha)e^{i\mathbf{k}\cdot\mathbf{r}_\ell^0},$$

$$p_i(\ell) = N^{-1/2}\sum_{\mathbf{k}\alpha}P_{\mathbf{k}\alpha}e_i(\mathbf{k}\alpha)e^{i\mathbf{k}\cdot\mathbf{r}_\ell^0}, \qquad (36)$$

where the summation is over all $\mathbf{k}$ and α. Using this transformation, the potential-energy term in Eq. (35) becomes

$$(1/2N)\sum_{\substack{\mathbf{k}'\alpha' \\ \mathbf{k}\alpha}}U_{\mathbf{k}\alpha}U_{\mathbf{k}'\alpha'}^+\sum_{ij}e_i(\mathbf{k}\alpha)e_j(\mathbf{k}'\alpha')\sum_{\ell\ell'}\phi_{ij}(\ell\ell')e^{i\mathbf{k}\cdot\mathbf{r}_\ell^0-i\mathbf{k}'\cdot\mathbf{r}_{\ell'}^0}.$$

Adding and subtracting $i\mathbf{k}'\cdot\mathbf{r}_\ell^0$ to the argument of the exponent gives

$$i(\mathbf{k}-\mathbf{k}')\cdot\mathbf{r}_\ell^0 + i\mathbf{k}'\cdot(\mathbf{r}_\ell^0-\mathbf{r}_{\ell'}^0).$$

With the definition of the dynamical matrix, [Eq. (19) on page 150],

$$\sum_{\ell\ell'}\phi_{ij}(\ell\ell')e^{i\mathbf{k}\cdot\mathbf{r}_\ell^0-i\mathbf{k}'\cdot\mathbf{r}_{\ell'}^0} = M\sum_\ell \mathcal{D}_{ij}(\mathbf{k}')e^{i(\mathbf{k}-\mathbf{k}')\cdot\mathbf{r}_\ell^0}.$$

The matrix $\mathcal{D}(\mathbf{k}')$ is independent of ℓ, and, as in Eq. (24), the sum of the exponential factor over ℓ becomes $N\delta_{\mathbf{k}\mathbf{k}'}$. Summing over $\mathbf{k}'$ and noting the property of the Hermitian adjoint, we now have for the potential energy,

$$(M/2) \sum_{\mathbf{k}\alpha\alpha'} U_{\mathbf{k}\alpha} U_{-\mathbf{k}\alpha'} \sum_{ij} e_i(\mathbf{k}\alpha) e_j(\mathbf{k}\alpha') \mathcal{D}_{ij}(\mathbf{k}).$$

Until now, we have not specified how the polarization vectors are to be determined. We see that the Hamiltonian is greatly simplified if the polarization vectors are chosen to be the classical eigenvectors determined by the diagonalization of the dynamical matrix,

$$\sum_j \mathcal{D}_{ij}(\mathbf{k}) e_j(\mathbf{k}\alpha') = \omega_{\mathbf{k}\alpha'}^2 e_i(\mathbf{k}\alpha'), \tag{37}$$

previously seen as Eq. (20) on page 151. The eigenvalue $\omega_{\mathbf{k}\alpha}^2$ is the square of the classical normal-mode frequency for wavevector $\mathbf{k}$ and polarization index α. In other words, these are the dispersion curves that we calculated in Chapter 6.

The final simplification of the potential-energy term comes with the realization that

$$\sum_i e_i(\mathbf{k}\alpha) e_i(\mathbf{k}\alpha') = \delta_{\alpha\alpha'}, \tag{38}$$

which states that the polarizations of the three modes with the same $\mathbf{k}$ are orthogonal.

The kinetic-energy term in the Hamiltonian is much more easily dealt with, and the final result including both kinetic and potential energies is

$$\mathcal{H} = \sum_{\mathbf{k}\alpha} \mathcal{H}_{\mathbf{k}\alpha}, \tag{39}$$

with the sub-Hamiltonian

$$\mathcal{H}_{\mathbf{k}\alpha} = (1/2M)(P_{\mathbf{k}\alpha} P_{-\mathbf{k}\alpha} + M^2 \omega_{\mathbf{k}\alpha}^2 U_{\mathbf{k}\alpha} U_{-\mathbf{k}\alpha}). \tag{40}$$

Again, we have succeeded in casting the Hamiltonian into the form of $3N$ uncoupled harmonic oscillators. Equations (27)–(34) describe the phonon quantization of this three-dimensional crystal with the simple generalization $k \Rightarrow \mathbf{k}\alpha$. For example, the operator $a_{\mathbf{k}\alpha}^+$ creates a phonon in a mode with wavevector $\mathbf{k}$ and polarization index α. For the periodic harmonic crystal, this wave is decoupled from all the other waves $\mathbf{k}'\alpha'$ in the crystal.

D.   Phonon scattering from isotopic defects

When an impurity atom substitutes for a host atom, or when the crystal is composed of atoms with different isotopic masses, the periodicity of the crystal is perturbed. Such imperfections may be regarded as mass defects, although the impurity case also involves a local change in the force constants. One effect of mass defects is to change the vibrational spectrum, or density of states, from that of the perfect crystal. This modification of the energy levels can be considerable for impurities (e.g., leading to local modes) but turns out to be rather small in the case of isotopic disorder, provided that the average atomic mass is used.[6] The principal effect of mass defects is their influence on the phonon transport.

The picture we espoused in the previous section treats eigenfunctions that uniformly fill the entire volume of the crystal. In practice, a phonon is created over some effective volume much less than that of the crystal. Such a localization requires a superposition of normal modes, centered around the wavevector $\mathbf{k}$ and frequency ω (Figure 18 on page 47). The time evolution of this packet involves a spatial transport of vibrational energy. A mass defect acts as a scattering center that interferes with this ballistic transport. If the distribution of scattering centers is sufficiently homogeneous, the phonons experience the same perturbations no matter where they are in the crystal. Phonon scattering from defects plays an important role in the thermal properties of crystals, especially at low temperatures.

We begin by considering a single isotopic impurity of mass M_1 in a Bravais lattice of host atoms with masses M. For isotopic impurities the mass difference $\Delta M = M - M_1$ is fairly small compared to M or M_1. The Hamiltonian of this hypothetical crystal is the same as Eq. (35) with one of the N kinetic-energy terms replaced by*

$$p_\ell^2/2M_1 = p_\ell^2/2M(1 - \Delta M/M) \simeq p_\ell^2/2M + (p_\ell^2/2M)(\Delta M/M). \quad (41)$$

In this section we use the shorthand notation, $p_\ell^2 \equiv \mathbf{p}^+(\ell) \cdot \mathbf{p}(\ell)$. The new Hamiltonian is

$$\mathcal{H} = \mathcal{H}_0 + K, \quad (42)$$

where $\mathcal{H}_0$ is the Hamiltonian of the perfect crystal [Eq. (35)] and K is the small perturbation in kinetic energy,

$$K = (p_\ell^2/2M)(\Delta M/M), \quad (43)$$

associated with the isotopic defect at the site labeled ℓ. The potential-energy terms in $\mathcal{H}_0$ are unchanged because the defect atom has the same bonding as the host atoms.

Imagine a single phonon in the perfect lattice with wavevector $\mathbf{k}_i$ and mode α_i, and "turn on" the constant perturbation K. Quantum mechanics tells us that at some time later there is a finite probability that the initial phonon

---

* The need for higher-order terms in this expansion in $\Delta M/M$ is considered in Ref. 8. For isotopic impurities, the first-order term is quite sufficient, but this assumption breaks down for chemical impurities with $\Delta M/M \geq 1$, regardless of their density.

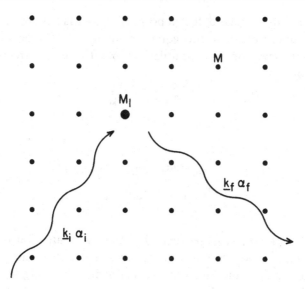

Figure 3 Schematic picture of a phonon scattering from a mass defect in the crystal.

disappears, leaving a phonon in another state $k_f \alpha_f$ (ignoring fission into two phonons for the moment). This process is illustrated in Figure 3. The scattering of the initial phonon can occur into a near continuum of final states. The standard result of time-dependent perturbation theory predicts that the coherent sum of these processes gives a scattering probability that increases linearly with time after the perturbation is turned on. The scattering rate for the initial phonon is

$$\tau_i^{-1} = (2\pi/\hbar) \sum_f |\langle f|K|i\rangle|^2 \delta(\omega_i - \omega_f). \tag{44}$$

The general form of an eigenstate of $\mathcal{H}_0$ is

$$|\ldots n_{k\alpha} \ldots n_{k'\alpha'} \ldots\rangle, \tag{45}$$

so the initial state described above has the form

$$|i\rangle = |\ldots 0\,0\,1\,0\,0\ldots\rangle, \tag{46a}$$

with $n_{k_i \alpha_i} = 1$ and all other $n_{k\alpha} = 0$. One possible final state is

$$\langle f| = \langle\ldots 0\,1\,0\,0\,0\ldots|, \tag{46b}$$

with $n_{k_f \alpha_f} = 1$ and all other $n_{k\alpha} = 0$.

To compute the matrix element in Eq. (44), we write the perturbation K in terms of normal momenta [Eq. (36)],

$$p_\ell^2 = N^{-1} \sum_{\substack{k'\alpha' \\ k\alpha}} P_{-k\alpha} P_{k'\alpha'} (\mathbf{e}_{k\alpha} \cdot \mathbf{e}_{k'\alpha}) e^{i(k'-k)\cdot r_\ell^0}. \tag{47}$$

For the case of a single mass defect at position $\mathbf{r}_\ell^0$ we may choose $\mathbf{r}_\ell^0 = 0$, which makes the exponent factor in this equation equal to 1. The normal momenta are related to the creation and annihilation operators by subtracting Eqs. (30) and solving for

$$P_{\mathbf{k}\alpha} = i[M\hbar\omega_{\mathbf{k}\alpha}/2]^{1/2}(a_{\mathbf{k}\alpha}^+ - a_{-\mathbf{k}\alpha}), \tag{48}$$

which gives

$$
\begin{aligned}
&P_{-\mathbf{k}\alpha} P_{\mathbf{k}'\alpha'} \\
&= -M\hbar(\omega_{\mathbf{k}\alpha}\omega_{\mathbf{k}'\alpha'})^{1/2}(a_{-\mathbf{k}\alpha}^+ - a_{\mathbf{k}\alpha})(a_{\mathbf{k}'\alpha'}^+ - a_{-\mathbf{k}'\alpha'})/2 \\
&= -M\hbar(\omega_{\mathbf{k}\alpha}\omega_{\mathbf{k}'\alpha'})^{1/2}(a_{-\mathbf{k}\alpha}^+ a_{\mathbf{k}'\alpha'}^+ + a_{\mathbf{k}\alpha}a_{-\mathbf{k}'\alpha'} - a_{-\mathbf{k}\alpha}^+ a_{-\mathbf{k}'\alpha'} - a_{\mathbf{k}\alpha}a_{\mathbf{k}'\alpha'}^+)/2.
\end{aligned}
\tag{49}
$$

The quadratic form of this perturbation energy means that in a particular scattering event the occupation numbers of at most two states are affected. For a nonzero matrix element in the sum shown in Eq. (44), a phonon must be removed from the state $\mathbf{k}_i\alpha_i$, requiring an annihilation factor $a_{\mathbf{k}_i\alpha_i}$. The last two terms of Eq. (49) each contribute one factor of this form because the components of $\mathbf{k}$ and $\mathbf{k}'$ range over both positive and negative values. The a^+ factors accompanying the $a_{\mathbf{k}_i\alpha_i}$ create a single phonon in a final state $\mathbf{k}_f\alpha_f$, so that

$$\langle f|a_{\mathbf{k}_f\alpha_f}^+ a_{\mathbf{k}_i\alpha_i}|i\rangle = \langle f|a_{\mathbf{k}_i\alpha_i} a_{\mathbf{k}_f\alpha_f}^+|i\rangle = 1$$

for the states (46a) and (46b), and all other terms vanish. We now see that final states involving more than one phonon (e.g., $\langle\ldots 0200000\ldots|$ and $\langle\ldots 0101000\ldots|$) do not contribute to τ_i^{-1}, due to the quadratic form of the perturbation. In other words, mass defect scattering is an elastic-scattering process, in which the initial and scattered phonons have the same energy ($\omega_i = \omega_f$).

The complete matrix element is then

$$\langle f|K|i\rangle = (2MN)^{-1}(\Delta M/M)[M\hbar(\omega_i\omega_f)^{1/2}](\mathbf{e}_i \cdot \mathbf{e}_f), \tag{50}$$

and the scattering rate is

$$\tau_i^{-1} = (\pi/2N^2)(\Delta M/M)^2\omega_i^2 \sum_f (\mathbf{e}_i \cdot \mathbf{e}_f)^2\delta(\omega_f - \omega_i). \tag{51}$$

This equation can be further simplified for cubic crystals, which, for an arbitrary function $F(\omega_i, \omega_f)$, have the property[7]

$$\sum_f (\mathbf{e}_i \cdot \mathbf{e}_f)^2 F(\omega_i, \omega_f) = (1/3)\sum_f F(\omega_i, \omega_f). \tag{52a}$$

Now, converting the sum over k space to an integral over frequency,

$$\sum_f \delta(\omega_i - \omega_f) = \int D(\omega_f)\delta(\omega_i - \omega_f)\,d\omega_f = D(\omega_i), \tag{52b}$$

where $D(\omega)$ is the density of phonon states for the crystal. With these manipulations, Eq. (51) becomes

$$\tau_i^{-1} = (\pi/6N^2)(\Delta M/M)^2 \omega_i^2 D(\omega_i). \tag{53}$$

It is useful to simplify this formula for the case of an isotropic simple-cubic solid of volume $L^3 = Na^3$, where a is the lattice constant. The density of states in $\mathbf{k}$ space is $1/(2\pi/L)^3$. In the continuum limit, the number of $\mathbf{k}$ states (with polarization index α and phase velocity v_α) that fall within a $\mathbf{k}$-space sphere of radius k is

$$\mathcal{N}_\alpha(k) = (4\pi/3)k^3/(2\pi/L)^3 = Na^3\omega^3/6\pi^2 v_\alpha^3, \tag{54}$$

giving the density of states,

$$D(\omega) \equiv \sum_\alpha d\mathcal{N}_\alpha/d\omega = 3Na^3\omega^2/2\pi^2 v^3, \tag{55}$$

with the average phase velocity v defined by

$$3/v^3 = \sum_\alpha 1/v_\alpha^3 = 1/v_L^3 + 2/v_T^3, \tag{56}$$

for isotropic longitudinal (L) and transverse (T) modes.

Equation (53) is the scattering rate from a single isotopic defect in the crystal. We now make the simple assumptions that a fraction $f \ll 1$ of the atoms in the crystal are isotopes with mass M_1, and that they incoherently scatter the phonon in question, a consequence of the random placement of isotopic defects, as discussed below. With these assumptions, the total scattering rate is fN times that in Eq. (53). Using Eq. (55), we have

$$\tau_i^{-1} = (1/4\pi) f (\Delta M/M)^2 (a/v)^3 \omega_i^4, \tag{57}$$

a useful result originally derived by Klemens.[7] It contains the well-known ω^4 dependence associated with the Rayleigh scattering of photons. From the above derivation, we see that a factor of ω^2 comes from the matrix element containing the kinetic-energy perturbation and another factor of ω^2 comes from the density of phonon states. The density of states in one and two dimensions has ω^0 and ω^1 dependencies, so the Rayleigh scattering rate in one and two dimensions has ω^2 and ω^3 dependencies, respectively.

To estimate the magnitude of this scattering rate, we normalize the phonon frequency to the average cutoff frequency of the lattice $\omega_0 = 2v/a$. [This relation comes from $\omega \cong \omega_0 \sin(ka/2)$, which implies that $\omega_0 ka/2 \cong vk$, in terms of an average sound velocity.] Equation (57) becomes

$$\tau_i^{-1} \cong (2/\pi) f (\Delta M/M)^2 \omega_0 (\omega_i/\omega_0)^4. \tag{58}$$

This says that, for a phonon with frequency near ω_0, the scattering rate is approximately ω_0 reduced by the factor $f(\Delta M/M)^2$, which reflects the isotopic disorder of the crystal. To get an idea of the magnitude of the isotopic scattering times, we substitute some representative numbers into Eq. (58): for $v_0 = \omega_0/2\pi = 3.2$ THz and $f(\Delta M/M)^2 = 10^{-1}(10^{-4})$, the scattering time is

$$\tau_i \cong [1\,\mu\text{s}](1\,\text{THz}/v_i)^4, \qquad (59)$$

with $v_i = \omega_i/2\pi$. The values of the coefficient in square brackets for selected crystals, calculated with less restrictive assumptions than above (see below), are listed as τ in Table 1 on page 172.

Calculations of isotope scattering rates, including elastic anisotropy, phonon dispersion, and randomly occurring isotopes, have been performed by Tamura[8,9] for crystals with diamond and zinc-blend structures. To account for randomly occurring isotopes in the scattering problem, one must perform a statistical sum of

$$\Delta M_\ell e^{i(\mathbf{k}-\mathbf{k}')\cdot\mathbf{r}_\ell^0} \qquad (60)$$

over all lattice sites [see Eqs. (43) and (47)]. Here, one defines the mass deviation of atom ℓ from the average mass as

$$\Delta M_\ell = M_\ell - \bar{M}. \qquad (61)$$

The net result of the isotopic disorder is to replace the quantity $f(\Delta M/M)^2$ in Eq. (57) by a sum over isotope types,

$$g = \sum_i f_i(\bar{m} - m_i)^2/\bar{m}^2, \qquad (62)$$

where m_i is the mass of isotope type i, f_i is the fraction of atoms of this type, and $\bar{m} \equiv \sum_i f_i m_i = \bar{M} \equiv 1/N \sum_\ell M_\ell$ is the average isotopic mass. Thus, Ge, with five different isotopes, has the disorder factor[8]

$$g = \sum_{i=1}^{5} f_i(1 - m_i/\bar{m})^2 = 5.87 \times 10^{-4}. \qquad (63)$$

In the limit of two isotopes with $f_1 \ll f_2$, it is obvious from Eq. (62) that $g \cong f_1(\Delta m/m_2)^2$, as previously assumed.

An important result of this calculation [Eq. (53) or (58)] is that the scattering rate depends only on the frequency of the initial phonon, not its wavevector direction or polarization. This result holds for isotope scattering in a Bravais lattice. The generalization of Eq. (53) for a Bravais lattice is[8]

$$\tau^{-1} = (\pi/6)V_0 g\omega^2 \mathbf{D}(\omega), \qquad (64)$$

where we have dropped the index i for the reason stated above. V_0 is the volume per atom, and $\mathbf{D}(\omega)$ is the density of states per unit volume [$=D(\omega)/Na^3$, above]. For compound crystals composed of more than one type of atom, each having a different set of isotopes, the scattering calculation includes a sum over the basis atoms labeled by s and having masses $M(s)$. The scattering rate still

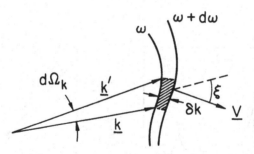

Figure 4 *Diagram for deriving the formula for the density of states. The total number of k states in the region between the two constant-frequency surfaces is $D(\omega)d\omega = D(\omega)(\nabla_k\omega) \cdot \delta k = D(\omega)V\delta k$. The number of states in the shaded region is $(k^2 d\Omega\delta k)/(2\pi/L)^3 \cos\xi$, with $\cos\xi = \mathbf{V} \cdot \mathbf{k}/Vk$. Integrating over all regions between the sheets, we have $D(\omega) = (2\pi/L)^{-3} \int_\omega (k^3 d\Omega)/\mathbf{V} \cdot \mathbf{k}$. In the long-wavelength limit $\mathbf{V} \cdot \mathbf{k} = \omega = vk$, which yields $D(\omega) = L^3/(2\pi)^3\omega^2 \int_\omega d\Omega/v^3 = L^3 \mathbf{D}(\omega)$.*

remains independent of the initial wavevector direction and polarization and is given by Eq. (64), provided that one replaces g in the above equation with an effective $g \equiv g_{eff}$ that includes all types of atoms.[9]

It is useful to know the scattering rate for an elastically anisotropic solid in the long-wavelength approximation. From Figure 4, the density of states per unit volume in the continuum limit is written

$$\mathbf{D}(\omega) = (2\pi)^{-3}\omega^2 \sum_\alpha \int_\omega [d\Omega_k/v_\alpha^3(\mathbf{k})], \tag{65}$$

$$\equiv 3\omega^2/2\pi^2 v^3, \tag{66}$$

where $d\Omega_k$ is an element of solid angle in $\mathbf{k}$ space and the integral is performed over the constant-frequency surface, $\omega(\mathbf{k}\alpha) = \omega$. The integral can be determined by numerical integration from the eigenvalues of the Christoffel equation [Eq. (18) on page 43] for the phase velocities $v_\alpha(\mathbf{k})$. As indicated by Eq. (66), the computed integral corresponds to an average velocity defined by

$$3/v^3 = (1/4\pi) \sum_\alpha \int_\omega [d\Omega_k/v_\alpha^3(\mathbf{k})], \tag{67}$$

which allows one to write the isotope-scattering formula for an anisotropic crystal in the long-wavelength limit in a simplified form,

$$\tau^{-1} = (1/4\pi)(V_0 g/v^3)\omega^4. \tag{68a}$$

or

$$\tau^{-1} = A_0 \nu^4, \tag{68b}$$

with $\omega = 2\pi\nu$. The scattering rate constants, $A_0 = 4\pi^3 V_0 g/v^3$, have been calculated for several crystals by Tamura and are listed in Table 1 of Chapter 7.

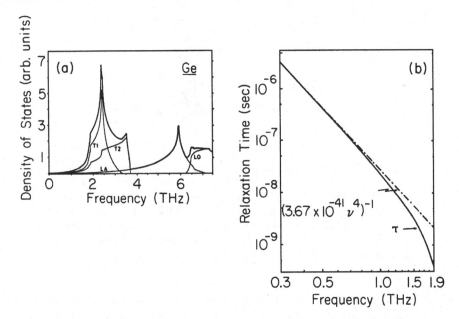

Figure 5 (a) Numerically computed density of states for Ge, showing the contributions from the various acoustic and optic modes. A rigid-ion lattice-dynamics model incorporating interatomic forces up to eighth nearest neighbors is fit to inelastic neutron scattering curves. (b) Frequency dependence of the isotope scattering time including the full lattice dynamics. The dotted-dashed line is the long-wavelength result given by Eq. (64). (From Tamura.[8])

For short-wavelength phonons the density of states [Eq. (65)] can be computed in a straightforward, though computer-intensive, manner: A uniform grid of wavevectors is chosen to cover a symmetry-reduced section of the Brillouin zone. With a given lattice-dynamics model, $\omega(k\alpha)$ are computed for each point on the grid and a histogram is built up of these frequencies, which is $D(\omega)$. The scattering rate for a given phonon is then determined from Eq. (64).

The results of Tamura's calculation[8] for Ge are shown in Figure 5. Figure 5(a) is the density of states for a rigid-ion model of the lattice dynamics. The computed isotope scattering time is plotted in Figure 5(b), showing the contributions from each of the modes. The continuum limit result is shown as the dotted-dashed line in Figure 5(b). This long-wavelength formula works quite well up to approximately 1 THz, but the scattering rate with dispersion included (solid line) becomes significantly larger above 1 THz. This is a consequence of the flattening of the dispersion curves near the zone boundary, which causes a large increase in the density of final states for high-frequency phonons. Calculations of isotope scattering rates using lattice dynamics models have been performed for GaAs,[9] InSb,[9] and Si.[10] Apart from the magnitude of the coefficients A_0, the same qualitative effects as seen in Figure 5(b) are found, regardless of the lattice-dynamics model chosen.

E. Anisotropy in mass-defect scattering

The isotope scattering rate given by Eq. (64) has no angular dependencies. That is, the scattering rate from mass defects in cubic crystals is independent of the wavevector and polarization directions of the *initial* phonon. However, a closer look shows that the distribution of *scattered* phonons from an initial $\mathbf{k}_i \alpha_i$ is far from isotropic. To see this, we refer to Eq. (51) (rewritten to reflect a random distribution of isotopes with disorder factor g),

$$\tau_i^{-1} = (\pi/2N)g\omega^2 \sum_f (\mathbf{e}_i \cdot \mathbf{e}_f)^2 \delta(\omega_f - \omega_i). \qquad (69)$$

The sum over final states involves all phonons with frequency equal to that of the initial phonon. We change the $\mathbf{k}$-space sum to an integral using Eqs. (52b) and (65),

$$\sum_f \delta(\omega_f - \omega_i) \Rightarrow (L/2\pi)^3 \omega^2 \sum_\alpha \int_\omega [d\Omega_k/v_\alpha^3(\mathbf{k})]. \qquad (70)$$

Incorporating this into Eq. (69), we see that the scattering rate of phonon $i = \mathbf{k}_i \alpha_i$ into all phonons $f = \mathbf{k}_f \alpha_f$ within the small solid angle $\Delta\Omega_f$ is[11]

$$\tau_{i \Rightarrow f}^{-1} = (1/16\pi^2) V_0 g \omega^4 (\mathbf{e}_i \cdot \mathbf{e}_f)^2 (\Delta\Omega_f/v_f^3), \qquad (71)$$

with $V_0 = L^3/N$. This process is quite anisotropic due to the dot product of initial and final phonon polarizations. If there is no mode conversion in the scattering event ($\alpha_i = \alpha_f$), the phonon prefers to scatter forward or backward. With mode conversion, the phonon prefers to scatter at right angles. The density-of-states factor ($1/v_f^3$) implies that most of the scattering occurs into the slow transverse (ST) mode. Typically, a phonon will have only a 10% chance of scattering into the L mode.

Is there any way to experimentally observe the microscopic selection rule governed by the polarization dot product? We shall see in Chapter 9 that the spatial distribution of scattered phonons can be measured by phonon-imaging techniques. The scattered phonons are experimentally isolated from the ballistic phonons by selecting their generally longer times of flight, and also by using special sample geometries that block the direct line of sight (ballistic path) between the phonon source and generator.

Detailed predictions of the scattered phonon flux originating from a point source have been made by Ramsbey et al.[11] using Monte Carlo techniques.

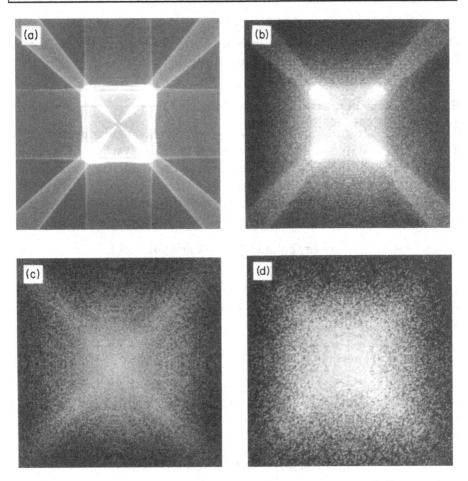

Figure 6 Monte Carlo calculations of phonons in GaAs: (a) ballistic, (b) once scattered, (c) twice scattered, (d) thrice scattered.

Their calculations are for a 1.82-mm-thick GaAs crystal, assuming an initial Planck distribution of phonons ($T = 22.5$ K) and a detector with threshold frequency of 0.7 THz. Phonons are injected into the crystal with an isotropic k-space distribution and a relative mode population of 0.565:0.351:0.084, given by the density of states for the ST, FT, and L modes, respectively. The time for the first scattering event is randomly generated with the scattering rate given by Eq. (68). If the initial phonon reaches the detector surface before scattering, it is collected as a ballistic phonon. For scattering events inside the crystal, Eq. (71) is used to calculate the mode and propagation direction of the scattered phonon. Phonons crossing the excitation surface are disregarded, and those crossing the detection surface are collected according to their history, as described below.

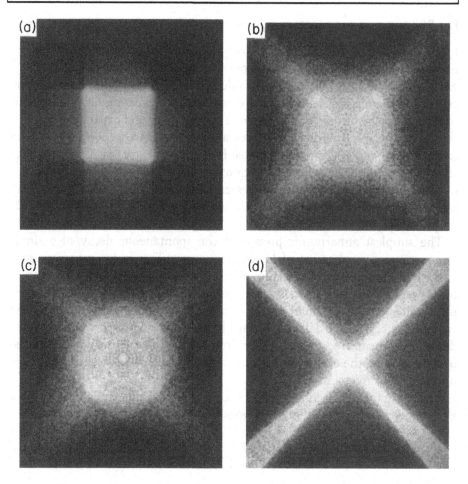

Figure 7 Transmitted flux of once-scattered phonons in GaAs classified according to the following mode-conversion processes: (a) ST ⇒ ST, (b) ST ⇒ FT, (c) FT ⇒ ST, and (d) FT ⇒ FT.

Figure 6(a) shows the calculated distribution of the ballistic phonons. Images (b), (c), and (d) are the spatial distribution of phonons that have scattered once, twice, and thrice, respectively. It is clear that the scattered phonons retain some of the ballistic anisotropy through several scattering events. This series of images depicts an evolution from ballistic to diffusive regimes of heat propagation.

Further information is gained by selecting phonons that have undergone specific mode-conversion processes. Figure 7 shows the spatial distribution of once-scattered phonons for the processes (a) ST ⇒ ST, (b) ST ⇒ FT, (c) FT ⇒ ST, and (d) FT ⇒ FT. Part of the anisotropy is due to the phonon focusing of the initial and final phonons, and part is due to the $(\mathbf{e}_i \cdot \mathbf{e}_f)$ selection rule. The relative contributions will be further discussed in Chapter 9.

F. Phonon-phonon scattering

In the harmonic approximation there are no intrinsic interactions between phonons because the normal modes of the crystal vibrate independently. Anharmonic corrections are required to describe the interatomic forces of real crystals with greater accuracy and are essential in predicting thermodynamic properties of the lattice such as thermal conductivity and expansion.

Classically, anharmonicity provides a coupling between normal modes of frequency ω_1 and ω_2, producing mixed frequencies $|\omega_1 \pm \omega_2|$. Quantum mechanically, the anharmonic perturbation induces transitions between phonons of different frequencies. Such processes allow the phonon bath to reach an internal thermal equilibrium, as described by the Planck distribution [Eq. (1) on page 21].

The simplest anharmonic process is the spontaneous decay of a single phonon into two phonons of lower frequency. This is an important process when a highly nonequilibrium distribution of phonons is injected into an otherwise cold crystal. Initially, the average phonon frequency may be quite high, and, as we have seen in the last section, high-frequency phonons scatter easily from crystalline defects. As the phonons downconvert by spontaneous decay, their mean free path increases dramatically; so the form of heat transport away from the excitation region (ballistic or diffusive) depends critically on the downconversion rate.

The spontaneous decay process is shown schematically in Figure 8. In the isotropic approximation, there are two possible decay routes for L phonons:

$$L \Rightarrow L + T,$$
$$L \Rightarrow T + T. \tag{72}$$

Both of these processes require conservation of energy and momentum,

$$\hbar\omega_1 = \hbar\omega_2 + \hbar\omega_3,$$
$$k_1 = k_2 + k_3. \tag{73}$$

In the isotropic case, the transverse modes are degenerate and T phonons cannot undergo anharmonic decay while conserving energy and momentum unless k_2 and k_3 are both collinear with k_1. This implies a negligible final-state density, which makes this process highly unlikely. A small amount of dispersion makes even this decay route impossible.

In the long-wavelength approximation, the harmonic energy of the strained lattice has the quadratic form, $(1/2)(\text{stress}) \cdot (\text{strain}) = (1/2)C_{ijlm}\varepsilon_{ij}\varepsilon_{lm}$, in terms of the elastic constants and the local strain tensor ε_{ij} (see Ref. 3 on page 58). Anharmonicity creates third-order elastic constants that provide a perturbation to the energy of the form $C_{ijlmno}\varepsilon_{ij}\varepsilon_{lm}\varepsilon_{no}$. (The C_{ijlmno} tensor contains six independent constants for a cubic crystal.) The strain at the

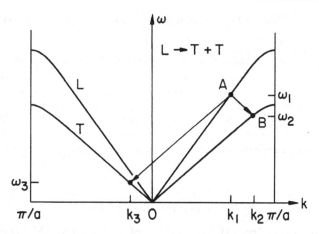

Figure 8 Schematic picture of the spontaneous decay of a longitudinal phonon (at point A) into two transverse phonons, conserving momentum and energy. Elastic isotropy is assumed, so the ST and FT modes are degenerate. The lowest-mode phonon (ST, in general) at point B can only decay by splitting into two ST phonons collinear with the first – a process highly improbable because of the small joint density of final states and impossible if the phonon at B is in the nonlinear region of the dispersion curve with elastic anisotropy.

position $\mathbf{r}$ is given by the spatial derivative of displacement in Eq. (36),

$$\varepsilon_{xy} \simeq \partial u_x/\partial y = iN^{-1/2} \sum_{\mathbf{k}} U_{\mathbf{k}} e_x(\mathbf{k}) k_y e^{i\mathbf{k}\cdot\mathbf{r}}, \tag{74}$$

where, for simplicity, the polarization indices that accompany each $\mathbf{k}$ have been suppressed. The $U_{\mathbf{k}}$ are proportional to $(a_{\mathbf{k}}^+ + a_{-\mathbf{k}})/\omega_{\mathbf{k}}^{1/2}$, by the inverse of Eqs. (30). Cubing ε_{ij} gives terms of the form $a_{\mathbf{k}_1} a_{\mathbf{k}_2}^+ a_{\mathbf{k}_3}^+/(\omega_1\omega_2\omega_3)^{1/2}$, which annihilate one phonon and create two others. Therefore, the transition rate takes the form

$$\tau_a^{-1} \cong \sum_{\mathbf{k}_2\mathbf{k}_3} (|\mathcal{M}|^2/\omega_1\omega_2\omega_3)\delta_{\mathbf{k}_1,\mathbf{k}_2+\mathbf{k}_3}\delta(\omega_1 - \omega_2 - \omega_3). \tag{75}$$

$|\mathcal{M}|^2$ is a messy sum of terms, each with six powers of k_i's and e_j's, preceded by combinations of third-order elastic constants.[12] A two-phonon density of states can be defined by

$$D_2(\mathbf{k}_1) = \sum_{\mathbf{k}_2\mathbf{k}_3} \delta_{\mathbf{k}_1,\mathbf{k}_2+\mathbf{k}_3}\delta(\omega_1 - \omega_2 - \omega_3), \tag{76}$$

summing over the combinations of $\mathbf{k}_2$ and $\mathbf{k}_3$ that satisfy momentum and energy conservation with $\mathbf{k}_1$. The frequency dependence of the decay rate can be seen by making the approximation $\omega_2 \cong \omega_3 \cong \omega_1/2$. (In the isotropic case depicted in Figure 8, ω_2 and ω_3 are fixed proportions of ω_1.) With this

Table 1. *Calculated anharmonic decay rates at $v = 1$ THz. Adapted from Tamura.*[12]

	Si	Ge	GaAs	LiF	NaF
τ_a^{-1} (10^5/s)	0.73	6.43	10.5 ± 3.0	5.14	28.0
$L \to L + T$	20%	26%	24%	33%	29%
$L \to T + T$	80%	74%	76%	67%	71%

simplification,[12]

$$\tau_a^{-1} = \text{constant} \cdot (\omega^3/v^2)\mathbf{D}_2(\mathbf{k}). \tag{77}$$

The ω^3 factor comes from the six powers of $k_i \cong \omega/v$ in $|\mathcal{M}|^2$ divided by the three powers of ω shown in Eq. (75). Okubo and Tamura[13] have computed the two-phonon density of states for Ge, using a lattice-dynamics model. In the low-frequency limit, $\mathbf{D}(\mathbf{k})$ scales as ω^2, leading to the dependence[14-17]

$$\tau_a^{-1} = B_0\omega^5. \tag{78}$$

At frequencies approaching the zone-boundary frequencies, the rate increases faster than ω^5 due to the higher density of states. Tamura[12] has calculated the spontaneous decay rates of L phonons for several crystals, assuming isotropic velocities and the long-wavelength approximation. Table 1 shows the calculated decay rates (τ_a^{-1}) for 1-THz phonons and indicates the relative contributions from the two processes in Eq. (72). These calculations indicate that $L \Rightarrow T + T$, depicted in Figure 8, is the dominant decay channel. Experimental verification of the ω^5 dependence has been published by Baumgartner et al.[18] and by Happek et al.[19]

The ω^5 dependence shown in Eq. (78) holds in the long-wavelength approximation, but the prefactor B_0 is a function of the direction and polarization of the initial phonon, unlike for the elastic-scattering case. The anisotropy of the spontaneous decay rate has been investigated in some detail by Tamura and Maris[20] and by Berke et al.[21] One of the interesting consequences of elastic anisotropy is that even the FT and ST modes can spontaneously decay. For some wavevector directions, the FT decay rate is comparable to that of the L mode!

To see how elastic anisotropy can destroy the stability of even ST phonons against anharmonic decay, we can examine Figure 9. Here the decay of a phonon with wavevector $\mathbf{k}_1$ into two phonons with wavevectors $\mathbf{k}_2$ and $\mathbf{k}_3$ is described by a Herring construction, as discussed by Maris and Tamura.[22] The constant-ω (slowness) surfaces of the daughter phonons are labeled S_1 and S_2 in the figure. For perfect isotropy and linear dispersion, the conservation of energy and momentum [Eqs. (73)] constrain the two spherical surfaces to touch tangentially, and only collinear decay is possible [Figure 9(a)]. If dispersion is accounted for, even collinear decay is impossible, as indicated in Figure 9(b).

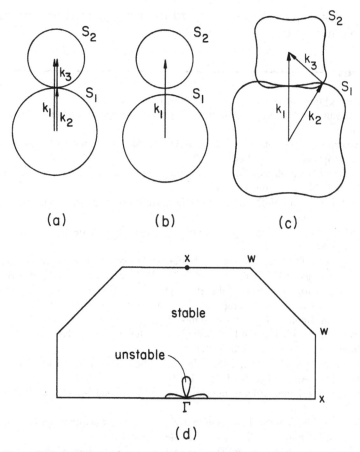

Figure 9 *Spontaneous decay of phonons in the lowest mode, ST $\Rightarrow$ ST + ST. (a) For an isotropic solid only collinear decay $k_1 \Rightarrow k_2 + k_3$ is possible, and only in the continuum limit. (b) For an isotropic solid with dispersion, a phonon in the transverse mode cannot decay. (c) For an anisotropic solid (even with mild dispersion as shown), noncollinear decays can conserve momentum and energy, due to the nonspherical constant-ω surfaces. Note that energy conservation requires the surfaces to touch along the direction of k_1. (d) Calculated region in the Brillouin zone of Si where ST phonons are unstable against anharmonic decay. (From Maris and Tamura.[22])*

For nonspherical constant-ω surfaces, however, other decay routes become accessible even in the dispersive regime, as shown in Figure 9(c). Nevertheless, as previously argued, dispersion tends to destroy anharmonic decay within a given mode. Consequently, there is a rather limited region in the Brillouin zone – whose shape depends on the particular type of crystal – in which the ST phonons are unstable. The calculated regions of instability for ST phonons with k_1 in the (001) plane in Si are shown in Figure 9(d).[22]

From the results of Figures 7 and 9, we conclude that even qualitative descriptions of elastic and inelastic scattering must include the effects of elastic anisotropy. We now turn to the experimental measurements of acoustic phonon scattering.

References

1. C. Kittel, *Quantum Theory of Solids* (Wiley, New York, 1963).
2. B. Donovan and J. F. Angress, *Lattice Vibrations* (Chapman and Hall, London, 1971), Chapter 6.
3. P. Brüesch, *Phonons: Theory and Experiments I*, Vol. 34 of Springer Series in Solid State Sciences (Springer-Verlag, Berlin, 1982).
4. W. Jones and N. H. March, *Theoretical Solid State Physics, Volume I*, Interscience Monographs 27 (Wiley, London, 1973).
5. N. W. Ashcroft and N. D. Merman, *Solid State Physics* (Holt, Rinehart and Winston, Philadelphia, 1976).
6. H. J. Maris, The vibrational spectrum of an isotopically disordered crystal, *Philos. Mag.* **13**, 465 (1966).
7. P. G. Klemens, The Scattering of low-frequency lattice waves by static imperfections, in *Solid State Physics, Vol. 7*, F. Seitz and D. Turnbull, eds. (Academic, New York, 1957).
8. S. Tamura, Isotope scattering of dispersive phonons in Ge, *Phys. Rev. B* **27**, 858 (1983).
9. S. Tamura, Isotope scattering of large-wavevector phonons in GaAs and InSb: deformation-dipole and overlap-shell models, *Phys. Rev. B* **30**, 849 (1989).
10. S. Tamura, J. A. Shields, and J. P. Wolfe, Lattice dynamics and elastic phonon scattering in silicon, *Phys. Rev. B* **44**, 3001 (1991).
11. M. T. Ramsbey, J. P. Wolfe, and S. Tamura, Phonon focusing of elastically scattered phonons in GaAs, *Z. Phys. B* **73**, 167 (1988).
12. S. Tamura, Spontaneous decay rates of LA phonons in quasi-isotropic solids, *Phys. Rev. B* **31**, 2574 (1985).
13. K. Okubo and S. Tamura, Two-phonon density of states and anharmonic decay of large-wavevector LA phonons, *Phys. Rev. B* **28**, 4847 (1983).
14. R. Orbach and L. A. Vredevoe, The attenuation of high frequency phonons at low temperatures, *Physics (USA)* **1**, 91 (1964).
15. P. G. Klemens, Decay of high-frequency longitudinal phonons, *J. Appl. Phys.* **38**, 4573 (1967).
16. H. J. Maris, Inelastic scattering of neutrons by an anharmonic crystal at low temperatures, *Phys. Lett.* **17**, 228 (1965).
17. P. F. Tua and G. D. Mahan, Lifetime of high-frequency longitudinal-acoustic phonons in CaF_2 at low crystal temperatures, *Phys. Rev. B* **26**, 2208 (1982).
18. R. Baumgartner, M. Engelhardt, and K. F. Rank, Spontaneous decay of high frequency acoustic phonons in CaF_2, *Phys. Rev. Lett.* **47**, 1403 (1981).
19. U. Happek, W. W. Fischer, and K. F. Renk, Decay of terahertz phonons in $Ca_{1-x}Sr_xF_2$ mixed crystals, in *Phonons 89*, S. Hunklinger, W. Ludwig, and G. Weiss, eds. (World Scientific, Singapore, 1990), p. 1248.
20. S. Tamura and H. J. Maris, Spontaneous decay of TA phonons, *Phys. Rev. B* **31**, 2595 (1985).
21. A. Berke, A. P. Mayer, and R. K. Wehner, Spontaneous decay of long-wavelength acoustic phonons, *J. Phys. C: Solid State Phys.* **21**, 2503 (1988); M. T. Labrot, A. P. Mayer, and R. K. Wehner, Decay cascades of acoustic phonons in calcium fluoride, *J. Phys. Cond. Matter* **1**, 8809 (1989).
22. H. J. Maris and S. Tamura, Anharmonic decay and the propagation of phonons in an isotopically pure crystal at low temperatures: application to dark-matter detection, *Phys. Rev. B* **47**, 727 (1993).

9

Bulk scattering
of phonons – Experiments

A. Phonon scattering and thermal conduction

We have seen that the mean free path of high-frequency phonons is greatly limited by scattering in the bulk of the crystal. The experiments discussed in Chapter 7 were concerned with the relatively small fraction of large-k phonons that traverse the crystal ballistically, as identified by their sharp focusing pattern. In those experiments, the bulk scattering of high-frequency phonons by mass defects acted as a low-pass filter, which, in combination with a high-pass detector, provided an effective means of frequency selection for the ballistic phonons. In Chapter 8, we developed a theoretical basis for phonon scattering. Now we will describe experiments that measure the phonon scattering processes in detail.

There are several different phonon-scattering processes to be considered. As discussed in Chapter 8, the simplest form of scattering is from atomic mass defects that differ from their neighbors only in isotopic mass. Because most atoms occur naturally with several isotopic masses, this type of defect is ubiquitous. Only a few crystals, such as NaF, occur naturally in an isotopically pure form. Because the natural isotopic abundance of atoms is generally well known, however, phonon scattering from randomly occurring isotopes (i.e., isotope scattering) is very predictable and serves as an important test case for phonon-scattering theories.

Another common form of defect scattering occurs when an impurity atom substitutes for a host atom in the crystal. In addition to being a mass defect, the impurity generally has a different bonding strength to its neighbors than the host atom it replaced. Thus, phonon scattering is induced by the local changes in both mass and elastic constant. Also, an enhanced scattering strength can occur if the impurity has low-lying electronic transitions (e.g., spin splittings) that resonate with some of the phonon frequencies. Other point defects include interstitial atoms and vacancies. Extended defects, in the form of dislocation lines, can be quite effective in scattering phonons. Essentially, anything that disrupts the periodicity of the crystal will scatter phonons.

Phonons also scatter from other phonons. This rather remarkable phenomenon arises because the phonon itself, by virtue of its displacement field, disturbs the translational symmetry of the crystal. The strength of the coupling between phonons depends upon the small anharmonic terms in the elastic energy (see Chapter 8, Section F). At large vibration amplitudes, which occur when local density of phonons is high, the lattice is no longer perfectly harmonic, allowing for transitions between modes of different frequencies. In the basic phonon–phonon scattering event, a single phonon of frequency ω_1 splits into two phonons such that $\omega_1 = \omega_2 + \omega_3$. This and the reverse process (fusion of two phonons into one) are important phonon-scattering mechanisms in the heat conduction of crystals at room temperature.[1]

Phonon scattering divides naturally into two categories: elastic and inelastic. Isotope and nonresonant impurity scattering are examples of elastic scattering, in which the incident phonon and scattered phonon have the same frequency because there is no mechanism to absorb the elastic energy. Phonon–phonon scattering, which involves three or more phonons, is defined as an inelastic process because the frequencies change. Resonant scattering from electronic impurities may be elastic or inelastic, depending on whether the excited electron re-emits a phonon of equal frequency or relaxes to intermediate states by emitting phonons of different frequency. Likewise, scattering from extended defects such as dislocations may be elastic or inelastic processes, depending on whether (a) the phonon is deflected by the local static strains or (b) it induces a dynamic motion of the dislocation line, which absorbs the incident phonon and re-emits the vibrational energy as other phonons. In general, if the object scattering the phonons has internal degrees of freedom, inelastic scattering can occur.

The most potent method for characterizing phonon scattering in crystalline and amorphous solids is the measurement of thermal conductivity.[2] The temperature dependence of the thermal conductivity is a type of phonon spectroscopy because the frequency at the peak of the equilibrium Planck distribution shifts linearly with sample temperature. Thus, measuring the thermal conductivity over a wide temperature range yields a frequency dependence of the phonon-scattering processes. This spectroscopic approach is valid when the mean frequency of the phonons is much smaller than the Debye frequency (or zone-boundary frequency) of the solid, i.e., for temperatures below approximately 100 K. Thermal-conductivity experiments have provided valuable information about phonon scattering from defects in solids.

Figure 1 illustrates these ideas in a classic thermal-conductivity experiment on Ge.[3] At temperatures below approximately 20 K, the phonon mean free

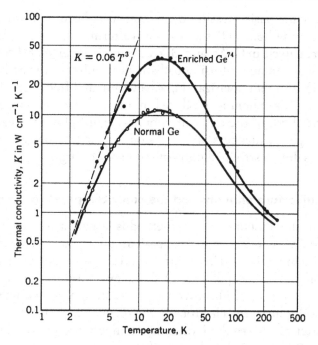

Figure 1 Isotope effect on the thermal conduction in Ge. (Data from Geballe and Hull.[9] This figure is adapted from one in Ref. 4.)

path, ℓ, is a constant roughly equal to the crystal dimension, and the lattice heat capacity, C, varies as T^3. Therefore, assuming an average phonon velocity v, the thermal conductivity, $\kappa = (1/3)Cv\ell$, approaches a T^3 dependence at low temperatures. This boundary-scattering regime is amenable to ballistic heat-pulse experiments. Above approximately 30 K, the thermal conductivity drops rapidly because phonon–phonon scattering (aided by the Umklapp process[4]) greatly reduces the mean free path.

Between the ballistic and phonon–phonon scattering regimes, the average phonon frequency is sufficiently high that defect scattering becomes important. For an undoped Ge crystal, mixed Ge isotopes provide the dominant mass defects. Figure 1 shows that for an isotopically enriched crystal the peak in the thermal-conductivity curve is enhanced by a factor of approximately three, indicating a reduced scattering from mixed isotopes.[5] Systematic isotope-scattering studies have been conducted on LiF, and recent experiments on isotopically enriched diamond suggest enhanced thermal conductivity even at 300 K.[6] Other experiments, notably those conducted with isotopically pure NaF, have shown similar changes in the thermal-conductivity curve with varying concentrations of impurities.[7]

An additional method for characterizing phonon-scattering processes in crystalline solids is provided by phonon-imaging techniques. In this chapter we examine some basic measurements that have been performed with this relatively new method: (1) the isotope scattering rate in Si, (2) scattering from residual defects in GaAs, and (3) scattering from extended defects in LiF. The essential advantages of using ballistic heat pulses is the ability to select phonons with different polarization, wavevector, and frequency. In addition, the nonequilibrium phonons produced in a heat pulse can have frequencies far exceeding those of the ambient phonons. This allows the experimenter to observe high-frequency scattering from crystalline defects without the complications due to strong phonon–phonon scattering.

B. Nonequilibrium phonons and the color temperature

The utility of the phonon-imaging method is largely attributable to the fact that a phonon has a longer mean free path at low crystal temperature than at high crystal temperature. This point deserves a short explanation. Imagine that the phonons in a heat pulse are produced by optically exciting a small spot of diameter d on a metal film covering the surface of a nonmetallic crystal. The laser locally heats the electrons in the metal, which, by strong electron–phonon scattering, raises the local lattice temperature in the film. We now suppose that this Planck distribution of nonequilibrium phonons is emitted into the cold crystal. The nonmetallic crystal does not have the high density of free electrons to quickly change this frequency distribution, so the emitted phonons form a heat pulse that may be envisioned as an expanding shell of thermal energy with roughly constant spectral distribution. In other words, we momentarily ignore frequency conversion.

What is the temperature of this heat pulse at a distance $r \gg d$ inside the crystal? The situation is analogous to the "color temperature" that describes the frequency distribution of photons arriving at the Earth's surface from the Sun. The phonons in the propagating heat pulse exhibit a frequency distribution similar to that of the heat source, but the occupation numbers of the phonons in a given mode are far fewer than that required for equilibrium. After all, temperature represents the kinetic energy per unit volume, and the volume occupied by the heat pulse in the crystal is far larger than the initial excited volume of metal. Consequently, the expanding region occupied by the heat pulse is not in equilibrium, yet the spectral distribution of the nonequilibrium phonons exhibits a color temperature equal to the equilibrium temperature of the metal source.*

The low occupation number of phonon modes in the cold crystal implies that the nonequilibrium high-frequency phonons interact with each other (i.e., interconvert) less frequently than if the crystal were raised to the color temperature corresponding to their frequency distribution. Spontaneous decay

* I thank Y. B. Levinson for pointing out this analogy.

of the phonons, however, does limit their lifetime, as discussed in Chapters 8 and 10. Nevertheless, the absence of stimulated processes makes it easier to isolate the scattering of high-frequency phonons from defects in the crystal. This is an advantage that the (nonequilibrium) heat-pulse methods have over (near-equilibrium) thermal-conductivity techniques.

At high excitation levels, however, phonon–phonon interactions do come into play near the excitation point. The full dynamics of phonons undergoing anharmonic decay and elastic scattering will be considered in Chapter 10. In that chapter we will also examine the interesting nonlinearities that occur at high excitation densities, in some cases involving the interaction of phonons and photoexcited electrons.

C. Elastic scattering of phonons in silicon

Phonons that have scattered in the bulk of a crystal have a pronounced effect on the detected heat-pulse signals: they produce an extended tail following the sharp ballistic pulses. This delayed signal due to the late-arriving scattered phonons is apparent, with varying degrees, in nearly all heat-pulse experiments. The greater the scattering rate, the larger the proportion of the phonons in the tail of the heat pulse. At very large scattering rates, the heat pulse becomes a broad pulse, with the arrival time of the peak much later than the ballistic arrival time. The InSb experiments described in Chapter 7 provide a good example of how the phonon scattering increases with increasing detection frequency. At the highest selected frequencies [Figure 7(a) on page 175], the broad heat pulse is due mainly to scattered phonons, although the pulse contains a small fraction of dispersive phonons that arrive ballistically with a distribution of velocities. The ballistic component can best be isolated by imaging the sharp spatial variations [Figure 8 on page 176] that lie on a broad diffusive background.

A determination of the scattering rate from the time traces is not easy, even if a sharp ballistic peak is distinct from the diffusive tail. First, there is the possible presence of late-arriving dispersive phonons, which contribute to the tail. Second, the ratio of the intensity of the ballistic peak to that of the tail depends strongly on the propagation direction, due to phonon focusing of the ballistic phonons. Finally, in most heat-pulse experiments, a broad distribution of phonon frequencies is detected, which implies that an even broader distribution of scattering rates are expected. (Recall that $\tau^{-1} \propto \nu^4$ for elastic scattering.)

To circumvent these difficulties, Shields et al.[8,9] have employed a sample geometry that provides *spatial* separation of the ballistic and scattered phonons.[10] A high-purity crystal of Si is slotted, as shown in Figure 2. A PbBi tunnel-junction detector is chosen for frequency selectivity ($2\Delta / h = 680$ GHz) and placed directly opposite the slot boundary, as shown in the figure. A

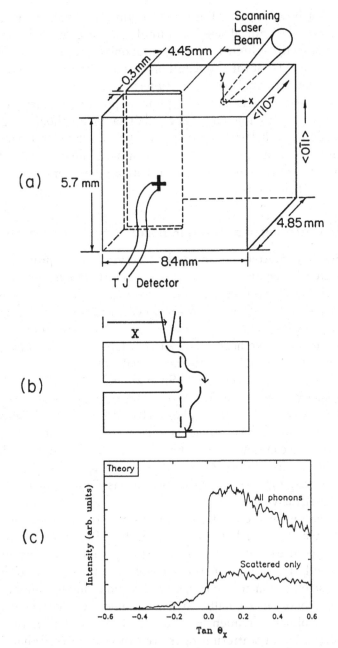

Figure 2 (a) Slotted sample geometry used for differentiating between scattered and ballistic phonons in a heat pulse. (b) Schematic of heat-pulse experiment showing that when the laser beam is positioned behind the slot, only scattered phonons can reach the detector. (c) Monte Carlo calculation of the detected phonon flux versus laser position for a Si crystal, assuming the theoretical isotope-scattering constant, A_0, and a source temperature $T = 10$ K. In the calculation it is possible to separately bin the scattered and ballistic phonons, as shown. (From Shields et al.[8])

pulsed laser beam is scanned across the opposing face, upon which a 2000 Å Cu film has been deposited. When the direct path between the excited spot and the detector is obstructed by the slot, only scattered phonons are detected. Otherwise, both scattered and ballistic components are present, as indicated in Figure 2(b).

A Monte Carlo calculation of the detected flux along a single scan line across the center of the crystal is plotted in Figure 2(c). As in the calculations described in Chapter 8 (page 206), the present calculation includes phonon focusing as well as the anisotropy in scattering associated with the phonon polarizations, [Eq. (71) on page 205]. The predicted isotope-scattering rate in Si is assumed. A phonon that intercepts the slot is discarded. The phonons collected at the detection surface are binned according to whether they scatter in the bulk or traverse the sample ballistically. Only phonons that arrive in a time between t_b and $1.5t_b$ are counted, where t_b is the shortest ballistic time of flight for transverse phonons across the sample. This time selection eliminates the need to account for sidewall reflections. A heater temperature of 10 K is assumed, although it is found that, with the time and frequency selection described above, the results are only weakly dependent on this parameter.

The calculation corresponds to the experiment shown in Figure 3. Figure 3(a) shows the focusing pattern of a Si sample with no slot, indicating the ballistic pattern for this crystal orientation. The dashes show the range of angles covered by the slotted sample, and Figure 3(b) is the phonon image of the slotted sample. A scan along the FT ridge [arrows in Figure 3(b)] is chosen to provide a simple ballistic structure for comparison to the scattered signal. The scan of intensity along the center of the FT ridge is shown in Figure 3(c). The measured intensity along this line is similar to that predicted in Figure 2(c).

A quantitative measure of the scattering rate in this crystal is obtained by measuring the intensity drop at the slot boundary. One can define a ledge ratio, $R = S/H$, that is related to the ratio of scattered to scattered-plus-ballistic phonons, as shown in Figure 3(c) for the line indicated by arrows in Figure 3(b). Of course, a different ledge ratio is obtained for a horizontal scan line off the FT ridge. Figure 4 shows a time trace on the FT ridge and intensity scans for a horizontal line slightly above the FT ridge in Figure 3(b), where the ballistic flux is not so intense. By sampling different times of flight [Figure 4(a)] one can experimentally select between the longitudinal and transverse phonons. Figure 4(b) shows that the faster L phonons have less chance of scattering in the crystal (i.e., smaller ledge ratio R) than the FT phonons, as expected from the dependence of mean free path on velocity. Of course, the ledge ratio also depends on focusing and channeling of the phonons.

The black dots in Figure 5 are the *predicted* ledge ratios along the FT ridge, assuming several values of A for the elastic scattering rate,

$$\tau^{-1} = Av^4. \qquad (1)$$

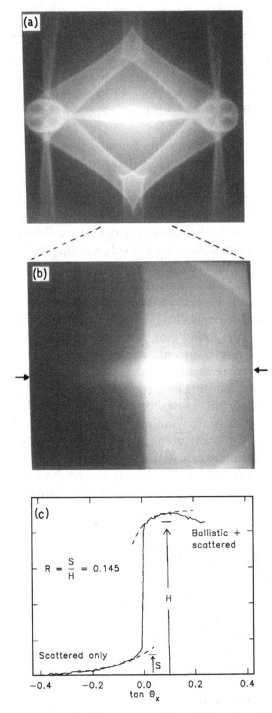

Figure 3 (a) Experimental phonon image for a (110) wafer of Si. (Tunnel-junction detector with onset frequency of 820 GHz.) The dashed region shows the range of angles scanned in the slot experiment. (b) Phonon image of the slotted sample drawn in Figure 2(a). (c) Phonon intensity along the line identified by arrows in (b). The ledge ratio is defined as R = S/H. (From Shields et al.[8])

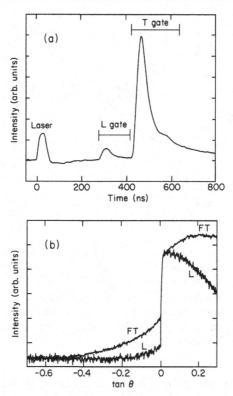

Figure 4 (a) Time trace of ballistic heat pulses in Si. To isolate L and T phonons, the time gates shown are used. (b) Phonon intensities along a line above the FT ridge in Figure 3(b). Less scattering is expected for the L phonons because their higher velocity produces a longer mean free path for a given scattering rate. (From Shields, unpublished.)

Physically, the variation in the constant A corresponds to a hypothetical variation in the concentration or scattering strength of mass defects, as described in Chapter 8. We see that there is a significant range over which the calculated ledge ratios depend approximately linearly on the scattering constant, indicated by the straight line in Figure 5. The horizontal line and the shaded region indicate the measured value of the ledge ratio and its uncertainty, for the 680-GHz detector and the 4.85-mm-thick Si crystal. This value corresponds to a mean scattering constant, $A = 1.06 A_0$, where $A_0 = 2.43 \times 10^{-42}$ sec^3 is the value calculated for isotope scattering in Si.[11] At the time of this writing, a frequency dependence of the ledge ratio in Si has not been performed, although a measurement at $2\Delta/h = 850$ GHz showed an increase in R commensurate with Eq. (1).[8]

The main conclusions of this study are that (1) imaging with a slotted geometry provides a quantitative measure of phonon scattering rates and (2)

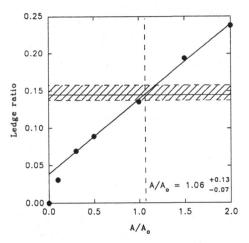

Figure 5 Determination of the elastic scattering constant in Si. The shaded region shows the experimental ledge ratio along the FT ridge [line F in Figure 3(b)] for a 680-GHz detector. The solid dots are the theoretically determined ledge ratios, obtained by analyzing data like Figure 2(b) for several elastic-scattering constants. The experimental value corresponds to a scattering rate very close to that predicted for isotope scattering, as given in the figure. (From Shields et al.[8])

the elastic scattering rate in high-purity Si agrees well with that expected for naturally occurring isotopes. One consequence of this observation is that the elastic scattering rate in Si may be reduced in isotopically enriched crystals, i.e., those in which one isotopic species dominates. It has been suggested that this would improve the characteristics of high-energy particle detectors based on ballistic-phonon detection.[12]

One of the key features of elastic scattering is the anisotropy associated with a single event [Eq. (71) on page 205],

$$\tau_{1\rightarrow2}^{-1} \propto |e_1 \cdot e_2|^2, \tag{2}$$

where e_1 and e_2 represent the polarizations of the incident and scattered phonons, respectively. This anisotropy averages out when one sums over all the possible scattered phonons, giving Eq. (1). However, with phonon imaging we have the ability to examine the regime where only a few scattering events have occurred. The calculations of Figure 6 on page 206 show that, in this regime between ballistic and diffusive propagation, significant remnants of the phonon-focusing anisotropies remain.

An example of how Eq. (2) affects the angular distribution of scattered phonons is illustrated in Figure 6. A portion of the FT phonons emitted from the source at point 0 are focused into the ridges centered on (100) planes and bounded by a pair of caustics. The polarization vectors of these phonons

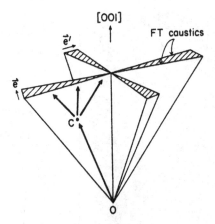

Figure 6 Schematic diagram of the FT phonon-focusing ridges. The phonons propagating in one of these high-intensity ridges has a polarization normal to the ridge, as shown. A phonon scattering at point C tends to stay in the ridge due to the $e_1 e_2$ factor in the scattering rate. (From Shields et al.[9])

are perpendicular to the ridge planes, as shown. The selection rule given by Eq. (2) implies that a phonon in one ridge will have a negligible probability of scattering into an orthogonal ridge. The greatest probability is for the phonon to scatter into the same ridge, where $e_1 \cdot e_2 = 1$. Thus, there is a tendency for the FT phonons to channel in the ridges.

This channeling effect is impressively demonstrated in the experiment shown in Figure 7.[9] The Si crystal is slotted as in Figure 2(a), except that the source, detector, and slot planes are {100}. The detector is placed such that the ST box along [100] is obscured, as indicated by the caustic pattern in Figure 7(b). The right side of the phonon image in Figure 7(a) shows the unobstructed ballistic pattern. The left side of the image contains only scattered phonons. The intensity of the scattered phonons is enhanced by approximately a factor of three in the printing process in order to make them visible. The channeling of the FT phonons is clearly seen as a continuation of the ballistic FT ridges.

The role of the $e_1 \cdot e_2$ factor in this channeling process is shown by the Monte Carlo calculations in Figure 8. Figure 8(a) is a pseudo–three-dimensional plot of the phonon intensities for the slot geometry of Figure 7, including the polarization-dependent scattering. Figure 7(b) is the same calculation except with *isotropic* scattering, i.e., setting $e_1 \cdot e_2 = $ constant throughout. The FT ridge channeling is obvious only when the polarization-dependent scattering is properly accounted for. This effect and the channeling of phonons in the other phonon modes were predicted in the calculations for GaAs shown in

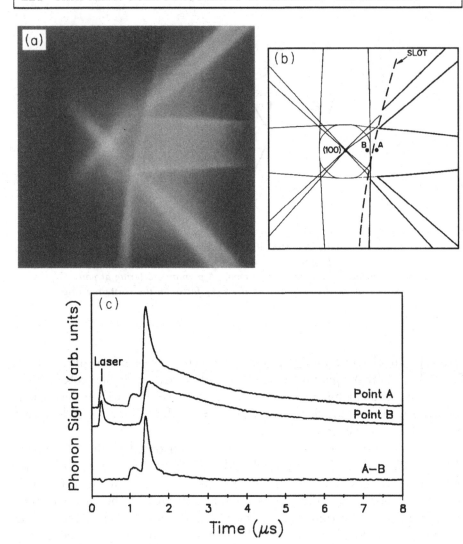

Figure 7 (a) Phonon image showing the channeling effect of scattered phonons; i.e., the tendency of phonons in the FT ridge to remain in the ridge. The cross structure at the left of the slot is an extension of the FT ridges, yet is solely due to scattered phonons. (b) Caustic map of Si, showing the position of the slot in the imaging experiment. (c) Time traces showing the heat-pulse signals at points A (ballistic-plus-scattered phonons) and B (only scattered phonons) in (b). The subtracted trace gives the ballistic-only phonons. (From Shields et al.[8])

Figure 7 on page 207.[13] A probable observation of phonon channeling was reported earlier by Stock et al.[14] They observed a peak in phonon intensity along the (100) axis in an unslotted GaAs crystal, which corresponds to the point at which the channeled ridges cross, but they mistakenly attributed the

(a)

(b)

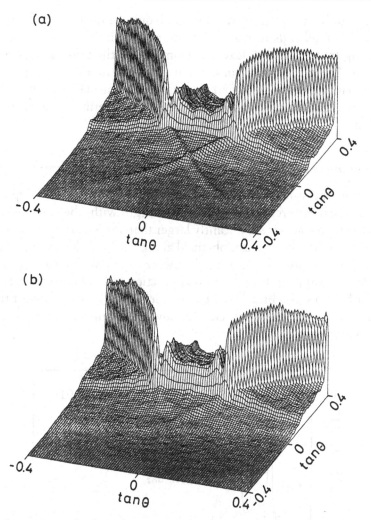

Figure 8 Pseudo–three-dimensional plots of the calculated phonon inten-sity for a slot geometry similar to that in Figure 7. (a) With the polarization dot-product included. (b) Assuming isotropic scattering. The channeling effect is much more pronounced with the polarization effect included, as seen in the experiment of Figure 7. (From Shields et al.[8])

directed flux to ballistic propagation of *dispersive* phonons[14-16] (see Chapter 7, Section A). Ramsbey et al.[13] studied the phonon-channeling effect in detail for GaAs. For a recent review of this controversy, see Ref. 17.

The slot geometry enables the experimenter to completely isolate the bal-listic and scattered components in the time-of-flight data.[8,10] Figure 7(c) is a plot of the heat pulses for the positions A and B marked in Figure 7(b). The time trace at position A includes both ballistic and scattered components, as is commonly measured in a heat-pulse experiment. The time trace at position

B contains only scattered phonons – the ubiquitous diffusive tail. Assuming that the *scattered* signals at A and B are nearly the same [see Figure 2(c)], subtracting trace B from A leaves predominantly the pure ballistic pulse for position A. The pure ballistic pulse contains information about the heat-source lifetime that cannot easily be extracted from the normal heat pulses, because of the long diffusive tail. In this case the source lifetime is found to be less than 50 ns.

D. Phonon scattering from impurities in GaAs

Similar slotted crystal experiments have been conducted for undoped GaAs, with somewhat different conclusions.[13] To begin with, the scattered component in the time traces is significantly larger than for Si at the same frequency. Figure 9 shows the heat pulses obtained along [110], under two detector conditions: (a) operating in the tunnel-junction mode with onset frequency of 690 GHz, and (b) operating the tunnel junction at high current in He vapor so that it behaves as a broadband bolometer, which is able to detect the low-frequency phonons. These data clearly show the strong frequency dependence of phonon scattering.

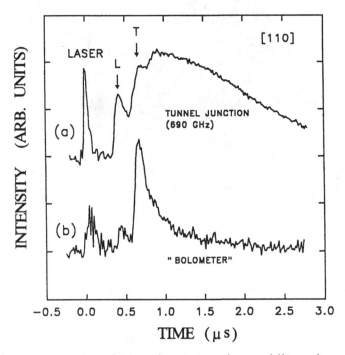

Figure 9 *Time traces of heat pulses in GaAs for two different detector conditions, as described in the text. The high-frequency detection shows much more phonon scattering. (From Ramsbey et al.[13])*

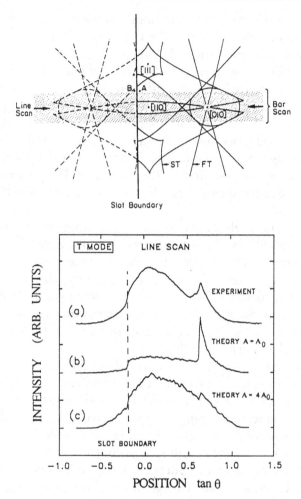

Figure 10 Experiment to determine the scattering rate of phonons in GaAs. The slot is positioned with respect to the caustic pattern, as shown at the top. The experimental line scan is seen to agree best with the Monte Carlo simulation, which assumes $A = 4A_0$, where A_0 is the elastic rate constant predicted for isotope scattering alone. The extra scattering is due to unidentified impurities or defects in this undoped crystal. (From Ramsbey et al.[13])

A quantitative estimate of the scattering rate is obtained from the spatial scans shown in Figure 10, which correspond to a tunnel-junction detector with 690-GHz onset and a time window between t_b and $1.5t_b$ for the transverse phonons. The bottom two traces are Monte Carlo calculations assuming a scattering constant of A_0 and $4A_0$, respectively. Clearly, the data [trace (a)] agree much better with the $4A_0$ calculation, indicating that impurities or structural defects are governing the transverse phonon scattering in this crystal.

Interestingly, a similar analysis indicates a scattering constant $A \leq 1.3 A_0$ for the longitudinal mode; that is, the L mode phonon scattering is limited principally by isotope scattering ($A = A_0$).

The different coupling strengths of L and T phonons to the defects, especially when they are compared for different propagation directions, can be useful in identifying the unknown defects in GaAs. Early experiments by Narayanamurti et al.[18] showed that Cr ions in GaAs had different effects on the L and T modes propagating along [100], [110], and [111] directions, indicating a tetragonal site symmetry. Further studies of phonon scattering from Cr impurities are described by Challis et al.[19] The undoped crystal described in the slot experiment above, however, contained no intended doping of Cr.

Another important defect in GaAs is the ubiquitous deep donor EL2, which is present at the 10^{16} cm^{-3} level and acts to compensate shallow acceptors, making the undoped crystal semi-insulating. One plausible identification of EL2 is an As atom on a Ga site (As_{Ga}), possibly with a neighboring As interstitial.[20] This defect produces a characteristic optical absorption below the GaAs band gap. Surprisingly, at low temperatures the optical absorption vanishes when the crystal is irradiated in the infrared (1.0–1.2 μm), but it reappears upon heating above 130 K.[21]

Culbertson et al.[22] examined the effect of optical radiation on heat pulses in GaAs, in an effort to characterize EL2. The differential effects on L and T modes along symmetry axes led them to conclude that EL2 is a point defect with trigonal symmetry, seemingly consistent with the above identification.

A phonon-imaging experiment by Shields and Wolfe[23] has produced the striking results on undoped (semi-insulating) GaAs shown in Figure 11. Employing the slotted sample geometry as in Figure 3(b) and a tunnel-junction detector, they found a marked change in the phonon-focusing pattern after the sample has been irradiated with 1.06-μm light at low temperature. After irradiation a broader FT ridge appears, superimposed on a narrower ridge observed both before and after irradiation. From Chapters 6 and 7, we know that the ridge width is directly correlated with the frequency of dispersive phonons. The appearance of the broader structure implies that the scattering of high-frequency phonons ($\nu > 600$ GHz) is greatly reduced when the defect (presumably EL2) is converted to its metastable state. A time trace taken at point D in Figure 11(c) before and after irradiation shows an immense change in the FT ballistic transmission (Figure 11(d)). These experiments are providing some interesting information about the changes in the spectra of ballistic phonons as the scattering center or centers are modified.

A recent paper by Xin et al.[24] suggests that the EL2 is only indirectly responsible for the phonon scattering in such experiments, presumably by changing the ionization state of more dilute, but strongly phonon-coupled, ions such as Cr^{3+}. The resolution of this issue remains a challenging problem.

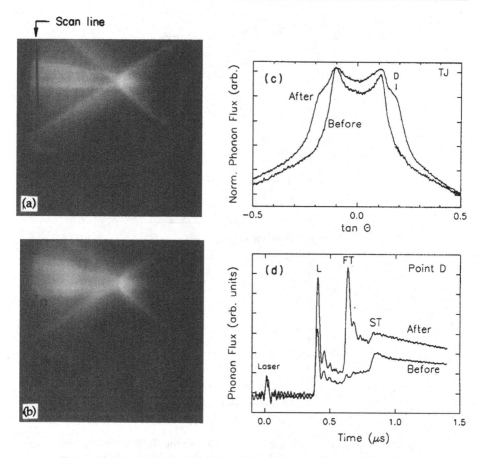

Figure 11 Focusing pattern of GaAs (FT ridge) (a) before and (b) after in situ irradiation of the crystal with 1.06-μm light. (c) Line scans across the FT ridge show the lessening of high-frequency phonon scattering after irradiation. (d) Striking enhancement of the FT flux is seen in the time trace recorded after optical excitation. (From Shields et al.[23])

Phonon-imaging experiments on semi-insulating GaAs wafers containing known amounts of impurities have been conducted by Held et al.[25] For a phonon source they used an electron beam in a low-temperature scanning electron microscope. Their detectors were tiny Al bolometers ($2 \times 2 \ \mu m^2$ to $10 \times 10 \ \mu m^2$). The frequency distribution of the bolometer-detected phonons is quite broad, with an estimated peak in the 450–500-GHz range. Figure 12 shows line scans across the [100] ST box. Wafer A, which contains a residual concentration of carbon atoms (2–3×10^{15}/cm^3) as well as EL2, displays sharp ballistic peaks with a smaller diffusive background signal. Wafer B, which contains Cr doping to compensate a higher residual impurity level of Si atoms (2–3×10^{16}/cm^3), displays a larger diffusive background signal with relatively

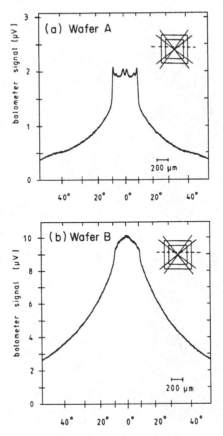

Figure 12 Line scans of phonon intensity across the [100] ST structure in GaAs. (a) For wafer A with low impurity content; (b) for wafer B with higher impurity content. A larger diffusive signal is observed in the latter case. (From Held et al.[25])

weaker ballistic peaks. Similar to the slot experiments, the intensity ratio B/A defined in the expanded line scan of Figure 13(a), is a measure of the diffusive component. Although the intensities A and B contain both ballistic and scattered phonons, this is a convenient method of gauging the phonon scattering when slotting the sample is not practical, such as in these thin wafers.

Monte Carlo simulations are used to predict the ratio B/A, using an average phonon mean free path, λ, as an adjustable parameter. Alternatively, a variable combination of pure ballistic and pure diffusive signals is matched to the data, yielding λ. With the simplifying assumption that the scattering processes are isotropic (i.e., no channeling), the results of these two algorithms are shown in Figure 13(b). Comparison with the data indicate that the average mean free path of phonons in wafer A is between 0.6 and 0.8 mm, and that

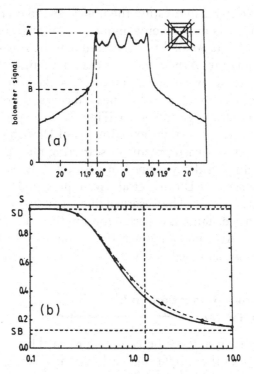

Figure 13 Method of quantifying the average mean free path of phonons in GaAs. (a) Experimental line scan and definition of the ratio S = A/B, which is related to the ratio of ballistic to scattered phonons. (b) Computed ratio S as a function of the assumed mean scattering length of the phonons in GaAs. Dots are from a Monte Carlo simulation (isotropic scattering), and the solid line is generated by adding pure diffusive and pure ballistic components. SD and SB are the values of S that correspond to the pure diffusive and pure ballistic limits. (From Held et al.[25])

in wafer B is between 0.35 and 0.42 mm. The scattering lengths are found to be correlated with the dislocation density, which varies smoothly across the wafers. Such experiments, in conjunction with electrical measurements, could have a practical value in characterizing and optimizing electronic-grade crystals.

Fieseler et al.[26] made a similar analysis of the phonon scattering in undoped GaAs, using a frequency-selective tunnel-junction detector ($2\Delta/h = 700$ GHz) and taking into account the channeling effect due to phonon focusing. By spatial scanning, they found a mean free path, $\lambda = 1.6$ mm, consistent with the predicted isotope scattering rate at 700 GHz in this crystal. Further analysis of the temporal pulse widths supports this conclusion for their undoped samples.[27]

The literature contains interesting theoretical speculation that the isotope scattering rate in GaAs may become anomalously small for zone-boundary acoustic phonons.[28] The possibility was raised that the zone-boundary lattice dynamics consisted only of the motion of the As atom, with the isotopically impure Ga being at the wave node. Tamura[29] looked for this effect, using lattice dynamics models, but found that the predicted scattering rate *increased*, rather than decreased, at high frequency [Figure 5(b) on page 204]. This issue was precipitated by time-of-flight experiments of Ulbrich et al.[30] In view of the recent quantitative measurements of scattering rates in GaAs, as well as the phonon-imaging results described in Chapter 7, it seems clear that the broad pulses observed by Ulbrich et al. were produced mainly by phonon scattering, not dispersion.[17] The observed linear scaling of the arrival time of the pulse *peaks* with distance is expected for purely diffusive propagation in a planar geometry.[31] Also, phonon downconversion has an effect on the shape of the heat pulses, as discussed in Chapter 10.

E. Scattering from dislocations in LiF – A phonon polarizer

The mechanical strength of a solid is often governed by the presence of dislocation lines, such as the simple edge dislocation depicted in Figure 14. The motion of dislocations under an applied stress allows plastic deformation of a solid at stresses considerably below the elastic yield limit of a perfect lattice. The introduction of impurity atoms and radiation-induced defects can pin dislocations and thereby increase the mechanical strength of the solid. The metallurgical implications of dislocation motion are great.

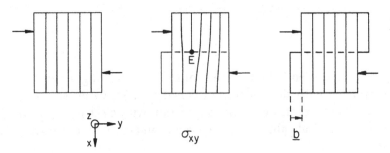

*Figure 14 Diagram showing the formation and motion of an edge dislocation (E) as a result of shear stress on the crystal. Vertical lines are atomic planes, and the horizontal dashed line is the slip plane of the dislocation, not necessarily the (100) basal plane of a cubic crystal. The dislocation line is perpendicular to the paper at point E. The Burgers vector, **b**, is along the slip direction and has a magnitude equal to the spacing between atomic planes. Dislocation motion allows plastic deformation of a crystal at moderate shear stresses.*

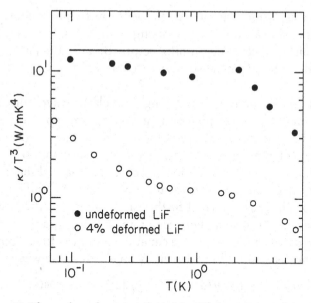

Figure 15 Thermal conductivity divided by T^3 for LiF as a function of temperature. The mean phonon frequency is proportional to temperature. Data before (dots) and after (circles) a 4% compressive deformation show increased phonon scattering due to the formation of dislocations. The solid line indicates the expected conductivity for boundary-limited scattering. (From Yang et al.[32])

Just as a static shear strain can cause nonelastic deformation of a solid by dislocation motion, an elastic wave of appropriate polarization and wavevector can couple to the dislocation line and scatter inelastically. Indeed, the attenuation of radio-frequency acoustic waves is a powerful probe of extended defects in both metallic and insulating materials. Similarly, high-frequency phonons scatter from dislocations, reducing the thermal conductivity, κ, of a solid. An example of this effect is seen in Figure 15, which contains measurements of the thermal conductivity of LiF by Yang et al.[32] The undeformed crystal displays a T^3 dependence of κ as expected when the phonon mean free path is limited by the boundaries. The thermal conductivity of a crystal plastically deformed by 4%, implying a large density of dislocations, is greatly decreased at low temperature. By irradiating the same sample, thereby inducing pinning centers, the scattering strength of the dislocations is significantly reduced and the conductivity rises.[32]

A physical picture that explains this remarkable behavior is the fluttering-string model of Granato,[33] originally devised to explain the acoustic attenuation in similarly deformed systems. The elastic wave induces a vibrational motion of the dislocation, which inelastically scatters the incident wave into other elastic waves. Unlike a single mass defect, the dislocation line is capable of undergoing vibrational motion at relatively low frequencies and can exhibit resonances like that of a fluttering string, with a fundamental wavelength equal to the distance between pinning points. Because the pinning points are statistically distributed, the absorption occurs over a broad range of phonon frequencies.

A classical calculation of the scattering of an elastic wave from a dislocation by this dynamic scattering process is given, for example, by Kneezel and Granato.[34] The essential ideas are outlined below. The elastic wave induces a force per unit length $b\sigma_0$ on the dislocation line, where $\mathbf{b}$ is the Burgers vector associated with the dislocation and σ_0 is the resolved shear stress – that component of stress that induces the dislocation motion. In Figure 14, where the dislocation line runs normal to the paper (along z), the Burgers vector is $\mathbf{b} = a\hat{y}$, with a the spacing between atomic planes, and the resolved shear stress is $\sigma_0 = \sigma_{xy}$. One can write the equation of motion for a damped, infinite string (i.e., the dislocation line) driven by an elastic plane wave as[34]

$$A(d^2y/dt^2) + B(dy/dt) - C(d^2y/dz^2) = b\sigma_0 \cos(kz - \omega t), \qquad (3)$$

where $y(z, t)$ is the displacement of the dislocation line, A is the linear mass density, B is the damping constant, C is the line tension, and k is the component of the phonon wavevector along the dislocation line. The solution of this equation, $y(z, t)$, is a traveling wave with components in and out of phase with the driving force. The out-of-phase component absorbs acoustic power from the elastic wave, and the resulting inverse mean free path, ℓ^{-1}, of the incident wave depends on the areal dislocation density, n_d, and the elastic stresses as,[34,35]

$$\ell^{-1} \propto n_d \Omega, \qquad (4)$$

where $\Omega = (\sigma_0/\sigma_p)^2$ is the resolved shear-stress factor for the incident phonon with total stress $\sigma_p = \text{Tr}(\sigma)$. Equation (4) assumes that all of the dislocations are oriented along the same direction in the crystal.

The resolved shear-stress factor, Ω, depends on the propagation direction of the phonon. In a phonon-imaging experiment, we measure the intensity of the heat pulses as a function of the (group-velocity) angles (θ, ϕ) between the source and the detector. Without scattering, the anisotropic phonon intensity (ballistic flux) due to phonon focusing may be written $I_0(\theta, \phi)$. With dislocation scattering, the detected intensity becomes

$$I(\theta, \phi) = I_0(\theta, \phi)e^{-\beta\Omega(\theta,\phi)}, \qquad (5)$$

where the scattering strength, β, contains the source-to-detector distance, the dislocation density, and the prefactors in Eq. (4).

Dislocations have been extensively characterized in the insulating crystal LiF.[36] The crystal structure is fcc with a two-atom basis (i.e., the NaCl

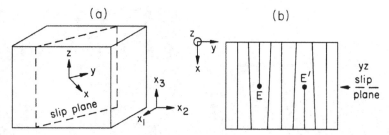

Figure 16 *Diagram showing the (110) slip plane in LiF. This plane contains the yz cartesian axes, as in Figure 14. Slip along these planes is induced by pressing the sample along x_1 while confining its x_3 dimension, thereby allowing it to expand along x_2. Dislocations form loops in the slip planes. A cross section of a loop showing its edge-type dislocation lines end on is illustrated in (b). Screw dislocations (not shown) connect the ends of the edge lines (E and E') to form the loop. The Burgers vector represents the atomic displacement associated with the section inside the the loop.*

structure). The slip planes are primarily {110}, as shown in Figure 16(a). In general, edge dislocations (E) do not extend entirely across the crystal; they may terminate on screw dislocations with the same Burgers vector and form dislocation loops.[37] A dislocation loop is described by a single Burgers vector and couples to elastic waves as indicated in Eq. (3). Shear stress tends to expand or shrink a dislocation loop, as may be understood by considering the cross section of the loop shown in Figure 16(b). A σ_{xy} stress causes the edge dislocations, E and E', to move apart or together, as reasoned from Figure 14.

By compressing the crystal illustrated in Figure 16(a) along its (100) faces (normal to the x_1 axis) and allowing it to expand only along the [010] direction, slip is possible along either (110) or ($1\bar{1}0$) planes. However, one of these slip directions may dominate, so that the crystal contains dislocations predominantly with a single Burgers vector direction.

Note that the phonon dislocation coupling is strongly dependent on the directions of the phonon wavevector and polarization. Imagine that the (100) faces in Figure 16(a) are the source and detector faces in a phonon-imaging experiment. Consider a phonon propagating in the horizontal FT ridge, i.e., the (001) plane. As seen in Figure 6, the polarization of an FT ridge phonon is perpendicular to the ridge plane, so the wave displacement in Figure 16(a) is

$$\mathbf{u} = u_0 \hat{z} e^{i(\mathbf{k} \cdot \mathbf{r} - \omega t)}, \tag{6}$$

where $\mathbf{k}$ lies in the xy plane. The strain accompanying this shear wave has no σ_{xy} component and, therefore, does not induce a motion of the dislocation line along the slip plane, as depicted in Figure 14. We show below that FT

Figure 17 At the left is a photograph of the two LiF crystals used in the experiments of Northrop et al.[35] On top is the undeformed crystal and below is the crystal deformed 10%, as in Figure 16(a). The front surfaces have bolometer detectors, and the rear surfaces have evaporated Cu films to absorb the incident laser light. The top right photo is a phonon image of the undeformed LiF crystal, using an Al bolometer for detection. The bottom right photo is a phonon image of the deformed LiF crystal.

phonons outside the horizontal ridge and ST phonons propagating in nearly all directions have significant σ_{xy} components. The net result is that the (vertically polarized) FT phonons in the horizontal ridge propagate much more freely through the sample than other phonons. A selectively dislocated crystal acts as a phonon polarizer!

This phonon-imaging experiment was performed by Northrop et al.,[35] and the results are displayed in Figure 17. A photo of the undeformed and deformed crystals with bolometer detectors is shown at the left. The top photo at the right is the phonon image for an undeformed LiF crystal, showing both the FT ridges and the ST structure centered around [100]. The bottom photo at the right is the phonon image for the crystal plastically deformed by 10% in the manner described in Figure 16(a). Figures 18(a) and 18(b) contain the

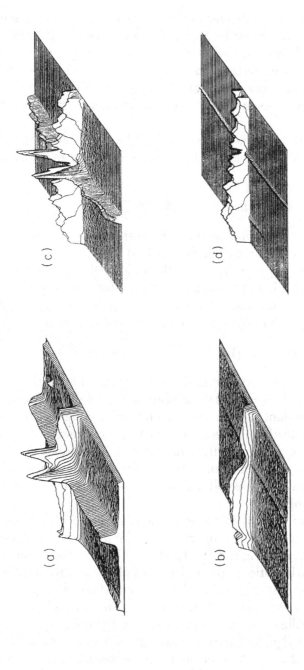

Figure 18 (a) Pseudo–three-dimensional representation of the experimental phonon image at the top right of Figure 17 (undeformed crystal). (b) Pseudo–three-dimensional representation of the phonon image at the bottom right of Figure 17 (deformed crystal). (c) Monte Carlo calculation of the ballistic phonon pattern in undeformed LiF. (d) Monte Carlo calculation of the transmitted ballistic flux using a scattering strength $\beta = 2800$, as discussed in the text. The ST mode is completely absorbed by dislocation scattering and the vertical FT ridge is greatly attenuated, but the horizontal ridge remains strong.

corresponding pseudo–three-dimensional plots of intensities for these two experimental images, compared to Monte Carlo calculations [Figures 18(c) and 18(d)] described below. As predicted, only phonons in the horizontal FT ridge (and a small remnant of those in the vertical FT ridge) are ballistically transmitted through the heavily dislocated crystal. The ST phonons contain sufficient shear components in the slip planes to be completely attenuated in this sample.

A more quantitative analysis of the phonon dislocation scattering follows. The strain associated with an elastic wave with wavevector $\mathbf{k}$ and polarization $\mathbf{e}$ is given by

$$\varepsilon_{lm} = \partial u_l / \partial x_m = i e_l k_m e^{i(\mathbf{k}\cdot\mathbf{r}-\omega t)}, \tag{7}$$

where the subscripts l, m refer to the cube axes, x_1, x_2, x_3, indicated in Figure 16(a). In terms of these axes, the resolved shear stress for the {110} slip planes is $\sigma_{xy} = (\sigma_{11} - \sigma_{22})/2$ (c.f. the rotational transformations in Figure 14 on page 39). Using the elasticity tensor for cubic crystals, given in Eq. (10) on page 41, one finds $(\sigma_{11} - \sigma_{22}) = (C_{11} - C_{12})(\varepsilon_{11} - \varepsilon_{22})$ and $\mathrm{Tr}(\sigma) = (C_{11} + 2C_{12})\mathrm{Tr}(\varepsilon)$. With Eq. (7), the resolved shear-stress factor is,

$$\Omega = |(\sigma_{11} - \sigma_{22})/2\, \mathrm{Tr}(\sigma)|^2 \propto (e_1 n_1 - e_2 n_2)^2, \tag{8}$$

where $n_i = k_i / k$ are the components of the wave normal. We see that, for the horizontal FT ridge, $e_1 = e_2 = 0$ and Ω vanishes, as argued above. Figure 19(a) shows contours of constant Ω in wavevector angle space, roughly corresponding to the angular scale of the experimental images. To construct this graph, it is necessary to solve the Christoffel equation [Eq. (18) on page 43] for the eigenvectors $\mathbf{e}$ over a grid of $\mathbf{n}$ values. The results of similar calculation for the ST mode is shown in Figure 19(b). Ω is seen to be greater than 10^{-2} over nearly the entire plane.

Equation (5) shows that ballistic transmission across the sample is significant when $\beta\Omega < 1$, where β is proportional to the number of dislocations. The cross-hatched areas in Figures 19(a) and 19(b) indicate the phonons that would be transmitted ballistically through the sample for a scattering strength of $\beta = 10$.[3] Even with this large value, there is a broad region of phase space around the horizontal (001) plane where phonons are not strongly attenuated. A smaller region of ballistic transmission is seen for FT phonons with wavevectors close to the vertical plane. For these phonons, $n_1 = e_2 = 1$ and $n_2 = e_1 = 0$ in Eq. (8), yielding $\Omega = 0$ exactly in the plane. However, n_2 increases linearly with increasing wavevector angle out of the vertical plane, bringing the value of Ω up rapidly. The broad transparency region for FT phonons near the horizontal plane is a consequence of phonon focusing, which keeps the polarization vector nearly parallel to $\hat{z}$ (i.e., e_1 and $e_2 \simeq 0$) for a significant range of $\mathbf{k}$ vectors out of this plane.

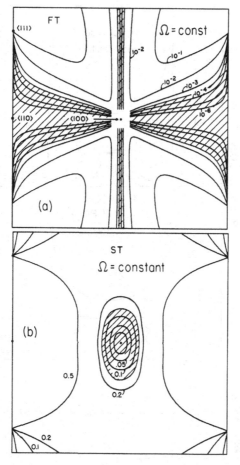

Figure 19 Contours of constant Ω – the resolved shear-stress factor – in wavevector angle space for (a) FT phonons and (b) ST phonons in a LiF crystal deformed as in Figure 16(a). The shaded region identifies phonons that should be transmitted ballistically through the 10% deformed crystal.

The different behavior of the horizontally and vertically polarized FT phonons provides a method for quantitatively determining the scattering strength β. Profiles of the vertical and horizontal ridges (taken at points 30° from the [100] axis) are generated by Monte Carlo calculations, as shown in Figures 20(a) and 20(b). As β increases, the ballistic flux in the ridges decreases – first in the vertical ridge, then in the horizontal ridge, as expected from Figure 19(a). The relative intensities of the two ridges are determined as a function of the scattering strength and are plotted in Figure 20(c). For the 10% deformed crystal, the experimental ratio of approximately 20 corresponds to $\beta = 2800$. The computed pseudo–three-dimensional phonon images plotted in Figures 18(c) and 18(d) for $\beta = 0$ and $\beta = 2800$ compare favorably with the corresponding experimental plots shown in that figure.

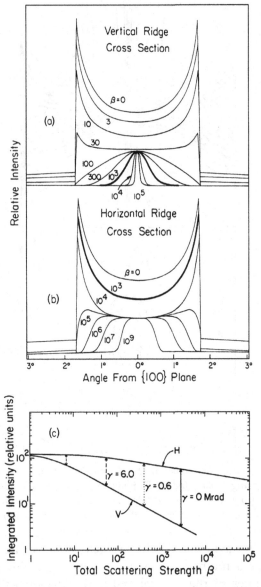

Figure 20 (a) and (b) Theoretical line scans across the FT ridges for several values of dislocation scattering strength, β. The experimental ratio of integrated intensities for the two ridges in the 10% deformed sample corresponds to a value $\beta = 2800$. The heavy lines show the FT profiles for $\beta = 10.^3$ (c) Integrated intensity of the horizontal (H) and vertical (V) FT ridges as a function of scattering strength, which depends on both the phonon coupling strength and the density of dislocations. The solid arrow shows the intensity ratio for the 10% deformed sample with no γ irradiation. Irradiation by 0.6 and 6.0 Mrad pin the dislocations and reduce the phonon scattering, as shown. The dotted-dashed arrow is for a 1.6% deformed sample with no irradiation. (From Northrop et al.[35])

Northrop et al.[35] further found that the scattering strength in the 10% deformed sample is reduced to $\beta = 400$ and 55 with exposures to γ irradiation of 0.6 and 6 Mrad, respectively. This treatment does not reduce the dislocation density but produces pinning centers that restrict the dislocation motion, providing further direct evidence that the phonon dislocation interaction is caused by the dynamics of dislocation motion rather than scattering by static strains around the dislocations.

Finally, the scattering strength can be independently estimated using an etch-pit analysis to measure the dislocation densities. It is found that the β determined by phonon imaging of a 1.6% deformed sample is consistent with this analysis, but the $\beta = 2800$ determined for the 10% deformed crystal is far higher than that expected from the etch-pit analysis. A similar discrepancy was found in the thermal-conductivity data.[34] One hypothesis is that the samples contain a high density of dislocation dipoles, which are not detected in the etch-pit count. Scattering from dipoles is also governed by the resolved shear-stress factor, Ω, and a dipole density $\simeq 10^{10}/cm^2$ would be required to explain the data.

We have seen in this chapter that anisotropies in the phonon scattering from point defects and extended defects are typically very large. By observing these anisotropies, phonon-imaging experiments can provide detailed information about the microscopic scattering processes and the objects responsible for phonon scattering.

References

1. P. Carruthers, Theory of thermal conductivity of solids at low temperatures, *Rev. Mod. Phys.* **33**, 92–138 (1961).

2. R. Berman, *Thermal Conduction in Solids* (Clarendon, Oxford, 1976).

3. T. H. Geballe and G. W. Hull, Isotopic and other types of thermal resistance in germanium, *Phys. Rev.* **110**, 773 (1958). An excellent review of recent studies on isotope effects in semiconductors has been written by E. E. Haller, Isotopically engineered semiconductors, *J. Appl. Phys.* **77**, 2857 (1995).

4. C. Kittel, *Introduction to Solid State Physics*, 5th ed. (Wiley, New York, 1976), p. 150. It was recently proposed that normal three-phonon processes can also affect thermal conductivity, due to elastic anisotropy: A. K. McCurdy, Energy-flow-reversing N processes in elastically anisotropic crystals, *Phys. Rev. B* **38**, 10335–10349 (1988).

5. G. A. Slack, Effect of isotopes on low-temperature thermal conductivity, *Phys. Rev.* **105**, 829 (1957); J. Callaway, Model for lattice thermal conductivity at low temperatures, *Phys. Rev.* **113**, 1046 (1959). See also R. Berman and J. C. F. Brock, *Proc. R. Soc. A* **289**, 46 (1965).

6. T. R. Anthony, W. F. Banholzer, J. F. Fleischer, L. Wei, P. K. Kuo, R. L. Thomas, and R. W. Pryor, Thermal diffusivity of isotopically enriched ^{12}C diamond, *Phys. Rev. B* **42**, 1104 (1990).

7. H. E. Jackson, C. T. Walker, and T. F. McNelly, Second sound in NaF, *Rev. Lett.* **25**, 26 (1970).

8. J. A. Shields, S. Tamura, and J. P. Wolfe, Elastic scattering acoustic phonons in Si, *Phys.*

Rev. B **43**, 4966 (1991); S. Tamura, J. A. Shields, and J. P. Wolfe, Lattice dynamics and elastic phonon scattering in silicon, *Phys. Rev. B* **44**, 3001 (1991).

9. J. A. Shields, J. P. Wolfe, and S. Tamura, Channeling of acoustic phonons in Si: the polarization dependence in elastic scattering, *Z. Phys. B* **76**, 295 (1989).

10. Slotted samples have also been used by B. Stock, R. G. Ulbrich, and M. Fieseler, Direct observation of ballistic large-wavevector phonon propagation in gallium arsenide, in *Phonon Scattering in Condensed Matter*, W. Eisenmenger, K. Lassmann, and S. Dottinger, eds. (Springer, Berlin, 1984), p. 100, and by R. J. von Gutfeld, Ref. 3 of Chapter 1.

11. S. Tamura, Spontaneous decay rates of TA phonons in quasi-isotropic solids, *Phys. Rev. B* **31**, 2574 (1985).

12. B. Cabrera, private communication. See *Proceedings of 5th International Workshop on Low Temperature Detectors* (Berkeley, CA), *J. Low Temp. Phys.* **93**, (1993).

13. M. T. Ramsbey, J. P. Wolfe, and S. Tamura, Phonon focusing of elastically scattered phonons in GaAs, *Z. Phys. B* **73**, 167–78 (1988).

14. B. Stock, M. Fieseler, and R. G. Ulbrich, in *Proceedings of the 17th International Conference on the Physics of Semiconductors*, J. D. Chadi and W. A. Harrison, eds. (Springer, Berlin, 1985), p. 1177.

15. R. G. Ulbrich, Generation, propagation and detection of terahertz phonons in GaAs, in *Nonequilibrium Phonon Dynamics*, W. E. Bron, ed. (Plenum, New York, 1985), p. 101.

16. M. Schreiber, M. Fieseler, A. Mazur, J. Pollman, B. Stock, and R. G. Ulbrich, Dispersive phonon focusing in gallium arsenide, in *Proceedings of the 18th International Conference on the Physics of Semiconductors*, O. Engström, ed. (World Scientific, Singapore, 1987), p. 1373.

17. J. P. Wolfe, Propagation of large-wavevector acoustic phonons: new perspectives from phonon imaging, *Festkörperprobleme* **29**, 75 (1989).

18. V. Narayanamurti, M. A. Chin, and R. A. Logan, Direct determination of symmetry of Cr ions in semi-insulating GaAs substrates through anisotropic ballistic-phonon propagation and attenuation, *Appl. Phys. Lett.* **33**, 481 (1978).

19. L. J. Challis, M. Locatelli, A. Ramdane, and B. Salce, Phonon scattering by Cr ions in GaAs, *J. Phys. C* **15**, 1419 (1982); L. J. Challis and A. M. de Goer, in *The Dynamical Jahn-Teller Effects in Localized Systems*, Yu E. Perlin and M. Wagner, eds. (North-Holland, Amsterdam, 1984), pp. 553–708.

20. H. J. von Bardeleben, D. Stiévenard, D. Deresmes, A. Huber, and J. C. Bourgoin, Identification of a defect in a semiconductor: EL2 in GaAs, *Phys. Rev. B* **34**, 7192 (1986); J. C. Bourgoin, H. J. von Bardelben, and D. Stievenard, Native defects in gallium arsenide, *J. Appl. Phys.* **64**, R65 (1988).

21. G. M. Martin, Optical assignment of the main electronic trap in bulk semi-insulating GaAs, *Appl. Phys. Lett.* **39**, 747 (1981).

22. J. C. Culbertson, U. Strom, and S. A. Wolf, Deep-level symmetry studies using ballistic-phonon transmission in undoped semi-insulating GaAs, *Phys. Rev. B* **36**, 2962 (1987); J. C. Culbertson, U. Strom, P. B. Klein, and S. A. Wolf, Phonon transport in photoexcited GaAs, *Phys. Rev. B* **29**, 7054 (1984).

23. J. A. Shields and J. P. Wolfe, Phonon scattering from residual defects in GaAs: observation of optically induced metastability by phonon imaging, *Phys. Rev. B* **50**, 8297 (1994).

24. Z. Xin, F. F. Ouali, L. J. Challis, B. Salce, and T. S. Cheng, Phonon scattering by impurities in semi-insulating GaAs wafers, in *Proceedings of 4th International Conference on Phonon Physics, Physica B* **219 & 220**, 56 (1996).

25. E. Held, W. Klein, and R. P. Huebener, Characterization of single-crystalline GaAs by imaging with ballistic phonons, *Z. Phys. B* **75**, 17–29 (1989).

26. M. Fieseler, M. Wenderoth, and R. G. Ulbrich, The mean free path of 0.7 THz acoustic

phonons in GaAs, in *Proceedings of the 19th International Conference on Physics of Semiconductors*, W. Zawadski, ed. (Institute of Physics, Polish Academy of Sciences, Warsaw, 1989), Vol. 2, p. 1477.

27. M. Fieseler, M. Schreiber, R. G. Ulbrich, M. Wenderoth, and R. Wichard, Numerical simulations and experimental data on transport of 0.7 THz acoustic phonons in GaAs, in *Phonons 89*, S. Hunklinker, W. Ludwig, and G. Weiss, eds. (World Scientific, Singapore, 1990), p. 1245.

28. M. Lax, P. Hu, and V. Narayanamurti, Spontaneous phonon decay selection rule: N and U processes, *Phys. Rev. B* **23**, 3095 (1981); M. Lax, V. Narayanamurti, P. Hu, and W. Weber, Lifetimes of high frequency phonons, *J. Phys.* **42**, 161 (1981).

29. S. Tamura, Isotope scattering of large-wavevector phonons in GaAs and InSb: deformation-dipole and overlap-shell models, *Phys. Rev. B* **30**, 849 (1984).

30. R. G. Ulbrich, V. Narayanamurti, and M. A. Chin, Properties of large-wavevector acoustic phonons in semiconductors, *Phys. Rev. Lett.* **45**, 1432 (1980).

31. M. Lax, V. Narayanamurti, and R. C. Fulton, in *Laser Optics of Condensed Matter*, H. Z. Cummins and A. A. Kaplyanskii, eds. (Plenum, New York, 1988), p. 229.

32. I.-S. Yang, A. C. Anderson, Y. S. Kim, and E. J. Cotts, Phonon dislocation interactions in deformed LiF, *Phys. Rev. B* **40**, 1297 (1989).

33. A. V. Granato, Thermal properties of mobile defects, *Phys. Rev.* **111**, 740 (1958).

34. G. A. Kneezel and A. V. Granato, Effect of independent and coupled vibrations of dislocations on low-temperature thermal conductivity in alkali halides, *Phys. Rev. B* **25**, 2851 (1982); G. A. Kneezel, Ph.D. Thesis, University of Illinois, 1980 (unpublished).

35. G. A. Northrop, E. J. Cotts, A. C. Anderson, and J. P. Wolfe, Anisotropic phonon dislocation scattering in deformed LiF, *Phys. Rev. B* **27**, 6395 (1983); *Phys. Rev. Lett.* **49**, 54 (1982).

36. J. J. Gilman and W. G. Johnston, in *Solid State Physics, Vol. 13*, F. Seitz and D. Turnbull, eds. (Academic, New York, 1962), p. 147; W. G. Johnston and J. J. Gilman, *J. Appl. Phys.* **30**, 129 (1959).

37. M. T. Sprakling, *The Plastic Deformation of Simple Ionic Crystals* (Academic, New York, 1976).

10

Quasidiffusion and the phonon source

This chapter attempts to characterize the acoustic phonons generated when a semiconductor crystal is excited by light. These phonons are a byproduct of the energy relaxation of photoexcited carriers. The problem naturally divides into two parts: (a) the distributions (in space, time, and frequency) of acoustic phonons *emanating* from the excitation region, i.e., the phonon source, and (b) the subsequent propagation and decay (downconversion) of acoustic phonons in the bulk of the sample. Because elastic scattering from mass defects is highly dependent on frequency, the instantaneous diffusion constant increases with time as phonons downconvert – a process commonly referred to as quasidiffusion. In principle, low-frequency phonon sources produce ballistically propagating phonons and high-frequency sources lead to quasidiffusion.

Previous chapters have dealt mainly with heat pulses generated by optical excitation of a metal film that was deposited on the nonmetallic crystal of interest. Nonequilibrium phonons are again produced by the relaxation of the carriers excited by the light. In the metal, however, there is a high density of free carriers (10^{22}–10^{23} cm^{-3}), and the interactions among carriers and between carriers and phonons tend to establish a single temperature. In that case, the phonons radiated from the metal film into the crystal have an approximately Planckian energy distribution, and the subsequent propagation in the crystal can be modeled by anharmonic decay, elastic scattering, and ballistic propagation (with frequency dispersion).

In a semiconductor at low temperatures a pulse of light produces free carriers at a density that is generally much smaller than the carrier density in a metal. Furthermore, the free carriers vanish within nanoseconds or microseconds by electron-hole recombination. At low densities the hot photoexcited carriers lose their excess kinetic energy mainly by emission of optical phonons. Thus, it is reasonable to expect that at low excitation levels nonequilibrium acoustic phonons are mainly produced by the decay of high-frequency optical phonons, plus some acoustic phonon emission by carriers with kinetic energies smaller than that of an optical phonon. This implies that the early acoustic-phonon spectrum will be rich in high-frequency phonons and is not likely to be described by a Planckian distribution. Indeed, this is the starting point of the quasidiffusion process.

This chapter is about the production and scattering of acoustic phonons in high-purity semiconductors. After a qualitative description of quasidiffusion, we concentrate on the particular indirect-gap semiconductor Si, for which the intrinsic elastic and anharmonic scattering rates have been measured or calculated. The character of heat pulses in this crystal depends markedly on the boundary conditions of the crystal and the excitation levels. Once the excitation surface is isolated from the liquid-helium coolant and the excitation density is reduced to low levels, quasidiffusive pulses are observed that agree quantitatively in shape and decay time to numerical and Monte Carlo models, also reviewed in this chapter.

Not so readily understood is the phonon production process at high excitation densities, where moderate densities of photoexcited carriers are present. The surprising result in Si is that high excitation density produces an abundance of low-frequency phonons. We discuss two quite different interpretations of this result – one involving the formation of a phonon hot spot and the other invoking the kinetics of phonons within electron-hole droplets. While many of the general ideas discussed here are expected to be applicable to other crystals and types of excitation (e.g., by photons, electrons, or high-energy particles), the specific electronic properties of the crystal are important to the phonon production process.

A. Qualitative description of quasidiffusion

At low temperatures, insulators and undoped semiconductors contain extremely low densities of free carriers (unpaired electrons and holes). For incident photon energies greater than the electronic band gap, energetic electron-hole pairs are created that lose most of their kinetic energy in less than a picosecond by phonon emission.[1] (We shall see later that this picture can be modified by carrier–carrier interactions at high carrier densities.) The generation of phonons produced by this electronic relaxation are zone center longitudinal-optical (LO) phonons.* After shedding their excess kinetic energy, the carriers eventually recombine radiatively or nonradiatively, generating more phonons.

The splitting of an LO phonon into two acoustic phonons generally takes place in less than a picosecond, as depicted by the heavy arrows in Figure 1.[2,3]

---

* This assumes that the incident photon energy, $h\nu$, is greater than the band gap, E_g, by more than $\hbar\Omega_{LO}$, generally a few tens of meV.

Table 1. *Frequency downconversion of a 7.5-THz longitudinal phonon in Si*

Frequency ν (THz)	Lifetime τ_a (ns)	Number of elastic scatterings τ_a/τ_e	Mean free path $V\tau_e$ (μm)	Diffusion length $(V^2\tau_e\tau_a/3)^{1/2}$ (μm)
7.5	0.0006	4	1.1	1.4
3.75	0.018	8–9	18	30
1.88	0.58	17–18	280	680
0.94	19	35	4400	15,000

Note: Hypothetically assuming no mode conversion to T phonons and $B = 1.2 \times 10^{-55}$ s^4, $A = 2.43 \times 10^{-42}$ s^3, and $V = 9 \times 10^5$ cm/s. (After Shields and Wolfe.[6])

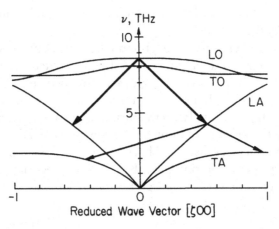

Figure 1 *The splitting of an LO phonon into two acoustic phonons and subsequent decay into lower-frequency phonons. The dispersion curves are for GaAs.*

In turn, the acoustic phonons split by anharmonic decay.[4] In Chapter 8 it was shown that the downconversion rate for acoustic phonons (neglecting dispersion) has a rapid frequency dependence,

$$\tau_a^{-1} = B\nu^5. \tag{1}$$

Therefore, each successive generation of phonons has roughly half the frequency and $2^5 = 32$ times the lifetime of its predecessors. To give an idea of the magnitude of this *inelastic* scattering for longitudinal phonons in Si, Table 1 shows that after approximately 1 μs, only phonons below approximately 1 THz remain, and these decay to smaller frequencies at a much slower rate.

Concurrent with the frequency decay of the phonon distribution, a *spatial* evolution of the thermal energy following electronic excitation is also occurring. We know from Chapter 8 that *elastic* scattering from defects in the crystal also has a steep dependence on frequency[5]:

$$\tau_e^{-1} = Av^4. \tag{2}$$

For each successive generation of phonons, the mean free path, $\ell = V\tau_e$, increases immensely: as a phonon bifurcates into two phonons of approximately half the frequency, the elastic scattering time increases by roughly $2^4 = 16$. One can define a diffusion length associated with a generation of mean frequency v that is equal to the distance the phonon elastically diffuses within its anharmonic lifetime. Representative values of τ_a, τ_a/τ_e, ℓ, and diffusion length for four generations of longitudinal phonons in Si are tabulated in Table 1. The ramifications of these numbers are discussed below.

Two theoretical approaches have been taken to describe this quasidiffusive transport process. The first is an analytic approach, introduced by Levinson and Kazakovtsev.[7-9] They have shown that the volume of the expanding phonon gas does not increase in the usual diffusive manner. Pure diffusion with no frequency conversion exhibits the $\mathcal{R} = (Dt)^{1/2}$ form, where $D = V^2\tau_e/3$ is the diffusion *constant* from the kinetic theory of gases. For the decaying phonon gas, the average frequency at time t is given by $v = (Bt)^{-1/5}$, so the average elastic scattering time takes on the time dependence, $\tau_e = v^{-4}/A = (Bt)^{4/5}/A$. Thus, the radius of the expanding phonon cloud is roughly described by

$$\mathcal{R} = (V^2 t\tau_e/3)^{1/2} = (V^2 B^{4/5}/3A)^{1/2}t^{9/10}, \tag{3}$$

which is approximately linear in time.* Spatially resolved phonon-spectroscopy experiments in sapphire and CaF_2 by Bron et al.[10] gave early qualitative support for this quasidiffusion description, based on these simple ideas.

The second theoretical approach to this problem is to use Monte Carlo techniques. Initially these studies have assumed isotropic scattering rates and simple collinear downconversion of phonons. Northrop and Wolfe[11] calculated the temporal arrival of phonon flux in GaAs assuming elastic isotropy, a v^5 spontaneous decay rate, and a v^4 elastic scattering rate. The shape of the predicted heat pulses yielded empirical agreement with data obtained for an optically excited crystal, supporting the quasidiffusion theory for this system. Happek et al.[12] used phonon-spectroscopy techniques to study the process in CaF_2. Their results indicated that the anharmonic decay rate in this crystal was more complicated than that given by the simple collinear process (see Section B below). Lee et al.[13] computed the temporal and spatial flux patterns in Si, incorporating phonon focusing but still employing isotropic elastic and inelastic scattering rates. At increasing times of flight the phonon caustics

---

* The range of validity of this type of scaling analysis has been considered by Maris and by Esipov and Tamura, as discussed in Section D.

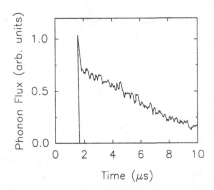

Figure 2 Theoretical time trace for quasidiffusive propagation through a 1-cm Si sample. (Calculated by Maris.[14])

are predicted to smear into a broader diffusive distribution. In contrast, early experiments showed sharp caustics in photoexcited Si – an anomaly that will be discussed in detail shortly.

In an effort to relate the analytical and statistical approaches, Maris[14] examined the range of validity of the scaling laws associated with quasidiffusive phonon propagation. He noted that the downconversion process produces a significant population of low-frequency phonons that are not described by the analytical quasidiffusion theory of Levinson and coworkers. Assuming elastic isotropy, Maris simulated heat pulses in Si and Ge and found agreement with these conclusions. The shapes of the Monte Carlo heat pulses display a component of flux arriving close to the ballistic time of flight, followed by a large component of late-arriving scattered phonons, as shown in Figure 2.

B. Improved models of anharmonic phonon decay

Early models of quasidiffusion generally assumed that the probability that a phonon at frequency ν_0 decays into two phonons at frequencies ν and $\nu_0 - \nu$ is given by

$$P(\nu_0, \nu) = C\nu^2(\nu_0 - \nu)^2. \tag{4}$$

This expression, plotted in Figure 3(a), results from the ν^2 density of phonon modes and assumes that the initial phonon decays into two phonons in the same nondispersive branch. To illustrate this process,[16] we choose an initial frequency $\nu_0 = 3.75$ THz phonons, corresponding to roughly the second generation of acoustic phonons in Si (Table 1). The daughters of these fast-relaxing phonons have measurable lifetimes and mean free paths on the scale of a typical heat-pulse experiment. Indeed a significant fraction of these third-generation phonons [shaded area in Figure (3a)] are predicted to propagate

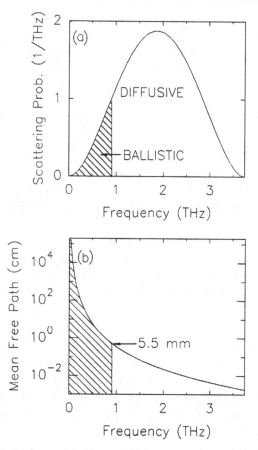

Figure 3 (a) Distribution of frequencies resulting from a single decay of 3.75-THz phonons, according to Eq. (4). (b) The mean free path, $V\tau_e$, with $V = 8.65$ mm/μs, an average longitudinal phonon velocity in Si, and τ_e given in Table 1. The shaded region indicates those phonons that travel ballistically through a 5.5-mm sample.

ballistically across a crystal with $d = 5.5$ mm thickness. (This is the crystal thickness used in an experiment described in the next section.) This can be seen by plotting the mean free path, $\ell = V\tau_e = V/(Av^4)$, versus frequency in Figure 3(b). Phonons with frequencies above 0.9 THz will contribute to the diffusive tail in the heat pulse (e.g., Figure 2).

This one-branch model of phonon decay is unphysical for the following reason: In order to conserve momentum and energy, the daughter phonons must have wavevectors collinear to the initial phonon, leaving essentially zero phase space for the final state. Indeed, a small amount of frequency dispersion can eliminate this process altogether, resulting in the prediction

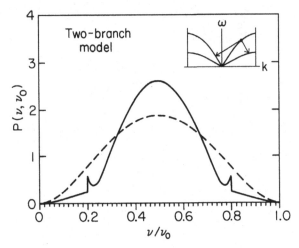

Figure 4 *The heavy curve is the downconversion probability in Si from the two-branch model of Tamura.[15] This model includes dispersive L and T branches. For comparison, the light curve is the isotropic single-branch model [Eq. (4)].*

that phonons in lowest (ST) branch should possess extremely long lifetimes.* Thus, any calculation of an average decay rate requires careful consideration (and detailed computation) of all possible decay channels. Commonly used values of B in the decay rate equation, $\tau_a^{-1} = B\nu^5$, are from the calculations by Tamura,[4] who considered the anharmonic decay of L phonons into L + T or T + T phonons.

Tamura[15] recently introduced a more realistic, two-branch model (with dispersion) into the calculation of quasidiffusive phonon propagation. The computed distribution of downconverted L phonons are shown in Figure 4 and compared to Eq. (4). We see that far fewer low-frequency phonons are produced than in the one-branch case. This has a significant effect on a heat pulse. The lower curve in Figure 5 represents the quasidiffusive Monte Carlo pulse calculated with this model, using the known elastic scattering rate in Si and assuming isotropic ballistic propagation. As expected, the early-arriving ballistic phonons are reduced from that computed in Maris' calculation (Figure 2). However, when phonon focusing from elastic anisotropy is included into the calculation, the early arriving (near-ballistic) portion of the heat pulse is restored.† Now we examine how these predictions compare with experimental observations.

---

* As shown in Figure 9 on page 211, however, *anisotropy* removes this restriction for a small subset of phonons and leads to nonzero decay times of the ST phonons. Also, ST phonons indirectly decay by converting to L or FT phonons through *elastic scattering*. The resulting L and FT phonons can decay anharmonically.

† Due to phonon focusing, the relative intensity of the early-arriving component depends upon the direction of propagation.

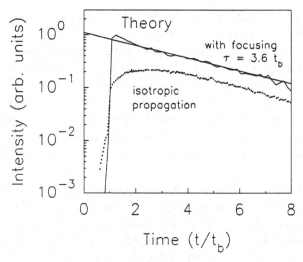

Figure 5 Monte Carlo simulation of the phonon signal versus time for two different models by Tamura[15] described in the text. Both calculations include one L and two T branches; the upper curve includes phonon focusing. The straight line is an exponential decay with a time constant equal to 3.6 times the ballistic time of flight.

C. Observation of quasidiffusion in Si

If the theoretically predicted downconversion rates are operative, it should be difficult to observe sharp phonon-focusing patterns in directly photoexcited Si. This startling prediction[16] can be understood by examining Table 1. Only phonons with frequencies below 1 THz have a significant probability of traveling ballistically across a 5-mm-thick crystal. The subterahertz phonons striking the detector will have evolved from a prior generation that diffused a significant distance before downconverting. In particular, the diffusion length of a 2-THz phonon is approximately 200 μm, which is then a prediction of the apparent size of the source of ballistic (<1 THz) phonons seen by the detector.[15] Of course, there are a few subterahertz phonons that are produced very near the excitation point and propagate ballistically across the crystal, giving rise to sharp caustics, but these phonons should be in the small minority.

In contrast to this prediction, experimental phonon images of directly photoexcited Si in a He bath display intense, sharp caustics [Figure 24(b) on page 56] and temporally sharp heat pulses,[6] as compared to those predicted in Figures 2 and 5. This seemingly anomalous behavior has only recently been understood, and it is caused mainly by two effects: (1) the loss of a large fraction of the high-frequency diffusive phonons into the He bath, and (2) the creation of a localized source of low-frequency phonons by free-carrier interactions at high excitation densities. The formation of a He bubble at a sharply focused excitation point can also affect the phonon source, as we have seen for metal-film excitation in Chapter 3.

In order to observe the predicted quasidiffusive propagation in Si, Shields et al.[17] and Msall et al.[18] isolated the excitation surface from the He bath and greatly reduced the photoexcitation density by defocusing the laser beam. The resulting heat pulses, shown as the heavy curve in Figure 6(a), indeed show a broad pulse very similar to that predicted by the quasidiffusive simulation by Tamura (upper curve in Figure 5). A much narrower pulse [light curve in Figure 6(a)] is obtained when liquid He is permitted to contact the excitation surface.

Why does the He bath have such a great effect on the heat pulses traversing the crystal? In essence, high-frequency phonons elastically scatter many times near the excitation surface, and when they encounter the surface they have a high probability of being transmitted into the bath (see also Chapter 11). A Monte Carlo simulation of this effect, assuming isotropy and one branch, is shown in Figure 6(b). The initial frequency in the simulation was chosen to be 4 THz. Good qualitative agreement is obtained with the experimental heat pulses of Figure 6(a). Figure 6(c) indicates the frequency distributions of those phonons lost into the bath and those detected. Also plotted as the solid lines in this figure is Eq. 4, normalized to the height of the data and using $\nu_0 = 4$ THz and 2 THz (solid lines). This equation gives a fairly good empirical representation of the resulting distributions.

Figure 7 shows the experimental heat pulse for Si with a vacuum interface, plotted on a semilogarithmic scale. The quasidiffusive tail is empirically characterized by a decay time of $t_0 = 3.6t_b$ for this 5.5-mm-thick crystal. This result compares favorably to the $t_0 = 3.6t_b$ extracted from Tamura's calculation. Although the calculated anharmonic decay is modeled without the full lattice dynamics theory (presently a prohibitive calculation), this remarkable agreement between theory and experiment gives us confidence that the basic processes of anharmonic decay and elastic scattering in Si are reasonably well understood at this time. A truly quantitative comparison between experiment and theory must also consider the actual surface conditions and the spectrum of phonons emitted from recombining carriers.

D. Quantitative description of quasidiffusion

Important insights are gained by seeking an analytical solution to the quasidiffusion process. The experiments and Monte Carlo calculations presented in the previous section displayed several interesting characteristics that suggest some simple underlying physics. The rapid rise in heat flux at the ballistic arrival time indicates that a significant number of subterahertz phonons are rather quickly produced by the decay of several-terahertz phonons. The intensity of the early-arriving component is expectedly dependent on

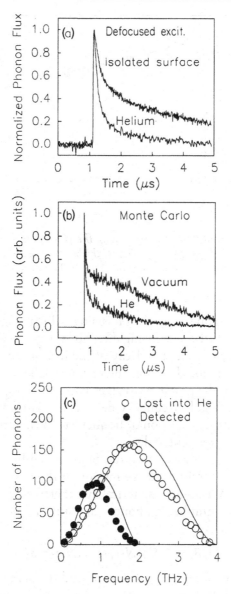

Figure 6 (a) Comparison of experimental time traces for weakly photoexcited Si, with and without liquid He in contact with the excitation
surface. Sample thickness = 5.5 mm. The traces are normalized to
the same peak height. Typically a factor of 3–7 reduction of the peak
height is observed when liquid He is in contact with the excitation surface.
(b) Calculation of time traces for isotropic phonon propagation through a
5.5-mm-thick Si sample in vacuum (heavy curve) or in a He bath (light
curve). The bath absorbs a large fraction of the diffusive phonons, which
produced the long tail in the vacuum trace, so that the He trace is a
sharp ballistic-like pulse. (c) Calculated frequency distribution of detected
phonons for both cases. (Data from Shields et al.[17])

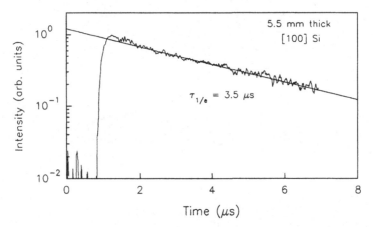

Figure 7 Experimental time trace along the [100] propagation direction in a photoexcited Si crystal of 5.5-mm thickness. The straight line is an exponential decay with time constant equal to 3.6 times the ballistic time of flight.

propagation direction due to phonon focusing, which magnifies the intensity of phonon flux along strong focusing directions (e.g., $\langle 100 \rangle$ for ST phonons). However, the nearly exponential decay of the heat pulse that is only weakly dependent on propagation direction and distance defies a simple explanation. In its simplest form, the problem involves two scattering times (anharmonic and elastic) that are highly frequency dependent. Thus, we are dealing with a broad and changing distribution of frequencies. What, then, is the physical origin of the single-exponential decay?

A partial answer to this question can be obtained by considering the detected signal from a purely diffusive motion of the phonons, i.e., neglecting anharmonic decay. We assume spherical symmetry and start N_0 phonons at $r = 0$ and $t = 0$. The number of phonons at point r at time t is then governed by the kinetic equation,*

$$(\partial/\partial t - D\nabla^2)N(r, t) = N_0\delta(r)\delta(t). \tag{5}$$

In the isotropic case, $\nabla^2 = (1/r)(\partial^2/\partial r^2)r$, and the standard solution to this diffusion equation is

$$N(r, t) = N_0(4\pi Dt)^{-3/2} \exp(-r^2/4Dt), \tag{6}$$

which states the well-known result that the characteristic radius of a diffusing cloud of particles increases as $(Dt)^{1/2}$.

If we detect (i.e., absorb) all the phonons that cross the surface at $r = R$, then a boundary condition, $N(R, t) = 0$ is imposed, and the solution to the

---

* This equation is a statement of local particle conservation, given the definition of D from Ficks law, $J = $ particle flux $= -D\nabla N$.

kinetic equation is conveniently written in terms of a Fourier series,

$$N(r, t) = \sum_{n=1}^{\infty} C(n, R, t)(n/r) \sin(n\pi r/R), \tag{7}$$

with the coefficient given by $C(n, R, t) = \exp(-n^2\pi^2 Dt/R^2)$. Using $D = V^2\tau_e/3$ and the ballistic time of flight, $t_b = R/V$, the factor in the exponential term equals $(\pi^2/3)(t\tau_e/t_b^2)$. If the detection time, t, and scattering time, τ_e, are somewhat larger than t_b, the first term in the Fourier series dominates. The detected signal is proportional to the particle flux at R,

$$J = -D\partial N/\partial r|_R = F(R)\exp(-t/t_0), \tag{8}$$

where the exponential decay constant is given by $t_0 = R^2/\pi^2 D^2 = 3t_b/\pi^2\tau_e$. Thus, in the purely diffusive case (i.e., no frequency conversion), the tail of the heat pulse is expected to exhibit a single-exponential decay time that scales as the square of the propagation distance. Using this result, the measured $t_0 = 3.6t_b$ for the 5.5-mm crystal implies a phonon frequency of approximately 1.4 THz, assuming no frequency conversion and Eq. (2) with $A = 2.43 \times 10^{-42}$ for Si.

Next consider the quasidiffusive case in which the phonons decay in frequency as they diffuse. Now the kinetic equation for the phonons of frequency v may be written in the form,[7]

$$(\partial/\partial t - D(v)\nabla^2)N(v, r, t) = -N(v, r, t)/\tau_a$$
$$+ \int_v^{\infty} dv' D(v')N(v', r, t)P(v', v)$$
$$+ S_0\delta(r)\delta(t)\delta(v - v_0), \tag{9}$$

where the terms on the right of this equation represent (a) the decay of phonons with frequency v, (b) the creation of phonons at frequency v due to downconversion of higher-frequency phonons, and (c) a point source of monochromatic phonons at frequency v_0. The systematic solution of this equation has been reported by Kazakovtsev and Levinson.[7] They show that in the case of spontaneous decay (no upconversion) in a single nondispersive mode, the solution takes on a scaling form,

$$N(v, r, t) = Av^{9.5} f[r/\mathcal{R}(v), t/\tau_a(v)], \tag{10}$$

where $\mathcal{R}(v) = [D(v)\tau_a(v)]^{1/2}$ is the diffusion length of phonons with frequency v and $\tau_a(v)$ is the anharmonic decay time. This result holds only for phonons that have frequencies much lower than the source frequency, v_0, and for observation times that are much longer than the ballistic time of flight. That is, the scaling analysis holds only after several generations of phonons decay and after many elastic scattering events.* Notice that with $D = V^2 \tau_e/3$, $\mathcal{R} \sim [\tau_e \tau_a]^{1/2} \sim \nu^{9/2} \sim t^{9/10}$ so radius of the expanding cloud predicted by Eq. (10) agrees with Eq. (3).

Esipov[19] has applied this analytic approach to understand the single-exponential decay of the phonons in both the Si experiments and the Monte Carlo calculations. For $t \gg t_b$ a solution of the form $N(\nu, r, t) = N(\nu, t)$ $\sin(\pi r/R)/r$ may be assumed for Eq. (9), which leads to the equation

$$[\partial/\partial t + \tau_e^{-1}(\nu) + \tau_D^{-1}(\nu)]N(\nu, t) = \int_0^\infty d\nu' N(\nu', t) P(\nu', \nu), \qquad (11)$$

where $\tau_D^{-1}(\nu) = \pi^2 V^2 \tau_e(\nu)/3R^2$. Esipov showed that below a certain bottleneck frequency the solution to this equation has the form $\exp(-t/t_0)$, where t_0^{-1} is the minimum of the function, $\tau_e^{-1}(\nu) + \tau_D^{-1}(\nu)$, with respect to ν. The value of ν at this minimum is the bottleneck frequency,

$$\nu_{BN} = (4\pi^2 V^2/15R^2 AB)^{1/9}, \qquad (12a)$$

and the resulting decay rate is

$$t_0^{-1} = (9/4)B\nu_{BN}^5. \qquad (12b)$$

The R dependence of t_0 is $R^{10/9}$, so this theory predicts that the ratio $t_0/t_b \simeq R^{1/9}$ is weakly dependent on R. For $R = 5.5$ mm in Si, one finds $\nu_{BN} = 1.68$ THz and $t_0 = 3.0t_b$, which is reasonably close to the value observed in the experiments. It should be noted, however, that the ratio t_0/t_b depends upon the geometry chosen (e.g., spherical or slab). Also, the use of a two-branch model will change this ratio somewhat. At present we can only say that there is a semiquantitative agreement between the analytic theories and the experiments.

Using Monte Carlo techniques, Tamura[20] has calculated the flux of phonons as a function of their frequency at arrival. As predicted by Esipov, the decay time of the signal, t_0, for phonons with frequency less than the bottleneck frequency is quite independent of the phonon frequency. Figure 8 shows the frequency distribution of the detected phonons, calculated by numerical solution to Eq. (11) and by Monte Carlo simulation. The agreement is quite good at high frequencies. As expected, at low frequencies the analytical model predicts a lower flux than that of the Monte Carlo calculation because it applies best to phonons that have scattered many times.

A pronounced peak in the distribution occurs near the bottleneck frequency, ν_{BN}, which corresponds to phonons whose diffusion length, $\mathcal{R} = (D\tau_a)^{1/2}$, is roughly equal to the sample dimension $\mathcal{R}$. As indicated by the analysis

---

* Maris[14] has pointed out that for real crystals under ordinary conditions the scaling analysis appears to break down because a significant number of low-frequency phonons are promptly produced by the downconversion process, Eq. (4). However, in analyzing the long decay time of the quasidiffusive heat pulse, we are considering phonons that have scattered many times and thus expect the analytical theory to be useful.

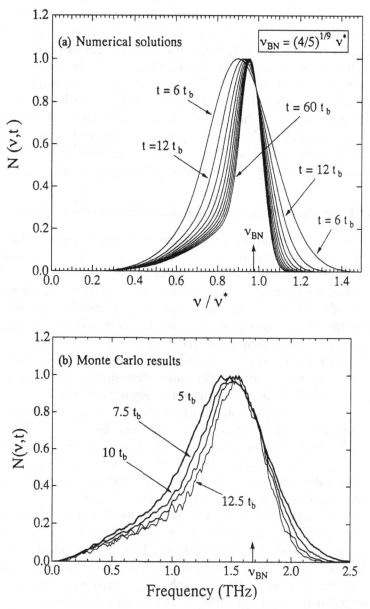

Figure 8 Comparison of analytic and statistical models of quasidiffusion by Tamura.[15] (a) Frequency distribution of detected phonons calculated by numerical solution of the diffusion equations for several detection times relative to the ballistic time of flight, t_b. (b) Monte Carlo results. Both calculations show a predominance of phonons near the bottleneck frequency for Si.

leading to Eq. (8), purely diffusing phonons display a heat pulse with a single exponential decay time. The analytical theory has given us an expression [Eq. (12a)] for the characteristic frequency of quasidiffusing phonons in terms of the scattering constants (A and B) and the propagation distance R. Those phonons that have frequency much lower than ν_{BN} leave the crystal and are not counted in the quasidiffusive tail. Those phonons with frequency much higher than ν_{BN} decay in the crystal before being detected. It is interesting to note that the peak of the detected phonon spectrum in the Monte Carlo calculation [Figure 12(b)] coincides quite well with the $\nu = 1.4$ THz that we estimated earlier [after Eq. (8)] for purely diffusing phonons.

E. Localized phonon source at high excitation density

A remarkable transformation in the character of the experimental heat pulses occurs when the excitation area is decreased at constant excitation power.[17,18] This is achieved by simply reducing the laser spot size on the sample. The optical arrangement and results are shown in Figure 9. Again, the excitation surface is isolated from the He bath. At low power density, a quasidiffusive pulse is obtained. At high power density, the early-arriving signal greatly increases at the expense of the tail signal [Figure 9(a)]. Apparently the fraction of low-frequency phonons (with near-ballistic times of flight) is larger when the optical excitation is more concentrated. Amazingly, with *high* excitation density a localized source of *low*-frequency phonons is created.

The peak intensity of the heat pulse (near $t \simeq 1.1~\mu s$) is plotted as a function of the computed power density in Figure 9(b). The data for several powers are normalized at low densities and appear to follow a common curve. The peak signal undergoes a transition in intensity that is centered around 20 W/mm^2. At low power densities the pulse has a quasidiffusive character, and at high power densities the pulse is primarily ballistic.

What is the physics behind this striking transition in the character of the heat pulses? The result cannot be understood solely in terms of quasidiffusion, which, in theory, does not depend on the local density of phonons. Two possibilities come to mind: (1) a local hot spot is formed due to phonon–phonon interactions, and (2) the initial phonon spectrum is somehow modified by the high density of photoexcited carriers. In what follows, we will consider each of these possibilities in detail. The conclusions reached will form a basis for examining the generation of acoustic phonons in other photoexcited semiconductors at low temperature.

F. A phonon hot spot?

So far, we have considered only the spontaneous *decay* of nonequilibrium phonons because the *fusion* process (two phonons merging into a single,

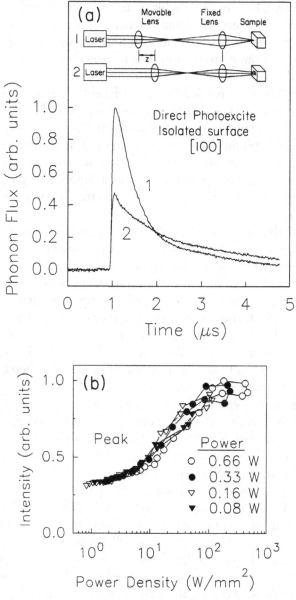

Figure 9 Experiment in which the power density is varied without chang-ing the total power deposited in the sample by defocusing the laser spot. (a) Comparison of heat pulses for a 5.5-mm-thick sample of Si for focused and defocused laser beams (200 and 2 W/mm², respectively). (b) The in-tensity of near-ballistic phonons arriving in the interval $1.05 t_b < t < 2 t_b$, as a function of power density, P_{abs}/area. The detected phonon signal is almost entirely ballistic at high power densities and quasidiffusive at low power densities. The transition between quasidiffusive and ballistic regimes occurs at a power density of approximately 20 W/mm². (Data from Shields et al.[17])

higher-frequency phonon) is less likely in the cold crystal. Of course, in thermal equilibrium there is a detailed balance of these downconversion and upconversion processes. As the nonequilibrium phonon population is increased, phonon–phonon interactions will increase the chances for fusion and, hence, lessen the phonon mean free path.

Hensel and Dynes[21] proposed that at sufficient excitation level a local hot spot is formed that possesses a temperature much higher than that of the bulk crystal. The mean free path of phonons in the hot-spot region is short, presumably due to phonon–phonon interactions. High-frequency phonons are trapped by their frequent scattering with other phonons. Low-frequency phonons, however, may escape from the heated region, giving rise to a pulse at the ballistic time of flight. The details of this picture have yet to be sharpened by experiment.

Evidence for the formation of a phonon hot spot in Ge at microjoule (i.e., high) excitation levels, corresponding to instantaneous power densities of 10^3–10^5 W/mm^2, seems to be well founded.[22-24] A localized source of ballistic phonons explains the phonon-driven transport of electron-hole droplets[25] in this crystal. Subsequent theories[26] and experiments[27] have supported the concept of hot-spot formation in Ge and other crystals. The presently discussed threshold for the source of low-frequency phonons in Si, however, occurs at much lower excitation densities (approximately 3 nJ energy and 50 W/mm^2 power density), so one must examine the likelihood of creating a phonon hot spot under these conditions.

What is the phonon energy density just after the 10-ns laser pulse? Referring to Figure 9(b), we see that for a peak power of 80 mW the half-point of the transition corresponds to an area of approximately 0.08 W/(20 W/mm^2) = 4×10^{-5} cm^2. The absorption depth of the green laser light in Si is approximately 1 μm. The average phonon frequency during the laser pulse can be estimated by setting the anharmonic decay time equal to the laser pulse length, 10 ns. The result is approximately 4 THz. The isotope-scattering formula [Eq. (2) and Table 1] implies that the elastic scattering time for phonons of this frequency is in the range of 1 ns. Phonons with this scattering time diffuse approximately 10 μm during the 10-ns pulse; therefore, the volume occupied by the 4-THz phonon gas is roughly 4×10^{-8} cm^3, which yields a thermal energy density during the pulse of approximately 2×10^{-2} J/cm^3.

The formation of a hot spot requires that the phonon occupation number at the average frequency be roughly 0.1.* In other words, there must be approximately $0.1h\nu$ thermal energy per unit cell, which has a volume $V_u = (0.34 \text{ nm})^3$ in Si. With an average frequency of 4 THz, this implies an

---

* The frequency at the peak of a Planckian distribution is $2.8k_B T/h$; therefore, the phonon occupation number at this average frequency is $(e^{2.8} - 1)^{-1} = 0.065$. A more rigorous analysis of power-density thresholds for hot-spot formation is given in Ref. 26.

energy-density threshold of approximately 7 J/cm^3. This required energy density for the formation of a hot spot is several hundred times greater than the actual experimental energy density estimated above. The conclusion is that the transition in Figure 9(b) occurs at a much lower energy density than that predicted for the establishment of a hot spot.

In view of this discrepancy, we feel compelled to ask: Is the phonon source really localized, and is it emitting mainly low-frequency phonons? Phonon-imaging experiments, taken both with and without liquid He contacting the excitation surface, provide strongly positive answers to both of those questions. [A transition curve similar though not identical to the one in Figure 9(b) is obtained in the He case.[17]] Figure 10(a) is a phonon image of directly photoexcited Si at excitation densities well above the threshold of Figure 9(b). A scan across one of the ST caustics at this power level shows a spatial width of approximately 30 μm – positive proof of subterahertz phonons originating from a localized spot with less than 30-μm dimension.

A further unusual result is that the width of the caustic [defined in Figure 10(b)] depends on the excitation power. Figure 10(c) shows spatial profiles of the caustic onset at increasing pulse energy. The width increases to approximately 100 μm at the highest level. A plot of the caustic width versus excitation energy in Figure 10(d) shows an approximate $P^{1/3}$ dependence. This empirical result implies that the volume of the phonon source increases linearly with excitation energy – an important clue about its origin.

G. Origin of the localized phonon source in Si

The lifetime of the localized phonon source has been determined by a method called spatial filtering.[28] This technique extracts from the heat pulse the contribution due to phonons scattered in the bulk, leaving only the signal originating from the phonon source. The diffusive component is removed by subtracting a time trace taken slightly off a phonon caustic from one taken directly on the caustic. The caustic is caused by ballistic phonons traveling directly from source to detector. Because the signal from scattered phonons is only weakly dependent on the laser position, the difference trace represents the decay curve for the phonon source.

Figure 11(a) shows the heat pulses on and off an ST caustic and the difference of these two traces.[17] The small component of long tail (due to scattered phonons) is largely removed in the difference trace. A semilogarithmic plot of the difference trace [Figure 11(b)] indicates a decay time of 170 ns over nearly two orders of magnitude in intensity. (A small amount of extended tail remains, probably due to incomplete subtraction of the scattered phonons.) By quadratically deconvolving the 60-ns response time of the detector (determined from direct optical excitation), we arrive at a localized source lifetime of $(170^2 - 60^2)^{1/2} = (160 \pm 20)$ μs. This value is relatively insensitive to power

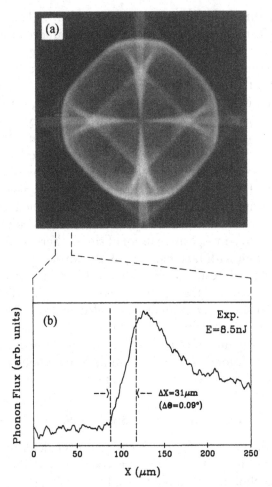

Figure 10 (a) High-resolution phonon image of photoexcited Si. The center of the image corresponds to the [001] propagation direction. The width of the image is 19.5 degrees left to right. (b) Line scan across the ST caustic as indicated by the white line in the image. (c) Line scans across the ST caustic at various incident laser energies. The width of the caustic varies significantly with incident energy, indicating an expansion of the ballistic phonon source. (d) Power dependence of the caustic width. The experimental resolution has been deconvolved from these data. (From Shields and Wolfe.[6])

density above the transition centered at 20 W/mm². What is the physics behind this phonon source?

The initial phonons resulting from photoexcitation of a semiconductor are produced by relaxation of the carriers. Those same carriers remain in the photoexcited region of the crystal for a time equal to the electron-hole recombination time. These carriers may continue to interact with the

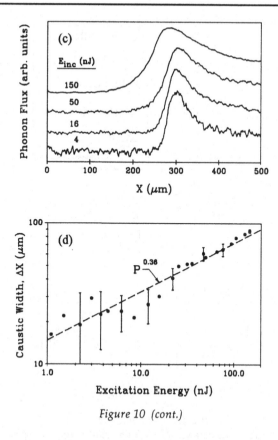

Figure 10 (cont.)

nonequilibrium phonons until they recombine at times ranging from approximately a nanosecond for direct-gap semiconductors to microseconds for indirect-gap semiconductors.

In relatively pure (undoped) semiconductors at low temperatures, the free electrons and holes produced by photoexcitation tend to bind pairwise into free excitons[29] – mobile electron-hole pairs (analogous to hydrogen atoms) that are bound by their coulomb attraction. In GaAs, the binding energy of a free exciton is approximately 4 meV and its Bohr radius is approximately 150 Å.[30] The lifetime due to radiative recombination of the free exciton in this *direct-gap* semiconductor is only about a nanosecond – not enough time for it to lose all of its excess kinetic energy acquired by photoproduction. (Photoluminescence experiments[31] show that the excitons in GaAs at liquid-helium temperatures reach an average temperature of approximately 15 K in their lifetime.) Under sufficiently intense pulsed excitation of GaAs, and most other semiconductors, the electrons and holes form a relatively dense electron-hole plasma that also decays within a few nanoseconds.[32]

In the *indirect-gap* semiconductors Si and Ge, new excitonic species come into the picture. Because the direct radiative recombination of an electron-hole

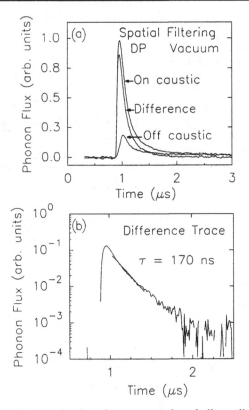

Figure 11 (a) Heat pulse for phonons traveling ballistically from the phonon source to the detector, obtained by subtracting time traces on and off a phonon caustic direction. (b) Semi-log plot of difference time trace shows an exponential decay of the ballistic pulse. Accounting for a detector response of 60 ns gives an estimate of the localized source lifetime, $\tau = 160 \pm 20$ ns. This lifetime is obtained using a 20-μm laser spot size and is valid over a range of incident total powers (30 mW < P_{abs} < 500 mW). (From Shields et al.[17])

pair is inhibited in the indirect-gap materials,* and because the electronic bands are multiply degenerate, excitons produced in sufficient densities can coalesce into droplets of electron-hole liquid, a degenerate Fermi fluid with many interesting properties.[25]

How does this all relate to the heat-pulse experiments we have been considering? It turns out that electron-hole droplets in Si have a measured lifetime of 140 ns, which is within the error bars for the measured decay time of the localized phonon source in the phonon experiments described above. In view of this coincidence, Esipov et al.[33] have suggested that the droplets themselves

---

* The conduction-band minimum and valence-band maximum are not at the same wavevector. To conserve momentum and energy, radiative recombination of an electron and a hole requires emission of a photon *plus* a phonon, which is a relatively weak transition.

can be the source of low-frequency acoustic phonons. The following is a short description of the hypothesized mechanism and some subsequent experiments that support the idea.

Photoexcitation of Si at $T < 23\,$K with visible light at moderate power densities is known to create a *cloud* of electron-hole droplets near the crystal surface. The pair density of each electron-hole droplet in Si has been spectroscopically measured to be $3 \times 10^{18}/\text{cm}^3$, and its radius is believed to be approximately 0.1 μm. The high pair density causes the carriers in the droplets to decay by nonradiative Auger recombination: an electron and hole recombine by giving their band-gap energy to the kinetic energy of a third carrier. This hot carrier must then shed its excess kinetic energy. In the usual single-particle relaxation process, the hot carrier cools by emitting optical phonons that subsequently decay into acoustic phonons. Such a phonon-generation process should lead to quasidiffusive heat pulses, just as with the optically excited hot carriers. But a carrier in a droplet can also lose its kinetic energy by collisions with other carriers, thereby heating the droplet as a whole. This latter process can bypass the generation of optical phonons and lead to the emission of lower-frequency acoustic phonons by the slightly heated droplet.

Thus the frequency spectrum of phonons emitted by electron-hole droplets depends on the relative rates of optical phonon emission and carrier-carrier scattering. Esipov's et al.[33] computed scattering rates for these processes as a function of carrier density are shown in Figure 12. This calculation shows the fraction of energy lost by a hot carrier due to carrier-carrier scattering as a function of the carrier density. We see that for the known pair density of a droplet (3×10^{18} cm^{-3} in Si), roughly 90% of the hot carrier energy goes

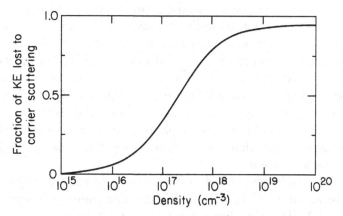

Figure 12 *Calculated fraction of the hot electron-hole pair energy losses transmitted to a large electron-hole droplet via intercarrier scattering as a function of carrier concentration. At low carrier densities, hot carrier cooling occurs by emission of optical phonons, but at higher densities this process is bypassed by rapid carrier collisions.*

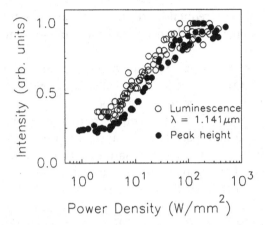

Figure 13 Detected photoluminescence from carrier recombination in electron-hole droplets in Si as a function of power density, normalized at the highest power density. The rise in luminescence intensity at approximately 20 W/mm² coincides with the phonon-signal transition shown in Figure 9(b). (From Msall et al.[18])

into raising the average kinetic energy of carriers in the droplet, rather than into optical phonons. Esipov et al. predict that Auger recombination and the subsequent carrier heating raise the temperature of the electron-hole liquid to between 5 and 10 K, a result that is in rough agreement with spectroscopic information.

The presence of electron-hole droplets under similar conditions to those used in the phonon experiments has been verified by Msall et al.[18] To observe photoluminescence signals at the relatively low excitation densities of Figure 9, the laser was mechanically chopped at 1 kHz and a lock-in detector provided the signal from a Ge photodiode at the exit slit of an optical spectrometer. The ratio of luminescence signals from electron-hole droplets and free excitons in Si was determined as a function of the excitation density with the same optical arrangement used in Figure 9. As shown in Figure 13, the droplet signal displays a distinct onset as a function of power density, corresponding quite well to the onset of the localized phonon source in photoexcited Si. This is pretty clear evidence that the droplets are instrumental in providing the acoustic phonon source in Si.

A critical issue is how the electron-hole droplets are distributed in the crystal. For continuous-wave excitation this question was answered earlier by Tamor and Wolfe,[34] who directly imaged the photoluminescence emitted by the electron-hole droplet cloud. The Si crystal was illuminated by a focused continuous-wave Ar laser beam, and an image of the glowing crystal was scanned across the entrance aperature of a spectrometer. The results are displayed in Figure 14. The size of the cloud of droplets increases with laser

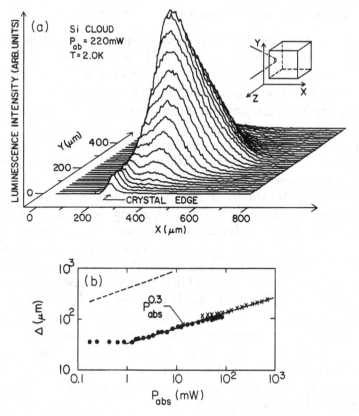

Figure 14 (a) Photoluminescence image of a cloud of droplets in Si, displayed as a pseudo–three-dimensional plot. Tamor and Wolfe[34] obtained these data using continuous excitation of an Ar laser. (b) Width of the cloud as a function of absorbed laser power.

power, as shown in Figure 14(b). The similarities between the expansion of the droplet cloud with increasing power, and that of the localized phonon source [Figure 10(c)] are striking, despite the different pumping conditions (pulsed versus continuous wave) in the two experiments. The line scans in the phonon-imaging experiments were integrated over time.

Further experiments showing the temporal expansion of the electron-hole droplet cloud under pulsed conditions ($10 \text{ K} < T < 25 \text{ K}$) were performed by Steranka and Wolfe.[35] These experiments indicated that phonons emitted from the droplets, themselves can be reabsorbed by other droplets, causing a transfer of momentum. This phonon-wind process, first theorized by Keldysh,[36] explained the anomalously high and power-dependent "diffusion constant" observed in such experiments. We conclude that the evolution of photoexcited carriers and phonons in the indirect-gap semiconductors are intricately intertwined.

H.   Extension to other semiconductors

The properties of the phonon source following photoexcitation in other semi-conductors are likely to be significantly different from those in Si. As noted before, the carrier lifetimes in GaAs at low temperature are much shorter than those in Si, and no electron-hole liquid has been clearly observed. The initial hot plasma at high excitation levels, however, may well affect the emitted acoustic phonon distribution in this crystal. A recent study[37] of heat pulses in photoexcited GaAs has been interpreted in terms of a phonon hot spot, although the authors note a considerable discrepancy in the theoretical threshold for its formation.

Germanium clearly exhibits electron-hole droplets, but the pair density is an order of magnitude smaller than that in Si, giving rise to much weaker Auger recombination and, presumably, a much less effective relaxation of hot carriers by carrier–carrier interactions. In addition, the droplet lifetime is much longer in Ge (40 μs). We might expect, then, that the phonon flux (power density per time) created by the relaxation of the initially photoexcited carriers would greatly exceed the flux from the slowly decaying droplets. Studies[35,38] indicate that both these thermalization – and recombination-generated phonons contribute to the cloud expansion. A particularly informative photoluminescence image of the cloud of electron-hole droplets in Ge, reported by Greenstein and Wolfe,[39] is reproduced in Figure 15(a). The sharp spikes in the cloud correspond to droplets pushed along the FT phonon-focusing ridges – clear evidence of a ballistic phonon wind.

This anisotropic cloud of droplets was produced with continuous-wave Ar-laser excitation of a high-purity Ge crystal. The time-resolved expansion of electron-hole droplets in Ge following intense pulsed excitation has also been observed [Figure 15(b)]. A rapidly expanding shell of droplets travels millimeters from the excitation point. The shell is punctured by the intense phonon fluxes along the FT ridges.[40] This time-resolved snapshot, taken 5 μs after a 20-μJ, 300-ns pulse from a Nd:YAG laser, contains evidence of a long-lived phonon hot spot, which provides the source of ballistic phonons that drive the shell of droplets at near sonic velocities. Similar experiments have been conducted for Si.[35] It must be emphasized that these time-resolved photoluminescence imaging studies employed much higher pulse energies than in the currently described phonon-imaging experiment, so the idea of a phonon hot spot is more relevant.

In view of these photoluminescence images, we should not be too surprised that the propagation of phonons in photoexcited semiconductors is inextricably linked to the photoexcited carriers. Heat-pulse experiments such as those described in this chapter provide complementary perspectives on these complicated processes.

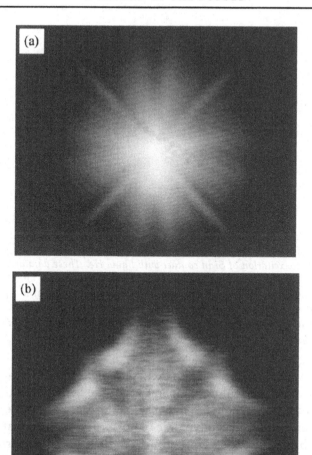

Figure 15 (a) Photoluminescence image of a cloud of electron-hole droplets in Ge produced by continuous focused with an Ar laser. Scale: 3 mm left to right. (b) Time-resolved image of droplet luminescence in Ge following an intense 300-ns pulse from a Nd : YAG laser. Scale: 8 mm left to right. (From Tamor et al.[40])

I. Return of the bubble

We have seen in this chapter that the He bath can be a significant sink for high-frequency phonons. A second consequence of He bath cooling is that at

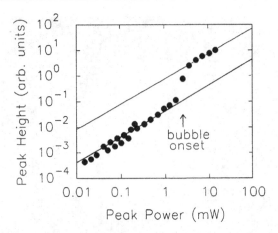

Figure 16 Peak phonon signal intensity as a function of absorbed power for photoexcitation of Si in contact with liquid He. These data are taken with a very long laser pulse (700 ns) and a correspondingly long boxcar gate (500 ns) in order to increase the total phonon signals at low power densities. Laser spot area = 200 μm². The jump in intensity at 2.5 mW is correlated with the formation of a He bubble at the excitation point.

a sufficient level of localized surface excitation, the surface temperature of the sample can exceed the boiling temperature of the bath, and a tiny bubble is formed at the excitation point. It may be that the crystal near the surface does not have to exhibit a thermodynamic temperature; merely that the density of nonequilibrium phonons emitted into the bath and converted to liquid vibrations is sufficient to vaporize a portion of the liquid.

This extremely nonlinear situation – undoubtedly present in a variety of low-temperature optical experiments – usually goes unnoticed and therefore has only been directly studied in a few instances. One manifestation of the bubble on heat pulses generated in a metal overlayer was described in Section C of Chapter 3. In that case, the metal film retained its elevated temperature for a longer time in the presence of a gas bubble.

The profound effect of the bubble to thermally isolate the surface near the excitation point from the superfluid He is shown in Figure 16. When the laser spot is tightly focused, the threshold power for bubble formation is extremely low, so, to observe the threshold, the laser is put into continuous mode and the beam is chopped at a low frequency.[17] The modulation of the bolometer by the phonons is detected with a lock-in detector, and its output is plotted in the figure. At low and high powers, the phonon signal increases linearly with power. At approximately 2 mW, an abrupt transition is observed that corresponds to an observable increase in scattered light at the excitation point, signaling a boiling of He at the excitation point. As seen in the figure, the presence of the bubble increases the transmitted signal by about an order of magnitude, which is qualitatively consistent with the heat-pulse study described in Figure 6.

The effects of the He bubble are not quantitatively understood at this time. As mentioned earlier, the important threshold curve given in Figure 9(b) is also observable when the excitation surface is immersed in liquid He. The onset of the bubble apparently accompanies the formation of the localized phonon source, and, therefore, the increase in peak height is approximately a factor of three larger than in the vacuum case. The transition from quasidiffusion to localized source is sharper in the He case, implying that the dynamics of the electron-hole liquid are also affected by the presence of the He. Such effects were previously noted in photoluminescence studies of Ge.[41]

J. Prospectus

Heat-pulse experiments have brought us to a better understanding of the physical processes that generate acoustic phonons in photoexcited semiconductors; however, a number of detailed issues remain to be addressed.

While the models of quasidiffusion appear to be in basic agreement with experiments, it would be interesting to know how the simulations depend on anisotropies in the anharmonic decay process and surface kinetics of the phonons. Details concerning specular or diffuse scattering at the interface, frequency conversion, and coupling to a He bath have not been explicitly included so far.

It would be useful to re-examine the conditions for hot-spot formation when nonequilibrium carriers are present. Existing models consider anharmonic and elastic scattering of phonons but not electron-phonon scattering. It is necessary to realistically treat the coupled electron-phonon system, where the nonequilibrium carriers are both the source of nonequilibrium phonons and are profoundly affected by them (e.g., via the phonon wind).

Finally, we would like to be able to understand the acoustic-phonon generation process when, instead of light, a high-energy particle is absorbed by the semiconductor. In particular, are excitons and/or electron-hole droplets involved? These issues take on an increasing importance as phonon-based particle detectors are developed.[13]

One of the principal results of the work described in this chapter is that the boundary conditions of a photoexcited semiconductor profoundly affect the frequency and spatial distributions of carriers and acoustic phonons. In the next two chapters, we consider in detail the interaction of phonons with interfaces.

References

1. A table of hot-carrier relaxation times in several semiconductors is given in V. F. Gantmakher and Y. B. Levinson, *Carrier Scattering in Metals and Semiconductors* (North-Holland, Amsterdam, 1987). For Ge, see E. M. Gershenson, Yu. A. Gurvich, E. N. Gusinskii, R. I. Rabinovich, and N. V. Soina, Interband transitions and cyclotron resonances of light holes in strong electric fields, *Sov. Phys. Solid State* **14**, 1493 (1972).

2. J. Menendez and M. Cardona, Temperature dependence of the first-order raman scattering by phonons in Si, Ge, and α-Sn: anharmonic effects, *Phys. Rev. B* **29**, 2051 (1984).

3. J. H. Parker, D. W. Feldman, and M. Ashkin, Raman scattering by silicon and germanium, *Phys. Rev.* **155**, 712 (1967).

4. S. Tamura, Spontaneous decay rates of LA phonons in quasi-isotropic solids, *Phys. Rev. B* **31**, 2574 (1985).

5. See, for example, S. Tamura, Isotope scattering of dispersive phonons in Ge, *Phys. Rev. B* **27**, 858 (1983).

6. J. A. Shields and J. P. Wolfe, Measurement of a phonon hot spot in photoexcited Si, *Z. Phys. B* **75**, 11 (1989).

7. D. V. Kazakovtsev and I. B. Levinson, Phonon diffusion with frequency downconversion, *Phys. Stat. Sol. B* **96**, 117 (1979).

8. Y. B. Levinson, Phonon propagation with frequency downconversion, in *Nonequilibrium Phonons in Nonmetallic Crystals*, W. Eisenmenger and A. A. Kaplyanski, eds. (North-Holland, Amsterdam, 1986), p. 91.

9. Y. B. Levinson, Phonon dynamics in highly nonequilibrium systems, in *Phonons 89*, S. Hunklinger, W. Ludwig, and G. Weiss, eds. (World Scientific, Singapore, 1990).

10. W. E. Bron, Y. B. Levinson, and J. M. O'Connor, Phonon propagation by quasidiffusion, *Phys. Rev. Lett.* **49**, 209 (1982).

11. G. A. Northrop and J. P. Wolfe, Search for large k-vector phonons in GaAs, in *Phonon Scattering in Condensed Matter IV*, W. Eisenmenger, K. Lassmann, and S. Döttinger, eds. (Springer-Verlag, Berlin, 1984).

12. U. Happek, Y. Ayant, R. Buisson, and K. F. Renk, Spontaneous decay cascade of high-frequency acoustic phonons, *Europhys. Lett.* **3** 1001 (1987).

13. A. T. Lee, Ph.D. Thesis, Stanford University (1992); A. T. Lee, B. Cabrera, B. L. Dougherty, and M. J. Penn, Measurement of ballistic phonon component resulting from nuclear recoils in crystalline silicon, *Phys. Rev. Lett.* **71**, 1395 (1993).

14. H. J. Maris, Phonon propagation with isotope scattering and spontaneous anharmonic decay, *Phys. Rev. B* **41**, 9736 (1990).

15. S. Tamura, Monte Carlo calculations of quasidiffusion in Si, *J. Low Temp. Phys.* **93**, 433 (1993); Numerical evidence for the bottleneck frequency of quasidiffusive acoustic phonons, *Phys. Rev. B*, **B56**, 13630 (1997).

16. H. J. Maris, Design of phonon detectors for neutrinos, in *Phonon Scattering in Condensed Matter V*, A. C. Anderson and J. P. Wolfe, eds., Vol. 68 of Springer Series in Solid State Sciences (Springer-Verlag, 1986), p. 404. Quantitative support for the idea that quasidiffusion broadens the source volume for ballistic phonons is given by S. Tamura.[15]

17. J. A. Shields, M. E. Msall, M. S. Carroll, and J. P. Wolfe, Propagation of optically generated acoustic phonons in Si, *Phys. Rev. B* **47**, 12510 (1993).

18. M. E. Msall, S. Tamura, S. E. Esipov, and J. P. Wolfe, Quasidiffusion and the localized phonon source in photoexcited Si, *Phys. Rev. Lett.* **70**, 3463 (1993); M. E. Msall, Photoproduced phonons in semiconductors, Ph.D. Thesis, University of Illinois at Urbana-Champaign (1994).

19. S. E. Esipov, Do phonons quasidiffuse in Si?, *Phys. Rev. B* **49**, 716 (1994).

20. S. Tamura, Quasidiffusive propagation of phonons in Si: Monte Carlo calculations, *Phys. Rev. B* **48**, 13502 (1993).

21. C. Hensel and R. C. Dynes, Interaction of electron-hole drops with ballistic phonons in heat pulses: the phonon wind, *Phys Rev. Lett.* **39**, 969 (1977).

22. M. Greenstein, M. A. Tamor, and J. P. Wolfe, Propagation of laser-generated heat pulses in crystals at low temperature: spatial filtering of ballistic phonons, *Phys. Rev. B* **26**, 5604 (1982); A. V. Akimov, A. A. Kaplyanskii, M. A. Pogarskii, and V. K. Tikhomirov, Kinetics of the emission of subterahertz acoustic phonons from the phonon hot spot in GaAs crystals, *JETP Lett.* **43**, 333 (1986); M. A. Tamor, M. Greenstein, and J. P. Wolfe, Time resolved studies of electron-hole-droplet transport in Ge, *Phys. Rev. B* **27**, 7353 (1983).

23. J. M. Hvam and I. Balslev, Exciton diffusion and motion of electron-hole drops in Ge, *Phys. Rev. B* **11**, 5053 (1975).

24. J. P. Wolfe, Transport of degenerate electron-hole plasmas in Si and Ge, *J. Lumin.* **30**, 82 (1985).

25. J. C. Hensel, T. G. Phillips, T. M. Rice, and G. A. Thomas, The electron-hole liquid in semiconductors, in *Solid State Physics, Vol 32*, H. Ehrenreich, F. Seitz, and D. Turnbull, eds. (Academic, New York, 1977); C. D. Jeffries and L. V. Keldysh, eds., *Electron-Hole Droplets in Semiconductors* (North-Holland, Amsterdam, 1983).

26. D. V. Kazakovtsev and Y. B. Levinson, Formation, dynamics and explosion of a phonon hot spot, *Zh. Eksp. Teor. Fiz.* **88**, 2228 (1985) [*Sov. Phys. JETP* **61**, 1318 (1985)]; The effect of phonon-phonon scattering in the substrate on the temperature dynamics of a phonon film injector, *Phys. Stat. Sol. B* **136**, 425 (1986).

27. N. M. Guseinov and G. S. Orudzhev, Phonon hot spot and emission of ballistic phonons, *Fiz. Tverd. Tela (Leningrad)* **29**, 2269 (1987) [*Sov. Phys. Solid State* **29**, 1308 (1987)]; V. I. Kozub, Phonon hot spot in pure substances, *Sov. Phys. JETP* **67**, 1191 (1988).

28. M. Greenstein, M. A. Tamor, and J. P. Wolfe, Propagation of laser-generated heat pulses in crystals at low temperature: spatial filtering of ballistic phonons, *Phys. Rev. B* **26**, 5604 (1982).

29. E. I. Rashba and M. D. Sturge, eds., *Excitons* (North-Holland, Amsterdam, 1982).

30. D. D. Sell, Resolved free-exciton transitions in the optical-absorption spectrum of GaAs, *Phys. Rev. B* **6**, 3750 (1972).

31. There are many optical experiments on carrier relaxation in semiconductors. A recent example with representative references is by H. W. Yoon, D. R. Wake, and J. P. Wolfe, Thermodynamics of excitons in GaAs quantum wells from their time resolved photoluminescence, submitted to *Phys. Rev. B.*

32. See, for example, Refs. 31 and 36. Lifetimes and spectra of photoexcited carriers in GaAs over a wide range of temperature and at low densities are reported by L. M. Smith, D. J. Wolford, J. Martinson, R. Venkatasubramanian, and S. K. Ghandhi, Photoexcited carrier lifetimes and spatial transport in surface-free GaAs homostructures, *J. Vac. Sci. Technol. B* **8**, 789 (1990).

33. S. E. Esipov, M. E. Msall, and J. P. Wolfe, Acoustic-phonon emission due to localized photoexcitation of Si: electron-hole droplets versus the phonon hot spot, *Phys. Rev. B* **7**, 13330 (1993).

34. M. A. Tamor and J. P. Wolfe, Spatial distribution of electron-hole droplets in silicon, *Phys. Rev. B* **21**, 739 (1980).

35. F. M. Steranka and J. P. Wolfe, Phonon-wind-driven electron-hole plasma in Si, *Phys. Rev. Lett.* **53**, 2181 (1984); Spatial expansion of electron-hole plasma in silicon, *Phys. Rev. B* **34**, 2561 (1986).

36. L. V. Keldysh, Phonon wind and dimensions of electron-hole drops in semiconductors, *JETP Lett.* **23**, 86 (1976).

37. B. A. Danil'chenko, D. V. Kazakovtsev, and M. I. Slutskii, Phonon hot spot formation in GaAs under optical excitation, *Phys. Lett. A* **138**, 77 (1989).

38. M. Greenstein and J. P. Wolfe, Formation of the electron-hole droplet cloud in germanium, *Sol. State Comm.* **33**, 1011 (1980).

39. M. Greenstein and J. P. Wolfe, Anisotropy in the shape of the electron-hole-droplet cloud in germanium, *Phys. Rev. Lett.* **41**, 526 (1878); M. Greenstein and J. P. Wolfe, Phonon-wind-induced anisotropy of the electron-hole droplet cloud in *Ge*, *Phys. Rev. B* **24**, 3318 (1981).

40. M. A. Tamor, M. Greenstein, and J. P. Wolfe, Time-resolved images of electron-hole droplets produced by intense pulsed-laser excitation of germanium, *Sol. State Comm.* **45**, 355 (1983).

41. References 50–52 in Chapter 3; M. Greenstein, Anisotropy and time evolution of the electron-hole droplet cloud in germanium, Ph.D. Thesis, University of Illinois at Urbana-Champaign (1981).

11

Phonon scattering at interfaces

A. The Kapitza problem

The transport of thermal energy across solid/solid and solid/liquid interfaces has occupied and perplexed many researchers since Kapitza's famous experiments[1] in 1941. He discovered the existence of a sharp temperature gradient at the surface of a heated solid that was immersed in liquid He. This surprisingly steep temperature drop was observed to occur within a few tens of micrometers of the surface, which was the spatial resolution of the experiment.

It was first suggested that the source of this discontinuity was a thick boundary layer of He that limited the thermal conductivity. However, such mechanisms proved unnecessary, as Khalatnakov[2] realized that the measured thermal boundary resistance between the solid and liquid was actually much *smaller* than predicted from a simple acoustic-mismatch theory. One defines the thermal boundary resistance as $R_K = \Delta T/\dot{Q}$, where ΔT is the temperature drop across the boundary for a given power flow, $\dot{Q}$.

Khalatnakov reasoned that the solid and the He have quite disparate acoustic impedances ($Z = \rho v$, with ρ the density and v the sound velocity), which implies a small phonon transmission coefficient across the interface. Modeling the solid and liquid as classical acoustic media, he predicted that the average transmission coefficient of an acoustic wave should be approximately 1%, whereas the measured thermal boundary resistances corresponded to a transmission coefficient of approximately 50%. This disparity between theory and experiment is known as the "Kapitza anomaly." Fifty years after Kapitza's measurements, this anomaly is not completely resolved. The status and extensive literature on this problem is the subject of several excellent reviews, listed in the references at the end of this chapter.[3-7]

Ballistic heat-pulse experiments have contributed to an understanding of the Kapitza conductance by resolving in time and space the phonons

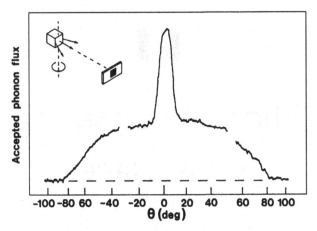

Figure 1 Angular distribution of phonons emitted from a NaF crystal into liquid He at 0.1 K and 18 bar. The sharp peak at the center represents the critical cone for specular refraction at the NaF/He boundary. The broad background is due to diffuse scattering at the boundary, which enhances the phonon transmission. (From Wyatt and Page.[8])

transmitted or reflected at an interface. A particularly informative experiment by Wyatt and Page[8] employed ballistic heat pulses and continuous scanning of the incident angle of phonons at the solid/liquid—[4]He interface to measure the *angular dependence* of the phonon transmission. Their data yield important clues about the surface scattering processes.

The experiment is illustrated in the inset of Figure 1. A small crystal of NaF is immersed in a bath of liquid [4]He at a temperature of 0.1 K and a pressure of 18 bar. Under these conditions, the phonon mean free path in the He is more than a centimeter. A bolometer is located approximately 1 cm from the crystal, and the surface of the crystal facing away from this detector has a metal film that may be electrically heated to produce the nonequilibrium phonons. The side surfaces of the crystal are coated with an amorphous material to prevent phonon emission. The crystal is rotated to change the angle between its surface normal and the direction of ballistic propagation to the detector.

The detected heat-pulse signal as a function of surface angle is plotted in Figure 1. The measured profile displays two components: (1) a narrow cone of high intensity within a half-angle of approximately 6° from the surface normal, and (2) a broad background signal. The narrow cone is just what one expects from the classical acoustic-mismatch model mentioned above. As we shall derive quantitatively later in this chapter, the phonons should refract from the NaF crystal only within a critical cone of half-angle given by

$$\sin \theta'_c = v'/v,$$

where v' and v are the sound velocities in liquid He and NaF, respectively. The ratio of He to NaF sound velocities is approximately 0.1, which implies $\theta'_c \cong 6°$, as measured in the experiment.

A striking observation in this experiment is that most of the phonons are leaking out of the crystal at angles beyond the critical angle predicted by acoustic-mismatch theory. This large-angle transmission is not possible for specular refraction obeying Snell's law. Yet experimental measurement shows that approximately 100 times as much thermal flux is emitted into the background channel as in the narrow cone! This can be seen by comparing the solid angles associated with specular and background channels in Figure 1; i.e., imagine rotating the graph about a vertical axis through $\theta = 0$ to get the transmission across the planar interface. The transmission through the background channel is large enough to account for the anomalous Kapitza conductance.

The background channel must be associated with diffuse, rather than specular, scattering at the interface. Simply put, phonon transmission outside the critical cone is due to phonons that are randomly scattered from imperfections at the surface and possibly converted to phonons of other frequencies. This idea is supported by other experiments that examine the internal *reflection* of phonons at an interface. Reflected heat pulses associated with diffuse scattering are found to diminish when liquid He or light-atom adsorbates are present at the surface.[9-12] The microscopic origins of diffuse scattering, however, are still open to question. Clearly inhomogeneities at the boundary are involved, but whether the inhomogeneities are associated with adsorbates on the surface, surface irregularities, or subsurface damage probably depends on the particular experiment and surface preparation.

Experiments on nearly perfect surfaces support this notion. Weber et al.[13] measured the ballistic phonons reflected from one surface of a NaF crystal by using heater and detector films evaporated on the face opposite the reflection surface. If the reflection surface was freshly cleaved in situ under high vacuum, the addition of He on the surface produced almost no change in the reflected heat-pulse intensities. That is, for a clean surface there appeared to be no anomalous transmission into the He. The phonon transmission increased significantly after the surface was exposed to air, suggesting that the anomalous Kapitza conductance in this case was due to diffuse scattering from adsorbates on the surface.

Other experiments have shown that surface damage has an effect on the phonon transmission and reflection.[3] For example, Kinder et al.[14] have observed an extinction of the Kapitza anomaly for longitudinal phonons normally incident on a Si/^{4}He boundary. In contrast, phonons that induce a strain along the surface (T phonons or obliquely incident L phonons) are anomalously transmitted. These results support Kinder's hypothesis[15] that the Kapitza anomaly in this case is due to surface defects that couple to the phonon strain field. Phonon reflection experiments on Si have demonstrated that in situ laser annealing, whereby a thin layer on the surface is melted and recrystallized, greatly reduces the diffuse scattering of phonons at the surface and provides a clean surface to study the effects of adsorbates.[10-12] As an example, the vibrational modes of rare-gas atoms and water molecules on Si have been studied.[16]

An important clue to the origin of the anomalous thermal boundary conductance is its temperature dependence.[17] Below temperatures of approximately 0.1 K, R_K is within a factor of three of that predicted by the acoustic-mismatch model, but the discrepancy increases by over an order of magnitude as the temperature is raised to 1 K and above. This is because the diffuse scattering rate increases with the average phonon frequency. Long-wavelength phonons do not "see" the surface inhomogeneities. This conclusion is supported, for example, by the phonon-spectroscopy experiments of Koblinger et al.,[18] which indicate an increased transmission through a metal/liquid−^{4}He interface above roughly 85 GHz.[19] Using a thermal-conductivity technique, Klitzner and Pohl[20] have observed 99% specular reflection of phonons from Si surfaces over the temperature range of 0.05 to 2 K, corresponding to 5–180 GHz. The surfaces were Syton polished, a standard preparation for Si wafers in the electronics industry.

A schematic plot of the transmission coefficient, $\bar{\alpha}$ (averaged over angle), at a solid/He interface is shown in Figure 2. The acoustic-mismatch model predicts a constant $\bar{\alpha}$ running across the bottom of this graph; i.e., the specular transmission coefficient is small and independent of frequency. The actual rise

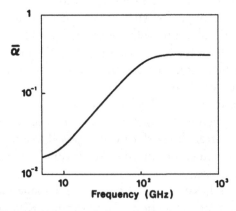

Figure 2 Schematic drawing of the typical transmission coefficient at a boundary as a function of phonon frequency. Due to the large acoustic mismatch between the solid and liquid He, specular transmission at the surface predicts $\bar{\alpha} \cong 10^{-2}$, independent of frequency. The large increase in $\bar{\alpha}$ at higher frequencies represents the Kapitza anomaly, which is believed to arise from diffuse scattering at the boundary. (From Wyatt.[3])

in $\bar{\alpha}$ at higher frequencies is due to diffuse scattering at the interface. The relation between R_K and $\bar{\alpha}$ is derived in Wyatt's review.[3]

B. Solid/solid interfaces

The situation for solid/solid interfaces is presently less well resolved than that for solid/liquid−[4]He interfaces. Again the acoustic-mismatch model fails to predict the thermal boundary resistance in many cases.[4,6] For example, using phonon frequencies in the range 280 GHz to 1 THz, Marx and Eisenmenger[21] found that acoustically well-matched metal films on a Si surface display *more* reflection than expected from acoustic-mismatch theory, and poorly matched inert gas layers show *less* reflection than expected. Indeed, all these films on polished and etched Si showed roughly the same reflectivity (30–60%), indicating to the authors that nearly 98% of the incident phonons were scattered diffusely at the surface.

This conclusion is similar to the one drawn from the He experiments described above. In those experiments, the key to resolving the specular and diffuse components was to measure the angular distribution of the phonons scattered at the surface. Thus, one may hope to learn more about the solid/solid interface scattering by using phonon-imaging techniques.

Indeed, informative imaging experiments have been conducted with solid/solid interfaces, which we will describe in Chapter 12. In particular, total internal reflection at specially prepared solid/solid interfaces has been observed, as well as evidence for frustrated total internal reflection from a thin film.[22] The observation of critical cones have provided a stringent test of specular transmission and reflection.[23] As in the He case, diffuse backgrounds can be quantitatively measured to give the ratio of diffuse to specular scattering.

Refraction and reflection at a solid/solid boundary is much more complicated (and interesting) for acoustic waves than for electromagnetic waves. Phonons can have longitudinal polarization, and thus mode conversion can occur between compression (L) and shear (T) waves. Hence, a phonon incident at an interface between two anisotropic materials produces as many as three transmitted waves and three reflected waves (trirefringence). Photon refraction is, at most, birefringent. In addition, phonon focusing reveals additional parameters of the problem, since the caustics allow one to positively identify phonons of given polarization and wavevector.

We shall see that the caustics survive the reflection and transmission process – if it is *specular* (i.e., obeying Snell's law). Indeed, new sets of caustics are created by mode conversion, allowing the experimenter to isolate particular mode-conversion events (e.g., L to ST or ST to FT). The mode conversion processes have been demonstrated graphically by phonon reflection imaging of sapphire[24] and Si[25] surfaces. In addition, as in the case of the solid/liquid interfaces, both specular and diffuse components have been identified by their differing angular distributions.

Phonon imaging has also shed some new light on the transmission across metal/crystal interfaces. Ordinary metal/crystal interfaces may have an intermediate layer of adsorbates, from the air or the surface treatment, which can affect the phonon transmission. Such an effect is seen in the phonon images produced with a metal-film heater on highly polished sapphire; these images display a sharp cone of increased phonon flux that cannot be attributed to phonon focusing in the bulk of the crystal.[23] The new structure is associated with an L mode propagating along the interface that couples to a T mode in the bulk, an effect termed critical-cone channeling.

To understand and appreciate these imaging measurements, one must have some familiarity with the problem of interface scattering between anisotropic media. Therefore, in this chapter we present a brief review of specular refraction at a boundary and discuss the phenomena of transmission coefficients, critical cones, and mode conversion. The elements of acoustic refraction at boundaries can be found in the texts by Auld[26] and by Aki and Richards.[27] We conclude this chapter with a discussion of imaging experiments on a solid/liquid interface. In Chapter 12 we will consider experimental measurements of phonon interactions at solid/solid interfaces.

C. Acoustic-mismatch model

We consider the refraction of acoustic waves at a perfectly bonded planar interface. The interface is assumed to be smooth, and continuum elasticity theory will be applied to the two media.

The principal boundary condition assumed at the planar interface is that the traction force (i.e., force per area applied to the surface) on one medium is equal to and opposite the traction force on the other. This is simply Newton's third law, assuming that the interface has no inertia. (Deviations from this assumption will be considered in Chapter 12.) By the definition of traction force [Eq. (6) on page 40], this means that, at the boundary,

$$\hat{n} \cdot \sigma = \hat{n} \cdot \sigma', \tag{1}$$

where $\hat{n}$ is a unit vector normal to the boundary and σ and σ' are the stress tensors in the two media. In the following examples, we will take $\hat{n}$ to be along the z axis and place the interface at $z = 0$. In cases where crystalline anisotropy is included, our examples will assume that this is the (001) symmetry plane of a cubic crystal. From the definition of the traction force [Eq. (6) and Figure 14 on page 40 and 39],

$$\hat{z} \cdot \sigma \equiv \hat{x}\sigma_{xz} + \hat{y}\sigma_{yz} + \hat{z}\sigma_{zz}, \tag{2}$$

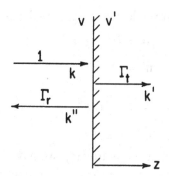

Figure 3 Transmission and reflection of an elastic wave that is normally incident at a boundary between two solids.

we see that σ_{xz}, σ_{yz}, and σ_{zz} must be continuous across the boundary. This provides three boundary-condition equations, associated with the gradient of the displacement, because $\sigma = C \cdot \varepsilon$.

Further boundary conditions that relate the displacements of the two media at the boundary may be assumed. For a perfectly rigid contact between the media,

$$\mathbf{u}(z = 0) = \mathbf{u}'(z = 0), \tag{3}$$

whose components provide three boundary conditions. In Section E we will consider a relaxation of this condition – i.e., a spongy interface. From the discussion on Kapitza conductance, the choice of boundary conditions should not be taken lightly. Nevertheless, we begin with these simple assumptions.

i) Isotropic solids

The stiffness tensor for an isotropic solid is the same as that for a cubic crystal [Eq. (10) on page 92], with the additional relation between the elastic constants, $C_{44} = (C_{11} - C_{12})/2$, that is, $\Delta = 0$, [Eq. (29) on page 96]. We separately consider refraction at a boundary with normal and oblique incidence.

Normal incidence: Figure 3 shows this simple case. A longitudinal wave with unit amplitude is incident on the boundary from the left. Interaction with the surface produces a transmitted longitudinal wave of amplitude Γ_t and a reflected longitudinal wave of amplitude Γ_r. The problem is to determine these amplitudes.

By symmetry, only the σ_{zz} element is nonzero. It is simply related to the strain component ε_{zz} by

$$\sigma_{zz} = C_{11}\varepsilon_{zz}. \tag{4}$$

The total displacement in the left and right media are

$$\mathbf{u} = \hat{z}e^{ikz} + \hat{z}\Gamma_r e^{-ik''z},$$

$$\mathbf{u}' = \hat{z}\Gamma_t e^{ik'z}, \qquad (5)$$

so

$$\varepsilon_{zz} = du_z/dz = ike^{ikz} - ik''\Gamma_r e^{-ik''z},$$

$$\varepsilon'_{zz} = du'_z/dz = ik'e^{ik'z}. \qquad (6)$$

Setting the stresses equal at the boundary, we have

$$C_{11}k - C_{11}k''\Gamma_r = C'_{11}k'\Gamma_t$$

or $\qquad (7)$

$$Z'\Gamma_t + Z\Gamma_r = Z,$$

with the definition of the longitudinal acoustic impedance,

$$Z \equiv \rho v = C_{11}/v = C_{11}k/\omega = C_{11}s, \qquad (8)$$

where $v^2 = C_{11}/\rho$ from Chapter 4 and $v = 1/s$. The second boundary condition, obtained by matching the displacement amplitudes at the boundary, is

$$1 + \Gamma_r = \Gamma_t. \qquad (9)$$

Combining Eqs. (7) and (9), we find,

$$\Gamma_t = 2Z/(Z + Z'),$$

$$\Gamma_r = (Z - Z')/(Z + Z'). \qquad (10)$$

From Eq. (35) on page 49, the power transmission associated with a plane acoustic wave with amplitude u_0 is

$$P = \mathbf{e} \cdot C \cdot \mathbf{ke}\,\omega\,u_0^2/2. \qquad (11)$$

In our case using the abbreviated subscripts introduced in Eqs. (9) and (10) of Chapter 2, where C is a 6×6 matrix [Eq. (10) on page 41] and σ and $\mathbf{ke}$ are six-element column vectors (see Appendix II),

$$\mathbf{e} = \begin{bmatrix} 0 \\ 0 \\ 1 \end{bmatrix}, \qquad \mathbf{k} = \begin{bmatrix} 0 \\ 0 \\ K \end{bmatrix} \qquad (12)$$

implying

$$\mathbf{ke} = \begin{bmatrix} 0 \\ 0 \\ k \\ 0 \\ 0 \\ 0 \end{bmatrix}, \qquad C \cdot \mathbf{ke} = \begin{bmatrix} C_{12} \\ C_{12} \\ C_{11} \\ 0 \\ 0 \\ 0 \end{bmatrix} k, \qquad (13)$$

and, with Eq. (5) of Appendix II,

$$\mathbf{e} \cdot C \cdot \mathbf{ke} = \hat{x}0 + \hat{y}0 + \hat{z}C_{11}k.$$

Therefore,

$$\mathbf{P} = C_{11}k\omega u_o^2 \hat{z}/2 = Z\omega^2 u_o^2 \hat{z}/2. \qquad (14)$$

If the wave amplitude in the first medium (impedance Z) is u_o, then the amplitude in the second medium (impedance Z') is $u_o\Gamma_t$. Using Eqs. (10) and (14), the ratio of transmitted power to incident power, known as the transmission coefficient, is found to be

$$\alpha \equiv (\mathbf{P}_t \cdot \hat{z})/(\mathbf{P}_{inc} \cdot \hat{z}) = (Z'/Z)\Gamma_t^2 = 4ZZ'/(Z + Z')^2. \qquad (15)$$

We can apply this result to the Kapitza problem already discussed. For an Al/He interface, the densities are 2.7 and 0.14 g/cm^3, respectively. The sound velocities are 6.4×10^5 and 0.24×10^5 cm/s; hence, the ratio of acoustic impedances, Z'/Z, is approximately 500, and $\alpha \cong 0.01$, as previously stated. By this model, only 1% of the longitudinal phonons normally incident on the interface from the Al will be transmitted into the liquid He. The commonly observed transmission coefficient of approximately 50% is indeed an anomaly.

If we had considered instead an incident shear wave with polarization along y, the solution to this problem would be formally the same as above, except the impedances would be those associated with a shear wave, $Z_s = C_{44}/v_s$.

Oblique incidence: The next simplest case involves a y-polarized shear wave incident in the (sagittal) x–z plane at an angle θ with respect to the surface normal, as shown in Figure 4. There is no coupling to longitudinal waves, or to transverse waves with polarization other than y. (In the parlance

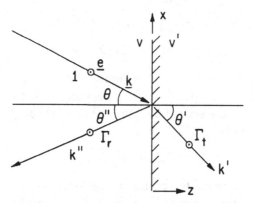

*Figure 4 Refraction (transmission at a deflected angle) and reflection of an elastic wave with wavevector **k** and polarization **e** parallel to the boundary between the solids. The case shown corresponds to $v' > v$.*

of acoustics, this is the horizontally polarized case.) The nonzero boundary condition is,

$$\sigma_{yz} = \sigma'_{yz}, \tag{16}$$

which implies

$$C_{44}\varepsilon_{yz} = C'_{44}\varepsilon'_{yz}. \tag{17}$$

Using $\varepsilon_{yz} = (\partial u_y/\partial z + \partial u_z/\partial y)/2$, we have

$$C_{44}k_z e^{i\mathbf{k}\cdot\mathbf{r}} + C_{44}k''_z \Gamma_r e^{i\mathbf{k}''\cdot\mathbf{r}} = C'_{44}k'_z \Gamma_t e^{i\mathbf{k}'\cdot\mathbf{r}} \tag{18}$$

at $z = 0$. Defining $\mathbf{k} = k_\parallel \hat{x} + k_\perp \hat{z}$, we find

$$C_{44}k_\perp e^{ik_\parallel x} + C_{44}k''_z \Gamma_r e^{ik''_\parallel x} = C'_{44}k'_\perp \Gamma_t e^{ik'_\parallel x}. \tag{19}$$

This equation can be satisfied for all x only if

$$k_\parallel = k'_\parallel = k''_\parallel, \tag{20}$$

which, with $k_\parallel = (\omega/v)\sin\theta$, is the basis for Snell's law of refraction,

$$\frac{\sin\theta}{v} = \frac{\sin\theta'}{v'}, \tag{21}$$

and $\theta = \theta''$. With $k_\perp = (\omega/v)\cos\theta$ and $C_{44} = vZ_s$, Eq. (19) becomes

$$Z'_s \Gamma_t \cos\theta' + Z_s \Gamma_r \cos\theta = Z_s \cos\theta. \tag{22}$$

Setting the left and right amplitudes equal at $z = 0$, we get

$$1 + \Gamma_r = \Gamma_t. \tag{23}$$

Solving Eqs. (22) and (23), we have

$$\Gamma_t = (2Z_s \cos\theta)/(Z_s \cos\theta + Z'_s \cos\theta')$$
$$\Gamma_r = (Z_s \cos\theta - Z'_s \cos\theta')/(Z_s \cos\theta + Z'_s \cos\theta'). \tag{24}$$

A plot of these results is shown in Figure 5(a) for the case $v' > v$. In this case, there is a critical angle of incidence beyond which it is not possible to transmit a wave into the second medium. The critical angle is determined by $\theta' = \pi/2$, which, with Eq. (21), leads to,

$$\sin\theta_c = v/v'. \tag{25}$$

For incident angles beyond the critical angle, there is total internal reflection, as illustrated with the constant-frequency surfaces in Figure 5(b). For the case of the slower medium on the right, we find $\sin\theta'_c = v'/v$, which gives a critical cone in the transmitted medium, as was described previously for phonon emission into liquid He.

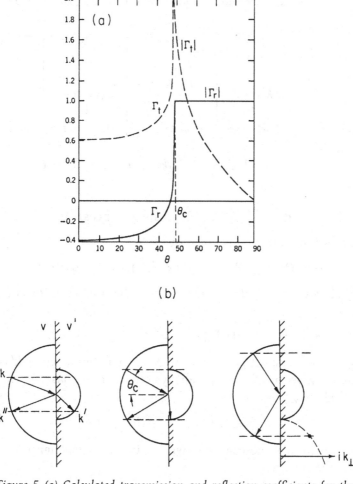

Figure 5 (a) Calculated transmission and reflection coefficients for the shear-horizontal case illustrated in Figure 4. (From Auld.[26]) (b) Explanation of the critical cone. Half-circles represent wavevectors of constant frequency in the two isotropic solids (slowness surfaces). By Snell's law, the components of wavevectors parallel to the surface are equal for incident, reflected, and transmitted waves. As the wavevector angle (with respect to the normal) is increased, it reaches a critical angle, θ_c, beyond which it is impossible to have $k_{\parallel}$ conservation with a bulk wave in the second medium. For $\theta > \theta_c$, the transmitted power goes to zero (total internal reflection), but an evanescent wave with exponentially decaying intensity, $\exp(-\beta z)$, exists in the second medium.

Remarkably, the transmission amplitude Γ_t does not go to zero at the critical angle. Equations (24) show that $\Gamma_t = 2$ and $\Gamma_r = 1$ at the critical angle. This does not mean that the transmitted power is greater than the incident power. Indeed, the power transmitted into the right medium is zero because the wave is propagating along the surface. Specifically, for an incident angle θ, we have

$\mathbf{k}' = -\hat{x}k'\sin\theta' + \hat{z}k'\cos\theta$ and $\mathbf{e}' = \hat{y}$, so with Eqs. (5), (10), and (11) on page 403-404,

$$k'e' = \begin{bmatrix} 0 \\ 0 \\ 0 \\ k'\cos\theta' \\ 0 \\ -k'\sin\theta' \end{bmatrix}, \quad C \cdot k'e' = \begin{bmatrix} 0 \\ 0 \\ 0 \\ C_{44}k'\cos\theta \\ 0 \\ -C_{44}k'\sin\theta \end{bmatrix}, \tag{26}$$

$$\mathbf{e}' \cdot C \cdot \mathbf{k}'e' = -\hat{x}C_{44}k'\sin\theta' + \hat{z}C_{44}k'\cos\theta,$$

and

$$\mathbf{P}_t = Z_s'\omega^2\Gamma_t^2 u_o^2/2(-\hat{x}\sin\theta' + \hat{z}\cos\theta'). \tag{27}$$

The transmission coefficient is given by

$$\alpha = (\mathbf{P}_t \cdot \hat{z})/(\mathbf{P}_{inc} \cdot \hat{z}) = (Z_s'/Z_s)(\cos\theta'/\cos\theta)\Gamma_t^2, \tag{28}$$

which equals zero at $\theta' = 90°$. One can show that the reflection coefficient is given by

$$R = [\mathbf{P}_r \cdot (-\hat{z})]/(\mathbf{P}_{inc} \cdot \hat{z}) = \Gamma_r^2 = 1 \quad \text{at } \theta' = 90°. \tag{29}$$

At angles $\theta > \theta_c$, Eq. (19) is satisfied if the normal component of $\mathbf{k}'$ is allowed to be imaginary, because $k_{||} = k_{||}'$ is always required for phase matching at the boundary. That is,

$$k_\perp' = [k_{||}'^2 - (\omega/v')^2]^{1/2} \equiv i\beta, \tag{30}$$

corresponding to an evanescent wave with an intensity exponentially decaying away from the surface,

$$\mathbf{u}' = \hat{y}\Gamma_t e^{ik_{||}y}e^{-\beta z}. \tag{31}$$

The extinction coefficient, $\beta = -ik_\perp'$, is a positive real number that depends on $k_{||}$ as given in Eq. (30). The hyperbolic function [Eq. (30)] is plotted as the dashed curve in the right diagram of Figure 5(b), which is a constant-frequency surface for the evanescent wave.

The amplitudes, Γ_r and Γ_t, of the reflected and transmitted waves at the surface are imaginary when $\theta > \theta_c$, implying a phase shift relative to the incident wave. The amplitudes of these waves, $|\Gamma_r|$ and $|\Gamma_t|$, are plotted in Figure 5(a), showing the total internal reflection, $|\Gamma_r| = 1$, and the finite evanescent wave intensity $|\Gamma_t|$, which decreases with increasing θ.

As we have seen, the evanescent wave has a large intensity at $z = 0$; however, it can extract energy away from the incident wave only if the second medium is dissipative. If, however, the second medium is a thin film covered by a softer third medium, an evanescent wave in the film can leak acoustic

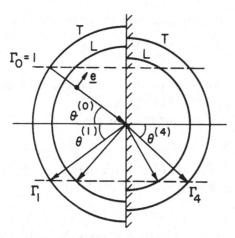

Figure 6 Transmission and reflection at the boundary between two isotropic solids when the incident wave has a polarization, e, in the saggital plane (the plane containing the normal to the surface and the incident k vector). The initial wave, shown here as transverse, creates reflected and transmitted waves of both polarizations, with wavevectors determined by the slowness surfaces and $k_{\parallel}$ conservation (Snell's law). This is an example of mode conversion at an interface.

energy into the third medium, a process known as frustrated total internal reflection.

Finally, we consider the case in which the polarization e of the incident shear wave is not parallel to the interface. An example in which e lies in the saggital plane containing the wavevectors is shown in Figure 6. In this vertically polarized case it is necessary to consider both L and T modes, because the incident wave can couple to both longitudinal and transverse waves in the two media. In the isotropic case where the transverse modes are degenerate, there will be two reflected waves and two transmitted waves, as shown. The total displacements in the left and right media are,

$$\mathbf{u} = \mathbf{e}^{(0)} e^{i\mathbf{k}^{(0)}\cdot\mathbf{r}} + \mathbf{e}^{(1)}\Gamma_1 e^{i\mathbf{k}^{(1)}\cdot\mathbf{r}} + \mathbf{e}^{(2)}\Gamma_2 e^{i\mathbf{k}^{(2)}\cdot\mathbf{r}},$$
$$\mathbf{u}' = \mathbf{e}^{(3)}\Gamma_3 e^{i\mathbf{k}^{(3)}\cdot\mathbf{r}} + \mathbf{e}^{(4)}\Gamma_4 e^{i\mathbf{k}^{(4)}\cdot\mathbf{r}}, \tag{32}$$

where the superscripts label the waves as indicated in Figure 6. For isotropic media, the polarization vectors are

$$\mathbf{e}^{(i)} = \mathbf{k}^{(i)}/k^{(i)} \qquad \text{(L modes)},$$
$$\mathbf{e}^{(i)} = \mathbf{k}^{(i)} \times \hat{y}/k^{(i)} \qquad \text{(T modes)}, \tag{33}$$

with

$$\mathbf{k}^{(i)} = k^{(i)}[\hat{x}\sin\theta^{(i)} + \hat{z}\cos\theta^{(i)}]. \tag{34}$$

The scattering angles in this equation are determined by Snell's law,

$$\frac{\sin \theta^{(o)}}{v^{(o)}} = \frac{\sin \theta^{(i)}}{v^{(i)}}, \tag{35}$$

where the velocities for these pure modes are given in terms of C_{11} and C_{44} for the two media, as in the previous cases.

The boundary conditions for this case are

$$\sigma_{zx} = \sigma'_{zx}, \quad \sigma_{yz} = \sigma'_{yz}, \quad u_x = u'_x, \quad u_y = u'_y \tag{36}$$

at $z = 0$, which are sufficient to determine the four reflection and transmission amplitudes, Γ_n. These four boundary condition equations take the form

$$A_{in}\Gamma_n = B_i, \tag{37}$$

where A_{in} and B_i contain the elastic constants C_{11} and C_{44} for both media, the four sound velocities, and the incident angle $\theta^{(o)}$. From the general form of the stiffness tensor, C_{ijlm} [related to the contracted C_{IJ}'s by Eq. (9) on page 41], one can write a general expression for the boundary condition coefficients,[28]

$$A_{in} = C_{zipq}e_p^{(n)}s_q^{(n)},$$
$$B_i = -C_{zipq}e_p^{(o)}s_q^{(o)}, \tag{38}$$

where $s = k/\omega$ and the summation over repeated indices $p, q = 1, \dots 3$ is implied. These factors have the dimension of impedance, analogous to Eqs. (8), (19), and (22).

The solution to the four linear equations (37) is given by[28]

$$\Gamma_n = \bar{A}_{nm}B_m/|A_{pq}|, \tag{39}$$

where $\bar{A}_{nm}$ is the cofactor of A_{mn} and $|A_{pq}|$ is the boundary condition determinant. Even for the isotropic case of Figure 6, the solution for the scattering amplitudes is best computed numerically on a computer. The ability to calculate the amplitudes of the reflected and transmitted waves is crucial to any predictions of the behavior of phonons near a boundary.

ii) Anisotropic solids

The formal solution for the scattering amplitudes in anisotropic media are also obtained from Eqs. (38) and (39). However, the boundary-condition matrix in this case is a 6×6 matrix because there are three possible transmitted and three possible reflected waves, as illustrated in Figure 7.

In the isotropic case with saggital plane polarization, we could simply write down the scattered slowness vectors $[s_q^{(n)}]$ and polarizations $[e_q^{(n)}]$, knowing the incident angle and Snell's law. In the anisotropic case, we are likewise given $s = k/\omega$ of the incident wave. By Snell's law, all the scattered slowness vectors have the same $s_\| = (s_x, s_y)$ as the incident wave. The determinant of the Christoffel equations [Eq. (22) on page 44] fixes the third component s_z

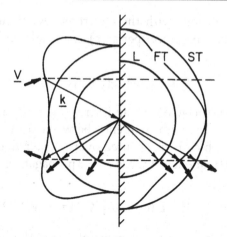

Figure 7 Transmission and reflection at a boundary between two anisotropic solids. In the case shown, the initial transverse wave creates three reflected and three transmitted waves. The light arrows are wavevectors, and the heavy arrows are group-velocity vectors.

for each scattered mode. (A practical algorithm for accomplishing this task is given by Kinder and Weiss.[29]) The polarization vectors $e_q^{(n)}$ for each mode are then determined by solving the Christoffel equations. Given these inputs, Eqs. (38) and (39) provide the amplitudes of the six scattered waves that are phase matched at the crystal surface.

D. Emission of phonons from a planar source

To compute a phonon-focusing pattern for a crystal, it is common to choose a uniform angular distribution of slowness vectors emanating from the heat source. This choice is equivalent to assuming a spherical heat source and has been assumed for most of the Monte Carlo images shown in this book. In reality, the experimental heat sources are planar metal films deposited on the crystal, although, at sufficient excitation densities, a hot spot may be formed in the crystal that approximates a spherical source.* The directions of the phonon-focusing caustics are the same for either spherical or planar phonon source geometry, but the relative intensities in a focusing pattern will depend on the shape of the source. In phonon refraction experiments, we wish to make quantitative measurements of the heat-pulse intensities as a function of angle, so theoretical simulations must choose an initial distribution of phonon wavevectors that best reflects the experimental conditions.

A polycrystalline metal film is generally approximated as an elastically isotropic medium. Phonons created by heating such a film by laser or electrical excitation are assumed to exhibit an isotropic distribution of phonons that

* Direct photoexcitation of a semiconductor may also produce a near-spherical source of acoustic phonons.

undergo frequent collisions with the electrons. As discussed in Chapter 2 [page 49], the energy flux (power per area) of a single acoustic wave is given by

$$\mathbf{P} = \mathbf{V} \cdot (\text{energy density}). \tag{40}$$

The energy density associated with an isotropic distribution of phonons in a given mode (L, ST, or FT) with frequency between ω and $\omega + d\omega$ is equal to $[(\hbar \omega D(\omega)\bar{n}(\omega)]d\omega$, which is distributed over 4π steradians. Therefore, the energy flux per unit frequency of those waves propagating within a solid angle $d\Omega$ is given by,

$$d\mathbf{P} = \mathbf{V}[\hbar \omega D(\omega)\bar{n}(\omega)]d\Omega/4\pi = \mathbf{V}\mathcal{U}d\Omega, \tag{41}$$

where $D(\omega) = \omega^2/2\pi^2 v^3$ is the density of states [Eq. (66) of Chapter 8], v is the phase velocity of the mode, $\bar{n}(\omega)$ is the phonon occupation number [Eq. (1) on page 21], and $\mathcal{U} = \hbar \omega^3 \bar{n}/8\pi^3 v^3$. We note that $V = v$ in the isotropic medium and consider a subset of waves striking the surface $\hat{n}A$ at (or near) a wavevector angle θ. The spectral power per unit solid angle that strikes this surface is

$$I = d\mathbf{P} \cdot \hat{n}A/d\Omega = v A \mathcal{U} \cos \theta, \tag{42}$$

where θ is the incident angle of the subset of phonons under consideration. This cosine law of illumination is the same factor that occurs for the flux of particles or photons escaping from a box with a small aperture. In this case, however, it represents the flux incident on the surface of the crystal. To determine the flux emitted into the crystal, we must take into account the effect of the interface. For a smooth interface with specular refraction, θ' and $\Delta\Omega'$ of the emitted beam are related to θ and $\Delta\Omega$ by Snell's law, $\sin \theta'/v' = \sin \theta/v$. The spectral power per (k-space) solid angle emitted into a particular phonon mode of the crystal is

$$I' = I\alpha(\Delta\Omega'/\Delta\Omega), \tag{43}$$

where the transmission coefficient (phonon transmission probability), $\alpha = \alpha(\theta)$, defined by Eq. (15), depends on the elasticities and densities of the two media* and may be expressed in terms of θ' using Snell's law.

Some insight into this problem is gained by assuming that the crystal is elastically isotropic. In that case, the slowness surfaces in both media are spheres with radii $s = 1/v$ and $s' = 1/v'$, as illustrated in Figure 5(b). We

---

* For example, we showed in Section C [Eq. (28)] that for oblique incidence of a y-polarized wave, $\alpha = (Z'_s/Z_s)(\cos \theta'/\cos \theta)\Gamma_t^2$, where Γ_t^2 is given by Eq. (24) and plotted in Figure 5(a). A general treatment of the transmission coefficient for anisotropic media is given by Weis.[31] His generalized transmission coefficient is denoted $t_{\sigma\tau}^{(10)}$, where 1 and 0 label the two media and σ and τ represent the modes in these media.

have

$$\Delta\Omega'/\Delta\Omega = \sin\theta' d\theta' d\phi' / \sin\theta d\theta d\phi = (v/v')^2 \cos\theta'/\cos\theta, \qquad (44)$$

where the last step is obtained by setting $d\phi' = d\phi$ and differentiating Snell's law. Combining this result with Eqs. (42) and (43), we notice that the cosine factor associated with the metal film cancels out, leaving

$$I' = \alpha(v/v')^2 Av\mathcal{U}\cos\theta'. \qquad (45)$$

So, for specular scattering at a perfect interface, a cosine factor in the phonon flux has carried over into the second medium.* The angular dependence of this transmitted k-space distribution, however, also includes the transmission coefficient $\alpha = \alpha(\theta)$. To obtain the transmitted intensity into a particular mode in the (isotropic) crystal, we must sum Eq. (45) over all three contributing modes in the metal. The transmission factors and phase velocities will differ for each incident mode, but the $\cos\theta'$ factor is common to all three.

Rösch and Weis[30] have published detailed calculations of the k-space flux transmitted into anisotropic crystals from metal-film radiators.† An example of their results is shown in Figure 8(a). At the left is depicted the flux transmitted from a constantan film into a sapphire crystal, assuming isotropic crystal velocities equal to the average L and T mode velocities. In this case, the $\cos\theta'$ factor dominates the angular distribution. The two polar plots at the right show the emitted flux when the anisotropy of the crystal is properly accounted for. Figure 8(b) contains renditions of the three-dimensional k-space flux emitted from a constantan film into a (100)-oriented Ge crystal. These distributions are not nearly as anisotropic as the phonon-focusing effect (e.g., there are no singularities), but they are a significant deviation from the simple $\cos\theta'$ dependence, especially at large angles.

In calculating a phonon image by the Monte Carlo method, it is not necessary to separately compute the k-space flux emitted into the crystal from a metal radiator. The cosine factor in Eq. (42) implies that the phonons incident on the crystal interface have a *uniform distribution of* $s_\parallel$ in the plane of the interface, because θ is also the projection angle of an element of the slowness surface onto the interface. Snell's law for a specular surface dictates that $s_\parallel$ is conserved across the interface; therefore one begins the calculation by assuming a uniform distribution of $s_\parallel$ in the plane. Given $s_\parallel$ and the slowness surface for the crystal, s for the transmitted phonon is determined.[29] The transmission coefficient is calculated by the formalism outlined in the last section. The group velocity $\mathbf{V}'$ of the wave with slowness s is determined by

---

* This is because the incident flux has a uniform distribution of $s_\parallel$'s and the slowness surfaces in the second medium are spherical, as discussed below.

† A detailed theoretical analysis of the mismatch problem is given by Weis,[31] and quantitative experiments – along symmetry axes – with metal-film radiators and bolometers have been reported by Müller and Weis.[32]

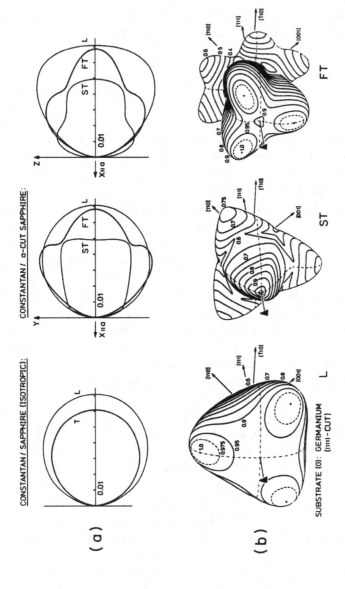

Figure 8 (a) Angular distribution of flux emitted from a metal film into a sapphire crystal, assuming specular transmission. An isotropic distribution of wavevectors in the heated film is converted by Snell's law into an anisotropic distribution of wavevectors in the crystal. Left: assuming isotropic L and T velocities equal to the average corresponding velocities in sapphire. Right: assuming the full elastic anisotropy of sapphire. (b) Calculated L and ST phonon flux emitted from a metal film into Ge. (From Rösch and Weis.[30])

the methods of Chapter 2, and the phonon is propagated along this direction to the opposite side of the crystal.

In a slab geometry, where the source and detector are located on opposite parallel faces of the crystal, the spectral power incident on the detector of area $\mathcal{A}'$ located a distance r from the source described above has the following dependence:

$$P_d = I'A(\mathcal{A}'\cos\theta'_v/r^2). \tag{46}$$

The quantity in parentheses is the solid angle subtended by the detector from the source, and the enhancement factor, $A = d\Omega'/d\Omega'_v$ [Eq. (42) on page 51], transforms the real-space solid angle to the k-space solid angle. Notice that the cosine factor contains the real-space angle, θ'_v, rather than the k-space angle, θ'. In a Monte Carlo calculation of the phonon flux, these factors are automatically taken care of by counting the phonons that propagate from the source with group velocity V'(along r to within a solid angle $\Delta\Omega'_v$) and intercept the area $\mathcal{A}'$. To determine the phonon power transmitted into the detector, one must calculate the transmission probability of the phonons into the detector film. This is straightforward, as $s_\parallel$ is already known.

If the surface between the metal-film generator and the crystal is rough on a scale larger than the phonon wavelength, specular transmission may still apply, but the distribution of surface normals will tend to average out the anisotropies in I' discussed above. However, it is known from optics that light *diffusely* scattered from a rough surface generally exhibits a cosine-like angular dependence, known as Lambert's law.[33] In that case, the microscopic processes include the multiple scattering of light with the surface or absorption and re-emission of photons, so the physical origin of the cosine dependence is different from that in Eq. (42). Nevertheless, it seems that a cosine factor in transmitted k-space flux may appear for both specular and diffuse scattering at the metal/crystal interface. In the specular case, additional anisotropies accrue due to the angular dependence of the transmission coefficient [e.g., $\alpha(\theta)$ in Eq. (43)].

E.   Model for the Kapitza anomaly in the helium/solid case

In the introduction of this chapter, we discussed the connection between the anomalous Kapitza conduction and diffuse scattering at the interface. Kinder and Weiss[29] have recently considered in detail the Kapitza anomaly in terms of a thin layer of lossy material separating the solid and liquid He. Their calculations, and the associated spatial imaging experiments of Kinder et al.,[34] provide important insights into this system and will be briefly summarized in this section.

The model assumes that the lossy film of atoms or molecules at the interface internally scatters the phonons – either elastically or inelastically –

randomizing their wavevectors. Thus, the diffusiveness of the surface arises from these scattering or absorption processes, not from roughness in the (two) interfaces, which are assumed to be smooth and obey acoustic-mismatch theory. Consequently, a uniform distribution of $s_\parallel$ is present in the three media: Si, lossy film, and liquid He.

It is assumed that an isotropic distribution of phonons is incident from the He. Recall from the experiment of Wyatt and Page (Figure 1) that only phonons within a small critical cone in He can cross a clean crystal boundary. A layer with losses relaxes this restriction because He phonons outside the critical cone can couple to evanescent waves in the solid, which are attenuated in the lossy medium.

By absorbing large-θ phonons from the He, the lossy layer is an effective channel of heat transfer. The angular distribution of the absorption of the film depends upon the loss mechanism. Two cases are considered within the context of classical continuum acoustics: (1) a complex elastic constant (arising from deformation-potential coupling to defects or force-constant fluctuations), and (2) complex mass density (arising from mass-defect scattering or density fluctuations).

Figure 9 shows the computed angular distribution of flux emitted from a point source of heat on the surface of interest and viewed on the far side of the (100)-oriented Si crystal. Figure 9(a) is for a perfectly lossy layer acting as a blackbody absorber and emitter (absorption independent of incident angle). The usual phonon-focusing pattern of Si is predicted. Figure 9(b) shows the opposite extreme of no intermediate layer, assuming specular refraction across the He/Si boundary. In this case, no phonons are transmitted into the FT ridges because the polarization of these phonons are in the plane of the boundary; phonons in the liquid are only longitudinally polarized, and the narrow critical cone means that there is little polarization component along the surface. Thus, the intensity of the transmitted FT mode is a sensitive indicator of the diffusive nature of the interface scattering.

Figure 9(c) assumes a layer with complex elastic constants. Compared to that in Figure 9(a), the central ST structure is reduced because the complex elastic constants have little effect on phonons with normal incidence. A ring of increased intensity (a halo) seen in the ST mode around the center indicates a critical cone channeling effect whereby the transverse phonons in the crystal couple to lossy longitudinal modes in the layer that propagate parallel to the interface. (See References 24 and 28 and Section B of Chapter 12.) Figure 9(d) is the flux pattern assuming a layer with complex mass density. This case resembles the blackbody case of Figure 9(a), but the critical cone is also present. The halo depends sensitively on film thickness.

Experiments performed by Kinder et al.[34] show some of these predictions. A small metal heater film was located 10–20 μm from the crystal surface in

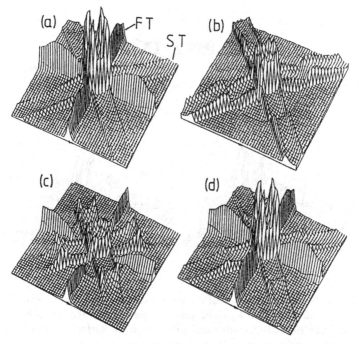

Figure 9 *Predicted phonon flux patterns from a point source of heat at the surface of a (100)-oriented Si crystal immersed in liquid He. (a) Assuming a perfectly absorbing and emitting layer between the Si and He. (b) Assuming no layer between Si and He and specular refraction. (c) Assuming elastic losses in the intervening layer. (d) Assuming inertial losses in the intervening layer. (From Kinder and Weiss.[29])*

the He bath and scanned laterally, as indicated in the schematic of Figure 10.* Figure 10(a) is the flux pattern observed for a normal surface, cleaned with acetone. The appearance of strong FT ridges and critical-cone halo shows that the scattering for this surface is diffuse. The pattern more nearly resembles the prediction for inertial losses than elastic losses. The surface in Figure 10(b) was specially etched and quickly isolated to minimize the contamination at the surface. In this case regions with negligible FT ridge transmission are observed, implying that there are clean regions that exhibit the predictions of a lossless surface.

Further transmission experiments have been performed on atomically clean (111) surfaces of Si by Wichert et al.[35] By examining the phonon reflection from such surfaces in both vacuum and liquid He, they estimated a transmission coefficient of only approximately 1%. Their transmission experiments

---

* The mean free path of phonons in the He at the chosen temperature and pressure is much smaller than $10\,\mu$m.

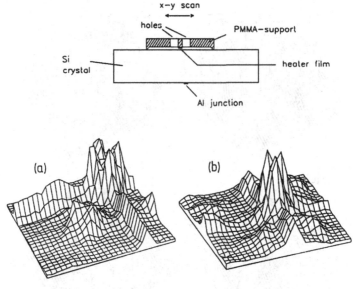

Figure 10 Top: Experimental setup for measuring phonon flux patterns in Si resulting from a localized source of phonons in the He bath. (a) Si surface cleaned with acetone. (b) Spin-etched surface mounted and evacuated within 10 min. (From Kinder et al.[34])

(like those described in Figure 10) show a clear transmission of L phonons normal to the surface, supporting the model for coupling by complex mass density. As expected, their experiments show extreme sensitivity of the anomalous Kapitza transmission to tiny amounts of adsorbates such as water vapor.

References

1. P. L. Kapitza, The study of heat transfer in helium II, *J. Phys. (USSR)* **4**, 181 (1941).

2. I. M. Khalatnakov, Heat exchange between a solid body and helium II, *Zh. Eksp. Theor. Fiz. (USSR)* **22**, 687 (1952) (in Russian); *An Introduction to the Theory of Superfluidity* (Benjamin, New York, 1965), Chapter 23.

3. A. F. G. Wyatt, Kapitza conductance of solid-liquid He interfaces, in *Nonequilibrium Superconductivity, Phonons and Kapitza Boundaries*, K. E. Gray, ed. (Plenum, New York, 1981).

4. A. C. Anderson, The thermal boundary resistance, in *Phonon Scattering in Solids*, L. J. Challis, V. W. Rampton, and A. F. G. Wyatt, eds. (Plenum, New York, 1976).

5. L. J. Challis, Kapitza conductance, in *The Helium Liquids*, J. G. M. Armitage, ed. (Academic, London, 1975); G. L. Pollack, Kapitza resistance, *Rev. Mod. Phys.* **41**, 48 (1969); T. Nakayama, Kapitza thermal boundary resistance and interactions of helium quasiparticles with surfaces, *Prog. Low Temp. Phys.* **12**, 115 (1989); E. T. Swartz and R. O. Pohl, *Rev. Mod. Phys.* **61**, 605 (1989).

6. A. C. Anderson, The Kapitza thermal boundary resistance between two solids, in *Nonequilibrium Superconductivity, Phonons, and Kapitza Boundaries*, K. E. Gray, ed. (Plenum, New York, 1981).

7. W. A. Little, The transport of heat between dissimilar solids at low temperatures, *Can. J. Phys.* **37**, 334 (1958).

8. A. F. G. Wyatt and G. J. Page, The transmission of phonons from liquid He to crystalline NaF, *J. Phys. C* **11**, 4927 (1978).

9. P. Taborek and D. Goodstein, Diffuse reflection of phonons and the anomalous Kapitza resistance, *Phys. Rev.* **22**, 1550 (1980).

10. H. C. Basso, W. Dietsche, and H. Kinder, Interaction of adsorbed atoms with phonon pulses, in *LT-17 (Contributed Papers)*, U. Eckern, A. Schmid, W. Weber, and H. Wühl, eds. (Elsevier, Amsterdam, 1984).

11. S. Burger, W. Eisenmenger, and K. Lassmann, Specular and diffusive scattering of high frequency phonons at sapphire surfaces related to surface treatment, in *LT-17 (Contributed Papers)*, U. Eckern, A. Schmid, W. Weber, and H. Wühl, eds. (Elsevier, Amsterdam, 1984).

12. E. Mok, S. Burger, S. Döttinger, K. Lassmann, and W. Eisenmenger, Effect of laser annealing on specular and diffuse scattering of 285 GHz phonons at polished silicon surfaces, *Phys. Lett.* **114A**, 473 (1986).

13. J. Weber, W. Sandmann, W. Dietsche, and H. Kinder, Absence of anomalous Kapitza conductance on freshly cleaved surfaces, *Phys. Rev. Lett.* **40**, 1469 (1978).

14. H. Kinder, A. De Ninno, D. Goodstein, G. Paternò, F. Scaramuzzi, and S. Cunsolo, Extinction of the Kapitza anomaly for phonons along the surface normal direction, *Phys. Rev. Lett.* **55**, 2441 (1985).

15. H. Kinder, Kapitza conductance of localized surface excitations, *Physica* **107B**, 549 (1981).

16. L. Koester, S. Wurdack, W. Dietsche, and H. Kinder, Resonant phonon scattering and Kapitza conductance induced by H_2O molecules on clean surfaces, in *Proceedings of the 18th International Conference on Low Temperature Physics*, Kyoto (1987).

17. R. E. Peterson and A. C. Anderson, The Kapitza thermal boundary resistance, *J. Low Temp. Phys.* **11**, 639 (1973); H. Haug and K. Weiss, *Phys. Lett.* **40A**, 19 (1972).

18. O. Koblinger, U. Heim, M. Welte, and W. Eisenmenger, Observation of phonon frequency thresholds in the anomalous Kapitza resistance, *Phys. Rev. Lett.* **51**, 284 (1983).

19. W. Klar and K. Lassmann, Investigation of the Kapitza anomaly by frequency-resolved phonon transport in silicon wafers, in *Proceedings of the 18th International Conference on Low Temperature Physics*, Kyoto (1987); *Jpn. J. Appl. Phys.* **26**, 369 (1987).

20. T. Klitsner and R. O. Pohl, Phonon scattering at silicon crystal surfaces, *Phys. Rev. B* **36**, 6551 (1987).

21. D. Marx and W. Eisenmenger, Phonon scattering at silicon crystal surfaces, *Z. Phys. B* **48**, 277 (1982).

22. C. Höss, J. P. Wolfe, and H. Kinder, Total internal reflection of high-frequency phonons: a test of specular refraction, *Phys. Rev. Lett.* **64**, 1134 (1990).

23. A. G. Every, G. L. Koos, and J. P. Wolfe, Ballistic phonon imaging in sapphire: bulk focusing and critical-cone channeling effects, *Phys. Rev. B* **29**, 2190 (1984).

24. G. A. Northrop and J. P. Wolfe, A determination of specular versus diffuse boundary scattering, *Phys. Rev. Lett.* **52**, 2156 (1984).

25. R. Wichard and W. Dietsche, Specular phonon reflection: refocusing and symmetry doubling, *Phys. Rev. B* **45**, 9705 (1992).

26. B. A. Auld, *Acoustic Fields and Waves in Solids, Volume II* (Wiley, New York, 1973).

27. K. Aki and P. G. Richards, *Quantitative Seismology: Theory and Methods* (Freeman, San Francisco, 1980).

28. A. G. Every, Pseudosurface wave structures in phonon imaging, *Phys. Rev. B* **33**, 2719 (1986); Phonon focusing in reflection and transmission, *Phys. Rev. B* **45**, 5270 (1992).

29. H. Kinder and K. Weiss, Kapitza conduction by thin lossy surface layers, *J. Phys. Condensed Matter* **5**, 2063 (1993).

30.   F. Rösch and O. Weis, Phonon transmission from incoherent radiators into quartz, sapphire, diamond, silicon and germanium within anisotropic continuum acoustics, Z. *Phys. B* **27**, 33 (1977).

31.   O. Weis, Phonon radiation across solid/solid interfaces within the acoustic mismatch model, in *Nonequilibrium Phonons in Nonmetallic Crystals*, W. Eisenmenger and A. A. Kaplyanskii, eds. (North-Holland, Amsterdam, 1986).

32.   G. Müller and O. Weis, Quantitative investigation of thermal phonon pulses in sapphire, silicon and quartz, Z. *Phys. B* **80**, 15, 25 (1990).

33.   M. Born and E. Wolf, *Principles of Optics*, 4th ed. (Pergamon, New York, 1970), p. 181.

34.   H. Kinder, K. H. Wichert, and C. Höss, Kapitza resistance: angular phonon distribution in the solid, in *Phonon Scattering in Condensed Matter VII*, M. Meissner and R. O. Pohl, eds. (Springer-Verlag, Berlin, 1993), p. 392.

35.   K. H. Wichert, C. Reimann, G. Blahusch, and H. Kinder, Kapitza Transmission through atomically clean silicon (111)-surfaces, in *Proceedings of 4th International Conference on Phonon Physics*, Sapporo, Japan (1995), *Physica B* **219** & **220**, 653 (1996).

12

Refraction and reflection at solid/solid interfaces – Experiment

Having examined the theoretical foundations for phonon scattering at an interface, we now turn to a variety of experiments that involve the interface between two solid materials. It is no surprise that the transmission and reflection of phonons at the boundary of two media have great practical significance, ranging from the dissipation of heat generated by an electrical device to the coupling of phonons into a thin-film detector.

As previously discussed, less-than-perfect interfaces can display more, or less, thermal transmission than that expected from the acoustic-mismatch theory. For the crystal/liquid-He and crystal/rare-gas-film cases, the thermal transmission is much greater than predicted. This is because *diffuse* phonon scattering at the interface increases the chances of transmission across the boundary.

In view of this, one might wonder if the detailed predictions developed in Chapter 11 for *specular* refraction and mode conversion at a solid/solid interface can be observed in practice. In particular, can one observe critical cones? ... total internal reflection? ... mode conversion? In this chapter, we examine phonon-imaging experiments that indeed show such effects of specular phonon scattering at a boundary. And, of course, we find that nature contains some surprises that our limited imaginations did not anticipate. Six different experimental systems are considered: (a) refraction at a specially prepared Ge/MgO interface, (b) transmission through an ordinary sapphire/metal-film interface, (c) internal reflection from high-quality sapphire and Si surfaces, with the crystals immersed in liquid He, and (d) two examples of transmission through *internal* interfaces: semiconductor superlattices and ferroelectric domains.

A. Phonon refraction at a solid/solid interface ──────────

For specular refraction of high-frequency phonons to occur at the interface between two solids, it seems reasonable to expect that the boundary must

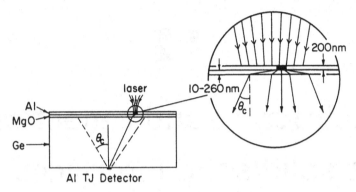

Figure 1 Schematic diagram illustrating the experiment for observing phonon refraction at a Ge/MgO interface.

be smooth on the scale of a phonon wavelength. Phonon wavelengths in a heat pulse are typically 10 to 30 nm, and, as previously discussed, imperfections at the boundary can easily cause diffuse scattering of these waves. For example, to reduce the damage and chemical contamination at the surface of a Si crystal, in situ laser annealing may be required. It has been observed that exposure of an annealed surface to air greatly increases the diffuse scattering.

One of the principal manifestations of specular scattering (or Snell's law) at an interface between two media is the existance of one or more critical cones. Just as in the case of NaF immersed in liquid He (Figure 1 on page 276), the transmitted phonons exhibit a critical cone when the phonons propagate from a fast medium into a slow medium. This condition is also satisfied for MgO films grown on Ge crystals, a system studied by Höss et al.[1] To remove the organic "dirt" usually present on the surface, they baked a highly polished Ge crystal at 700 K in vacuum for 4 h and deposited MgO films at a substrate temperature of 550 K. This procedure produces a polycrystalline film of MgO with approximately isotropic acoustic properties (C_{11} = 3.4 Mbar, C_{44} = 1.345 Mbar, and ρ = 3.595 g/cm^3). The elastic constants of Ge are given in Table 1 of Chapter 4.

A schematic diagram of the experiment is shown in Figure 1. A 200-nm film of Al is deposited on top of the MgO film to absorb the laser light and act as a Planckian source of phonons with average frequency of $\simeq$300 GHz (determined by the excitation intensity). An approximately uniform distribution of parallel slowness, $s_{\parallel}$, is created in the metal and transmitted into the MgO. These phonons propagate through the MgO film, are refracted at the MgO/Ge boundary, and propagate in the Ge. Phonons with frequency of $\leq$300 GHz have an isotope-scattering length that is longer than the 3-mm crystal thickness. Phonons are detected with an Al tunnel junction, which is sensitive to phonons above $2\Delta/h \simeq 140$ GHz.

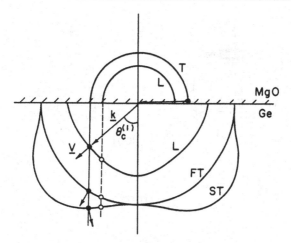

Figure 2 *Slowness curves of MgO and Ge in the (100) plane, showing refraction of an incident wave at the boundary and the resulting critical angles. (From Höss, et al.[1])*

What does one expect to observe in such an experiment? Slices of the MgO and Ge slowness surfaces are plotted in Figure 2. The L and T modes in the (isotropic) MgO film each produce three critical cones corresponding to the L, FT, and ST modes in Ge. The critical angle for T $\rightarrow$ L mode conversion in the (100) plane is denoted $\theta_c^{(1)}$. In this plane it is not possible to produce L phonons in Ge at an angle greater than this critical angle. Critical angles for other types of mode conversion in the (100) plane are defined by the dots and circles in Figure 2. In three dimensions the loci of critical angles form critical cones, as indicated in Figure 3 by solid lines on the Ge slowness surfaces. These k-space cones map into the group-velocity contours shown in Figure 4, which represent the intersections of critical-angle phonons (emanating from a point source) with the detection surface of the crystal. The real-space critical contours display loops due to folds in the group-velocity surface (i.e., phonon focusing). The dashed lines in this figure indicate the positions of the ST and FT caustics.

An experimental phonon image corresponding to Figure 1 with a 230-nm MgO film is shown in Figure 5(b). For comparison, the image for a crystal without the MgO film is shown in Figure 5(a). When the MgO film is present, both FT and ST modes show a rapid dropoff in intensity beyond the critical cones predicted in Figure 4.

A quantitative comparison between theory and experiment is made by performing a Monte Carlo calculation of the phonon flux, taking into account the specular refraction at Al/MgO and MgO/Ge interfaces. A uniform distribution of $s_\parallel$ in the Al film is chosen, and the transmission coefficients (Γ_n) of each phonon are computed as discussed in Section C of Chapter 11. The

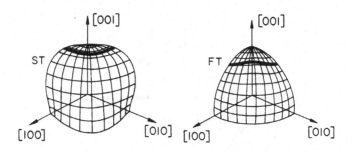

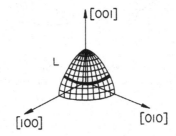

Figure 3 Slowness surfaces for the three modes in Ge, indicating with heavy lines three critical cones in k space for total internal reflection. (Adapted from Höss et al.[1])

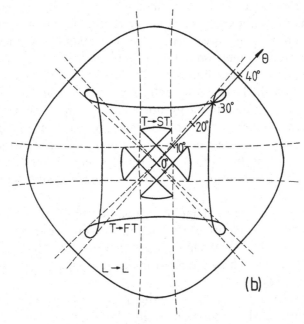

Figure 4 Heavy lines map three of the six critical cones of Ge/MgO onto the detection surface of the crystal. Dashed lines represent a superimposed caustic map of Ge. (Adapted from Höss et al.[1])

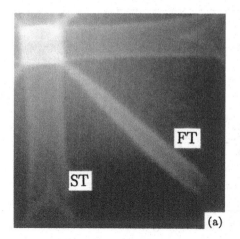 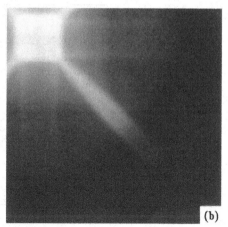

(a)   (b)

*Figure 5 (a) Phonon image of Ge without the MgO film. (b) Phonon image
of Ge with a 230-nm MgO film, showing an attenuation of heat flux beyond
the critical cones predicted in Figure 4. (From Höss et al.[1])*

phonons transmitted into the Ge crystal are propagated to the detection surface according to their computed group velocities.

Line scans of the computed flux along the FT ridge (arrow in Figure 4) are
plotted in Figure 6(b) and compared to the experimental data in Figure 6(a).
Because the intersection of the critical cone with the FT ridge produces a loop
in real space, the phonon intensity does not display an abrupt cutoff along the
ridge. Experiment and theory are in agreement for the 230- and 0-nm MgO
films, clearly demonstrating the phenomenon of total internal reflection at
propagation angles exceeding the critical angle. A small amount of ballistic
signal is observed beyond the critical angle and ascribed to a small component
of diffuse scattering at the MgO/Ge interface. It is expected that the diffuse
scattering would produce a distribution similar to the the no-film case. The
shaded profile plotted in Figure 6(a) represents the 0-nm profile multiplied
by a factor of 0.15. Thus, the phonon transmission through the MgO/Ge
interface appears to be approximately 85% specular.

If an MgO film with only 12-nm thickness is grown on the Ge crystal,
the transmission beyond the critical angle is greatly increased, as seen in
Figure 6(a). Höss et al.[1] interpret this phenomenon as frustrated total internal
reflection, in analogy with a similar effect in optics.* The Al film is softer than
MgO and contains phonons with $s_\parallel$ larger than those that can propagate in
bulk MgO. These large-$s_\parallel$ waves in Al produce evanescent waves in the MgO
with an extinction length, ℓ, given by [see Eqs. (30) and (31) and Figure 5 on

---

* In optics, frustrated total internal reflection, or optical barrier penetration, was discovered by Sir Isaac
Newton, *Opticks* (Dover, New York, 1952) pp. 194 and 205. A precise experimental verification was
reported by D. D. Coon, *Am. J. Phys.* **34**, 240 (1966).

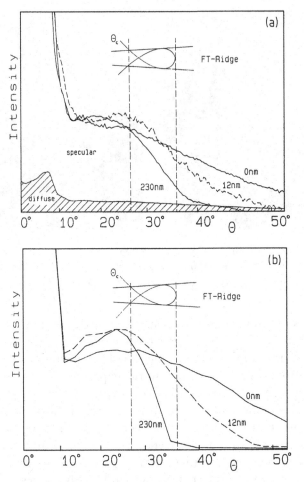

Figure 6 (a) Heat flux plotted along the FT ridge for 230-, 12-, and 0-nm films on Ge. The decrease in intensity near 30° for the thick film corresponds to the loop in the FT critical cone (Figure 4), beyond which phonons undergo total internal reflection. The increased transmission beyond 36° for the 12-nm film implies frustrated total internal reflection. The shaded profile is the 0-nm trace reduced to match the 230-nm trace between 40° and 43°, indicating that the transmission due to diffuse scattering is ≃15%. (b) Theoretical simulation of ballistic heat flux in the above experiment, assuming a phonon frequency of 140 GHz, corresponding to the onset of the Al tunnel junction. (From Höss et al.[1])

pages 285-286]

$$\ell^{-2} = \beta^2 = (\omega/v')^2 - (\omega/v)^2 \sin^2\theta, \tag{1}$$

where v, v', and v'' are the relevant velocities in Al, MgO, and Ge, respectively, and $\sin\theta/v = \sin\theta''/v''$ by Snell's law. If the MgO film thickness is comparable to the extinction length of the evanescent wave, then large-$s_\parallel$ waves can be produced in the Ge, giving rise to a transmitted intensity beyond the critical angle. A prediction of the transmitted phonon intensity for a 12-nm film, assuming a phonon frequency of 140 GHz, is shown in Figure 6(b). The agreement with experiment is quite good.

B. Critical-cone channeling at a loosely bonded surface

Metal films comprise the largest class of heating and detecting elements in heat-pulse experiments. Generally the experimenter does not go to great lengths in preparing the crystal surface on which the metal is deposited (such as baking or annealing). Nor is highly controlled growth of the film employed for most applications (such as heating the substrate or using molecular beam epitaxy methods). Of course, the experimenter rarely reports the case in which the film simply peels off the crystal before any results are obtained. Suffice it to say that, in most experiments, the boundary conditions between the metal film and the crystal are less ideal than assumed by Eqs. (1) and (3) on pages 280-281. Empirical techniques have been developed to characterize and improve the metal–insulator bond, and sometimes one cannot simply ignore this problem.

In this section, we consider the not-uncommon case in which a metal overlayer is weakly bonded to the crystal. The sensitivity of phonon imaging to this situation was discovered in the phonon-focusing pattern of sapphire (Al_2O_3). The discovery[2,3] has led to some informative insights about surface waves and their effect on the transmission across a loosely bonded interface.

The extreme limit of loose bonding, of course, is no bonding. First consider the case of a free surface, a crystal with a vacuum interface. The situation is illustrated by the isotropic slowness surfaces shown in the inset of Figure 7. A T phonon with wavevector $\mathbf{k}$ that strikes the interface at a small incident angle θ produces both L and T reflected waves, constrained in angle by $k_\parallel$ conservation. As the incident angle is increased, a critical angle is reached, beyond which it is impossible to produce a reflected L wave. All the reflected energy goes into the specularly reflected T wave, although an evanescent L-polarized wave, with extinction coefficient,

$$\beta = ik_\perp = [(\omega/v_L)^2 - (\omega/v_T)^2 \sin\theta]^{1/2}, \qquad (2)$$

is produced at the surface. Having a real wavevector component $k_\parallel$, this L component of the scattered wave may be thought of as a pseudo–surface wave that is coupled to the reflected bulk T wave with the same parallel wavevector.

The L evanescent wave is called a pseudo–surface wave because it exists only in concert with a bulk T wave. The boundary conditions that determine the reflected amplitudes Γ_L and Γ_T for a given $k_\parallel$ are

$$\sigma_{xz} = 0 \quad \text{and} \quad \sigma_{yz} = 0,$$

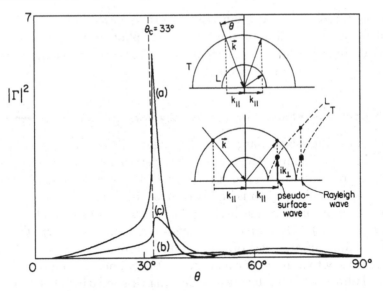

Figure 7 *Amplitude of the reflected or evanescent longitudinal wave in sapphire as a function of the angle of the incident transverse wave, assuming isotropic elastic constants appropriate to sapphire and Cu. The three curves correspond to (a) no bonding, (b) perfect bonding, and (c) loose bonding, as described by Every et al.[3]*

for the boundary at $z = 0$. The pseudo–surface wave differs from a true surface wave (a Rayleigh wave), which contains only evanescent components, as shown schematically* in the lower inset of Figure 7.[3] For a given $k_\parallel$ it is generally not possible to satisfy the boundary conditions with two evanescent waves; however, if $k_\parallel$ is allowed to vary because no incident bulk wave constrains it, a unique Rayleigh wave is obtainable. The resulting slowness of the Rayleigh wave is larger than that of any bulk wave.

Taking longitudinal and transverse velocities that simulate an isotropic sapphire crystal, the squared amplitude $|\Gamma_L|^2$ of the L component of the pseudo–surface wave is computed and plotted as curve (a) in Figure 7. This evanescent wave component is seen to have a large intensity at the surface for incident angles just beyond the critical angle,

$$\theta_c = \arcsin(C_{44}/C_{11})^{-1/2} \simeq 33°. \tag{3}$$

The pseudo–surface wave amplitude dies out rapidly at greater angles. The situation is similar to that depicted in Figure 5 of on page 285.

---

* As discussed in Chapter 14 with regard to surface waves, Eq. (2) is only an approximate condition for the evanescent wave that applies near the critical angle. For real solids the $ik_\perp$ curve of the L mode does not actually retain a hyperbolic form as it approaches the transverse modes (see Figure 7 on page 363), but there are indeed two evanescent wave branches beyond the maximum bulk wavevector of the T modes, as indicated schematically in Figure 7.

If, now, a softer solid such as Cu is weakly coupled to the sapphire crystal, the transmission of energy from sapphire into this solid is approximately proportional to the square of the total wave amplitude at the sapphire surface. Hence, the largest transfer of energy into the softer solid occurs for phonons in sapphire with incident angles just beyond the L–T critical cone; i.e., those which create a large-amplitude evanescent wave. Conversely, for an angular distribution of phonons in the Cu film, those with $s_\parallel$ roughly equal to that of the L–T critical cone in sapphire will couple most strongly to the sapphire.

The amplitude of the pseudo–surface wave depends on how heavily the sapphire surface is loaded by the Cu film. If there is a perfect contact, such that $u_{Cu} = u_{sapphire}$ at the surface, the evanescent wave amplitude is greatly reduced, as shown by curve (b) in Figure 7. Every et al.[3] considered also the more general case of variable bonding strength between the two solids, whereby the traction forces between the solids are proportional to the relative displacement of the surfaces, $\Delta u = u_{Cu} - u_{sapphire}$. One can postulate the force constants $K_\parallel$ and $K_\perp$ per unit area of the interface. The stresses in sapphire at the $z = 0$ interface are

$$\sigma_{xz} = K_\parallel \Delta u_x, \quad \sigma_{yz} = K_\parallel \Delta u_y, \quad \sigma_{zz} = K_\perp \Delta u_z, \tag{4}$$

and three additional equations apply to the Cu surface. These conditions reduce to those of two free surfaces in the limit $K_\parallel, K_\perp \to 0$ and to the perfectly bonded interface when $K_\parallel, K_\perp \to \infty$. The amplitudes of four partial waves (two reflected and two transmitted) can be determined using the procedures developed Section C of Chapter 11. Curve (c) in Figure 7 shows the squared amplitude of the reflected L wave in sapphire for intermediate values of the coupling constants, $K_\parallel$ and $K_\perp$.

By a natural extension of these calculations, Every et al.[3] computed the thermal emissivity of Cu into the sapphire crystal. As the bonding strength is reduced, the overall emission of phonons into sapphire decreases and a peak in emissivity emerges near the L–T critical cone of sapphire, due to the strong coupling of the Cu emitter with the pseudo–surface wave of a weakly loaded sapphire surface. This so-called critical-cone channeling is absent when the two solids are perfectly bonded.

Experiments by Koos, Every, and Wolfe[2,3] show this effect beautifully. The phonon-focusing pattern of a sapphire crystal, coated with a Cu film that is excited by a laser beam, is shown in Figure 14(b) on page 137. An approximately circular ring of increased intensity is clearly seen in this image. This structure is not present in a Monte Carlo calculation of the focusing pattern of sapphire [Figure 8], which does not take into account the crystal metal interface. Superimposed on the Monte Carlo calculation is a white line that is the calculated L–T critical contour. The folded shape of this contour agrees well with the bright ring in the experimental data.

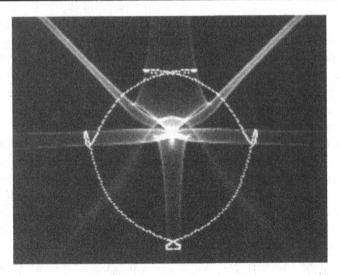

Figure 8 Monte Carlo calculation of heat flux in sapphire, corresponding to the phonon image of Figure 14(b) on page 137. The bright line is a plot of the ST critical cone. A similar structure in the experimental data indicates the increased transmission across the sapphire/metal-film interface at critical-cone angles, beyond which are generated pseudo–surface waves. (From Every et al.[3])

A quantitative comparison of the experiment to the calculations of emissivity indicates that the interface bond strength in the above experiment is about an order of magnitude weaker than the atomic bonding in sapphire. This suggests that there are several layers of soft matter between the Cu and sapphire. The critical-cone effect is greatly reduced if the sapphire surface is roughened before the metal is deposited, or if the metal film is deposited on a clean surface by molecular beam epitaxy. These observations suggest that critical-cone channeling could be a useful probe of the interface quality between two solids.

Extensive calculations of pseudo–surface wave structures have been performed for a variety of solids by Every.[4] An interesting range of structures are predicted, involving not only the L–T critical cone, but critical cones between the two transverse modes and even re-entrant cones involving a single mode. Experimental observations of critical-cone channeling are not so common, probably reflecting the recurring theme in the introduction to Chapter 11 that the average surface in an experiment diffusely scatters high-frequency phonons. However, striking critical-cone channeling effects have been observed in diamond by Hurley et al.[5] as shown in Figure 9. A close inspection of the pattern reveals two sets of fine rings – corresponding to heater and detector films, which are evaporated on slightly nonparallel surfaces. The extremely sharp nature of these critical-cone structures results from the nearly

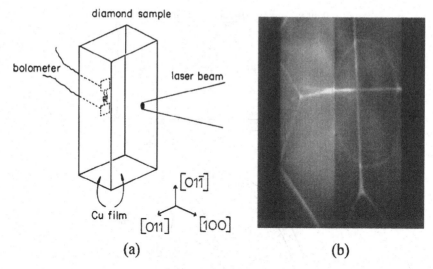

diamond sample

bolometer

laser beam

Cu film

$[01\bar{1}]$

$[011]$ $[\bar{1}00]$

(a)   (b)

Figure 9 (a) Experimental sample geometry. (b) Phonon image of a di-amond crystal, rotated to show both faces. In addition to the phonon-focusing caustics (straight lines), bright circles due to critical-cone chan-neling are observed. (From Hurley et al.[5])

isotropic elasticity of diamond. The critical angle, $\theta_c = \arcsin(v_T/v_L) \simeq 35°$, provides a nearly perfect coupling between the L and T waves.

C. Phonon reflection imaging

We have considered so far the *transmission* of phonons at an interface. Now let us examine the phonons that are reflected. The basic ideas are shown in Figure 10(a). The phonon source and detector are located on the same crystal surface, and, of course, we wish to scan the distance between the source and detector in order to get a picture of the reflected flux. Phonons will be detected that undergo either (a) specular scattering from the back surface, (b) diffuse scattering from the back surface, or (c) scattering in the bulk of the crystal.

Figure 10(a) naively suggests that the specularly scattered phonons undergo a mirror-like reflection, whereas the actual situation is far more interesting. Recall the general case for specular scattering of phonons at an interface, shown in Figure 7 of Chapter 11, which indicates the crucial role that mode conversion plays in this process. Snell's law ($k_{\parallel}$ conservation) provides a unique mapping of the single incident wave into as many as three transmitted and three reflected waves. The pattern of reflected flux is obviously going to be more complicated than the pattern of incident flux, which we already know as the ballistic phonon-focusing pattern.

Does the *reflected* flux pattern exhibit caustics? The answer depends on whether the scattering is specular or diffuse. Like ballistic focusing, the

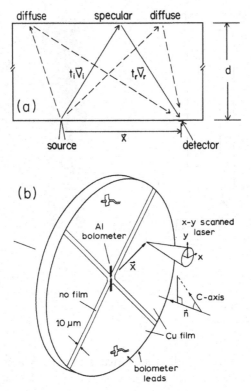

Figure 10 (a) Schematic of a reflection imaging experiment, which may sense both specularly and diffusively reflected phonons. (b) Sample configuration for a sapphire crystal that allows detection and generation of phonons on the same face. The Cu films double as generator films and detector contacts. (From Northrop and Wolfe.[6])

intensity of specularly scattered flux at the detection surface involves a deterministic mapping from the incident **k** vector to the reflected group-velocity vectors. Hence, specularly scattered flux is expected to exhibit caustics. On the other hand, diffuse scattering randomizes the wavevector at the reflecting surface; therefore, mathematical singularities in the detected flux are not expected to occur.

The actual geometry of a reflection imaging experiment, performed by Northrop and Wolfe,[6] is drawn in Figure 10(b). A highly polished sapphire crystal – actually, a standard cryostat window, commercially produced – is coated with a Cu film, except where masked by two wires in the shape of an X. Two of the four electrically isolated sections of Cu are connected with an Al bolometer strip, as shown. The Cu acts both as the heating element (when illuminated with a focused laser beam) and as the contacts for the detector. The purpose of this geometry is to allow a maximum scanning range

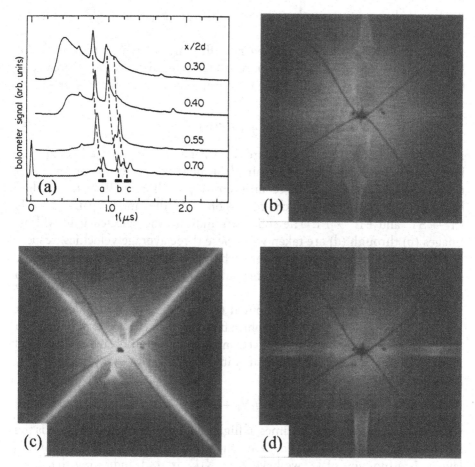

Figure 11 (a) Heat-pulse signals obtained for the experimental configuration of Figure 10(b) and several source-to-detector distances, X. Sharp
pulses are reflected from the back surface, and the broad underlying pulse
is due to scattering of phonons in the bulk. (b)–(d) Phonon images for time
gates selected as in (a). The various caustic patterns can be identified in
the caustic map of the next figure. (From Northrop and Wolfe.[6])

for the heater source. The reflection surface of the sapphire is uncoated, and
the crystal is immersed in superfluid He.

Typical heat pulses for several source-to-detector distances, X, are shown in
Figure 11(a). The early-arriving broad pulse seen in the small-angle ($X/2d$)
traces is due to phonons scattering in the bulk of the crystal. The sharp
pulses are due to phonons that have scattered off of the back surface. Earlier experiments by Taborek and Goodstein[7] and others[8-10] have shown that
sharp pulses can be produced by either specular or diffuse scattering, so the
time-of-flight data alone are difficult to interpret. Clearly, the intensities of

the heat pulses are extremely sensitive to the propagation angles, as seen in Figure 11(a).

The angular patterns of reflected heat flux are shown in the images of Figures 11(b)–11(d). Because the source-to-detector distance varies greatly as the source is scanned, it is desirable to continuously adjust the boxcar gate to select constant velocity, i.e.,

$$t_{\text{gate}} = 2[d^2 + (X/2)^2]^{1/2}/V_{\text{av}}, \tag{5}$$

referring to Figure 10(a) and Figure 3 on page 66. V_{av} is the average velocity estimated for a particular pair of incident and reflected modes. This time-gate dependence for three different V_{av} is illustrated as the dashed lines in Figure 11(a), which correspond to the mode pairs L $\rightarrow$ T, FT $\rightarrow$ ST, and FT $\rightarrow$ FT. The solid bars indicate the selected gate widths. Images (b) through (d) are taken with these three average velocities, respectively. Distinct patterns are seen for each of the mode conversion pairs. The occurrence of sharp caustic structures indicates that these patterns are due to specular scattering.

Northrop and Wolfe[6] also computed the spatial pattern of phonons that are specularly reflected from a sapphire boundary. An incident phonon with wavevector $\mathbf{k}_i$ and mode α_i that reflects into a phonon with wavevector $\mathbf{k}_r$ and mode α_r produces a net displacement $\mathbf{X}$ in the source–detector plane given by [see Figure 10(a)]

$$\mathbf{X} = t_i \mathbf{V}_i + t_r \mathbf{V}_r, \tag{6}$$

where $t_{i,r} = d/\mathbf{n} \cdot \mathbf{V}_{i,r}$ are the times of flight, d is the sample thickness, $\mathbf{n}$ is the surface normal, and $\mathbf{V}_{i,r}(\mathbf{k}_{i,r}, \alpha_{i,r})$ are the group velocities. Since, by Snell's law, $\mathbf{k}_r$ is a function of $\mathbf{k}_i$, we have $\mathbf{X} = \mathbf{X}(\mathbf{k}_i, \alpha_i, \alpha_r)$, indicating that for a specified pair of modes there is a unique mapping from the two-dimensional $\mathbf{k}_i$-space to the two-dimensional $\mathbf{X}$ space. Analogous to simple phonon focusing, the elastic anisotropy will produce caustics in the observation plane.

Monte Carlo images are calculated by mapping a Lambert source distribution of $\mathbf{k}_i$ into the observation plane ($\mathbf{X}$). A superposition of reflected flux patterns for all nine mode pairs (integrating over time) is shown in Figure 12(b). The caustics are drawn as solid lines in Figure 12(a) and identified with particular mode pairs. The dashed lines in this figure indicate an intense, but nonsingular, portion of the FT $\rightarrow$ FT structure. The caustic map shows some similarities to the direct phonon image of sapphire, shown in Figure 8. For example, somewhat distorted ST neckties are observed for mode pairs involving ST phonons. The lack of reflection symmetry about a vertical plane implies that the symmetry axis for this crystal is not exactly normal to the flat crystal surfaces, a fact used in generating the theoretical images. The qualitative agreement between experimental and theoretical images implies that the boundary scattering in this sapphire experiment is mainly specular.

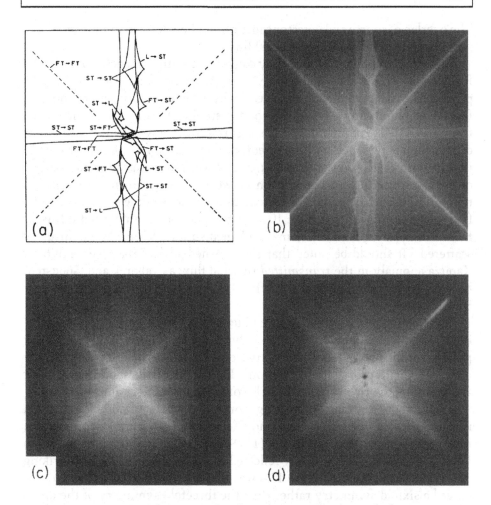

Figure 12 (a) Caustic map for the various mode-converted phonons. (b) Calculated phonon-focusing pattern, integrated over all time and assuming perfectly specular reflection. (c) Monte Carlo simulation for perfectly diffuse scattering. (d) Reflection image of a sapphire surface roughened by 1-μm diamond polish. (From Northrop and Wolfe.[6])

What flux pattern is expected for diffuse scattering? In this case there is no unique relation between $\mathbf{k}_i$ and $\mathbf{X}$. Diffuse scattering from the entire back surface can contribute to the flux at a given $\mathbf{X}$, as indicated in Figure 10(a). However, due to phonon focusing, certain regions of the back surface will be sampled more strongly than others, and rather sharp pulses can be observed. For example, Marx and Eisenmenger[10] and Lassman and coworkers[8,9] have observed and calculated temporally sharp reflected pulses in Si under diffusive surface conditions. Taborek and Goodstein[7] have pointed out that the diffusely scattered heat flux will be highly anisotropic, as it represents an overlap between phonon-focusing patterns centered on source and detector.

The overlap of source and detector caustic lines, however, occur only at points, producing no caustics in the reflected flux.

A Monte Carlo image produced by convolving source and detector phonon-focusing patterns is shown in Figure 12(c). As expected, no sharp caustics are produced. These predictions are tested by roughening the reflection surface of sapphire with a 1-μm diamond paste directly on a glass slide. The time-integrated experimental image, Figure 12(d), shows no signs of specular caustics and agrees well with the theoretical image. In fact, the experimental images with the highly polished surfaces contain a background component of diffuse scattering. A comparison between the experimental images and theoretical images composed of various combinations of the specular and diffuse patterns reveals that $20 \pm 10\%$ of the phonons are reflected diffusely from this surface. For the 1-μm-roughened surface, $97 \pm 3\%$ are diffusely scattered. It should be noted that the polished surface may well exhibit a Kapitza anomaly in the *transmitted* thermal flux; as Taborek and Goodstein have observed,[7] phonons that are diffusely scattered at the surface are largely transmitted into the He.

The experimental results of Figure 12 underscore the importance of employing a movable source or detector in studying the reflection of high-frequency phonons from a boundary.[11] Intensities of specularly reflected phonons vary immensely with propagation directions. Phonon images graphically show for the first time the rich variety of mode conversion processes at a boundary.

Finally, there are some interesting symmetry aspects of phonon reflection imaging pointed out by Wichard and Dietsche.[12] They performed reflection imaging experiments on (100)- and (111)-oriented crystals of Si, immersed in liquid He to reduce the diffusely reflected components.[7] One of their striking observations – shown in Figure 13 – was that the (111)-oriented sample displayed a sixfold symmetry rather than the threefold symmetry of the direct phonon image. The reason for the inversion symmetry in the reflection image (also evident in the sapphire reflection images above) can be seen from the diagram accompanying Figure 13. There is a pairwise symmetry between, for example, ST $\rightarrow$ L and L $\rightarrow$ ST conversion processes that contributes intensities at inversion symmetric points on the detection surface. Wichard and Dietche also pointed out that reflection from high symmetry surfaces [e.g., (100) in Si] can lead to interesting refocusing of reflected flux due to the mode conversion process.

D. Internal interfaces

i) Phonon transmission through superlattices

With the advent of epitaxial crystal growth techniques such as molecular beam epitaxy, it is possible to literally grow crystals one atomic layer at a time. By

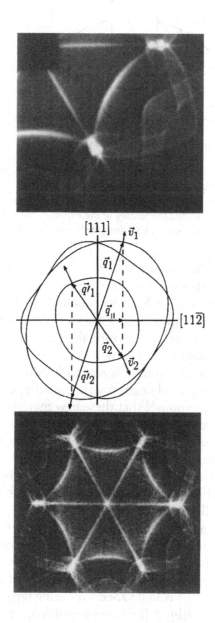

Figure 13 (a) Reflection image from the (111) surface of Si. Only part of the sixfold-symmetric pattern is shown. (b) Diagram of reflected waves showing how the reflection image is inversion symmetric (a general property), here leading to a pattern with sixfold symmetry. (c) Calculated flux pattern, which agrees with the data. (From Wichard and Dietsche.[12])

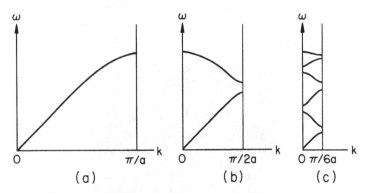

Figure 14 (a) Dispersion curve for compression waves (L) on a monatomic lattice with period a. (b) Dispersion curve for a diatomic lattice. (c) Dispersion curve for a superlattice (alternating layers of two types of atoms) with period D = 6a, showing the folding of the Brillouin zone. Frequency gaps, or stop bands, are formed due to Bragg reflection from the superlattice.

changing the atomic constituents abruptly, nearly perfect boundaries can be created between two different crystals. Carrying this process one step further, superlattice structures can be built by modulating the constituents with a periodicity larger than the interatomic spacing. These synthetic superlattices have opened up new avenues for tailoring the electrical, optical, and magnetic properties of semiconductors and metals.

The ability to fabricate multiple layers with differing acoustic properties has stimulated a variety of phonon experiments that explore this adjustable periodicity. Early suggestions[13] that these new materials could spawn a field of phonon optics – providing acoustic interference filters or gratings akin to those used in optical spectroscopy – have not yet come to pass; however, the theoretical and experimental understanding of phonon interactions with these structures has grown steadily over the last decade. It has been firmly established, for example, that acoustic phonons can exhibit a high degree of specular refraction at epitaxially produced interfaces.

The basic effect of a periodically modulated structure on phonon propagation is illustrated in Figure 14. Figure 14(a) shows the well-known dispersion curve for a compression wave on a monatomic lattice with spacing a between planes. Changing alternate planes to a different type of atom doubles the lattice period and halves the Brillouin zone, producing separate acoustic and optic branches, as in Figure 14(b). A forbidden gap of frequencies separates the two branches at the new zone boundary. Increasing the lattice period further by alternating two types of multiatom layers creates a folded zone with a large number of branches, as in Figure 14(c). Narrow stop bands are now produced both at the zone boundary and at $k = 0$. Physically, the phonons with these frequencies are Bragg scattered from the periodic structure, i.e., for normal

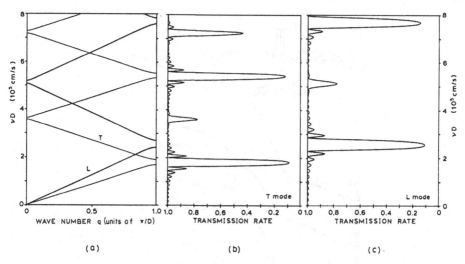

Figure 15 (a) Dispersion curves for normal incidence to a AlAs/GaAs superlattice with $d_A = d_B = D/2$. Transmission rate for T and L phonons is
shown in (b) and (c). (From Tamura et al.[16])

incidence they satisfy $m\lambda = 2D$, where m is an integer, $\lambda = 2\pi/k$, and D
is the period of the superlattice. The existence of the zone center modes has
been proven by light-scattering experiments.[14,15]

The lowest-frequency stop bands in Figure 14(c) can extend well into the
nondispersive phonon regime, where the dispersion curve is linear. Indeed,
by selecting an appropriate layer thickness, it is possible to create one or
more stop bands in the frequency range accessible to heat-pulse experiments.
Observation of the zone boundary stop bands using nonequilibrium phonons,
however, requires high-quality interfaces that exhibit specular refraction and
some means of selecting phonon wavelength. We expect that the ubiquitous
elastic anisotropy of crystals will make the story much more interesting than
the one-dimensional models of Figure 14 suggest.

The calculated phonon dispersion relation for L and T phonons normally
incident on a GaAs/AlAs superlattice is shown in Figure 15. Also shown is
the predicted transmission rate for both modes as a function of frequency.[16]
Strong dips in the transmission mark the aforementioned stop bands, and the
smaller satellite dips are due to the finite number of superlattice periods (15)
assumed in the calculation. The general methods for calculating the phonon
transmission through superlattices are described by Tamura et al.[16]

The first observation of a phonon stop band in a superlattice was reported
for GaAs/AlGaAs by Narayanamurti et al.[17] The result of their experiment,
using *normal* incidence and a phonon-spectroscopy technique, is reproduced
in Figure 16(a). A tunable, quasi-monochromatic source of phonons is

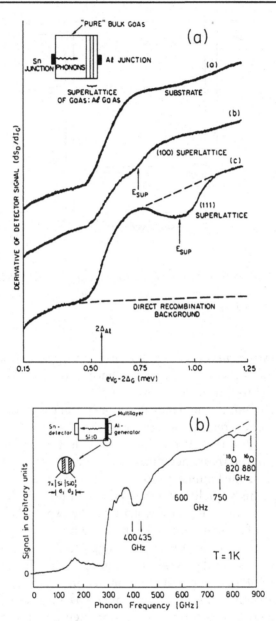

Figure 16 (a) Longitudinal phonon spectrum transmitted through a Ga/AlGaAs superlattice, indicated schematically in the inset. Two superlattice orientations are shown. The Al tunnel-junction detector onset is at approximately 0.5 meV. Phonons of a scannable frequency are modulated by the Sn tunnel-junction generator. A dip in the transmission occurs at the Bragg condition. (From Narayanamurti et al.[17]) (b) Similar phonon-spectroscopy experiment for an amorphous Si/SiO$_2$ superlattice. The observed stop bands at approximately 420 and 675 GHz correspond to $d_{Si} = 70$ Å and $d_{SiO_2} = 30$ Å, in fair agreement with values measured with a crystal thickness monitor. (From Koblinger et al.[18])

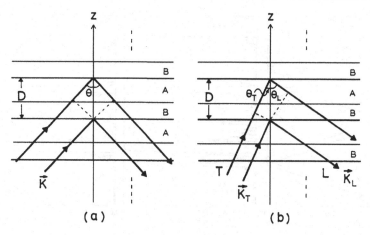

Figure 17 *Diagram of Bragg reflection from a superlattice (a) for a single phonon mode and (b) with mode conversion. (From Tamura et al.[16])*

produced by modulating a tunnel junction. (See Ref. 9 on page 19 for an explanation of this method.) A dip in the transmission indicates the Bragg-scattering frequency for longitudinal phonons. A similar phonon-spectroscopy experiment utilizing amorphous films has been reported by Koblinger et al.,[18] as shown in Figure 16(b). An interesting aspect of this experiment is that the phonons propagate ballistically through seven periods of amorphous Si/SiO_2 material, where each double layer is 100 Å thick. In other words, the mean free path of 400-GHz phonons in the *amorphous* material is greater than 700 Å.

Now we consider phonons that are incident on the superlattice at an *oblique* angle. Figure 17(a) illustrates the simplest process whereby the modes of the scattered and incident phonons are the same. The well-known Bragg condition, $m\lambda = 2D\cos\theta$, can be written

$$\cos\theta/v = m/2vD, \qquad (7)$$

where $v = v/\lambda$. Figure 17(b) illustrates the remaining possibility, in which the mode of the scattered phonon is different from that of the incident phonon (i.e., mode conversion occurs). Examination of this figure yields the Bragg condition,

$$\cos\theta_L/v_L + \cos\theta_{FT}/v_{FT} = m/vD, \qquad (8)$$

which, with the additional constraint of Snell's law,

$$v_L/\sin\theta_L = v_{FT}/\sin\theta_{FT}, \qquad (9)$$

determines both the incident and scattered angles, θ_L and θ_{FT}, for Bragg reflection. In principle, an incident phonon can scatter into any of the three

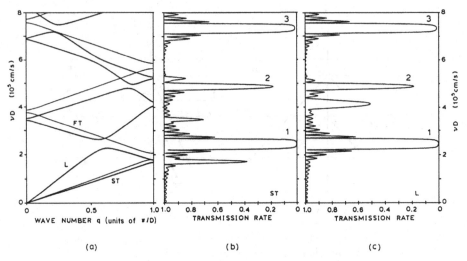

Figure 18 (a) Dispersion curves for phonons at oblique incidence to a (001) Al/Ga superlattice. Coupling between modes (e.g., L and FT) causes anti-crossings (gaps) to appear interior to the Brillouin zone boundaries. These gaps give rise to intramode stop bands, as indicated in the predicted transmission spectra for the ST and L modes [dips 1, 2, and 3 in (a) and (b)]. (From Tamura et al.[16])

phonon modes. Equation (8) corresponds to intermode (or coupled-mode) phonon scattering. If the incident and scattered modes are the same, this equation reduces to Eq. (7), which governs intramode scattering.

The coupled-mode scattering process was predicted by Tamura and Wolfe[19] and is illustrated in Figure 18. This process gives rise to a stop band (i.e., frequency gap) at a wavevector within the zone boundaries. This new gap is a result of an anticrossing of the dispersion curves for two modes, produced by the coupling between these modes at oblique propagation angles. A clear example of this in Figure 18 is the L and ST coupling, labeled "1," which gives rise to a strong intrazone stop band that can be observed when either L or ST phonons are directed at their respective Bragg-scattering angle, given by an equation analogous to Eq. (8).

Experimental control of the incident phonon direction is accomplished by phonon-imaging techniques. In a typical experiment, the superlattice is grown on a carefully prepared (100) surface of GaAs. A small detector is deposited on the superlattice, and a scanable laser beam incident on the opposite side of the GaAs substrate provides the heat source. Frequency selectivity is accomplished, as described in Figure 6(b) on page 173, by the combination of a high-pass frequency onset of a superconducting tunnel junction with the low-pass substrate transmission, due to mass-defect scattering. A rather narrow, fixed band of frequencies is detected, and the control parameter is the

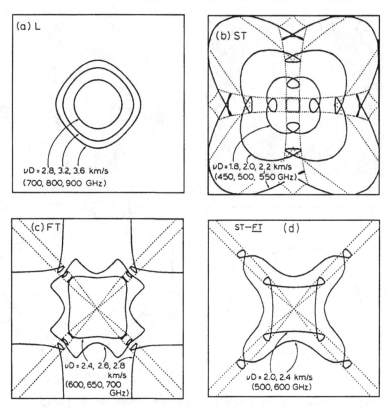

Figure 19 (a)–(c) Maps of the Bragg-scattering cones with no mode conversion for a (100) AlAs/GaAs superlattice. Dotted lines are caustic maps for reference. (d) Intermode Bragg cone for mode conversion between ST and FT phonons. (From Tamura et al.[16])

incident angle of the phonons on the superlattice. The Bragg condition will be satisfied for a series of cones, with $\cos\theta$ given by Eq. (7), with fixed v and D and integer values of m.

Of course, the Bragg cones of reduced transmission through the superlattice are not perfectly conical, due to the elastic anisotropies of the substrate and superlattice. The major deviation from conical arises from the phonon focusing in the substrate. For the transverse modes, a circular cone in k space contains folds in V space. Figures 19(a)–19(c) show the predicted intramode Bragg curves (heavy lines) superimposed on the caustics of a GaAs substrate (dotted lines).[16] Figure 19(d) shows one example of the Bragg contours for an intermode scattering process.

What do the experiments show? The simplest case is the intramode (zone boundary) scattering of L phonons from a superlattice (SL). This mode has no caustics, and a nearly circular ring of reduced intensity was observed by Hurley et al.[20] for a selected InGaAs/AlAs superlattice and $v \simeq 850$ GHz,

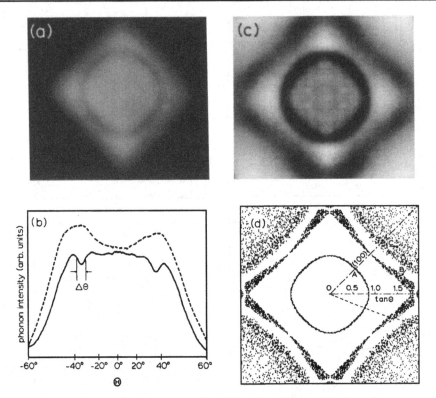

Figure 20 (a) Phonon image of L phonons propagating through an InGaAs/GaAs superlattice on a GaAs substrate. The dark circle is the L critical cone for Bragg reflection. (b) Line scan through the image. (c) Same image with a substrate-only image subtracted. Now two stop bands can be seen, the outer one due to intramode L-ST scattering, as predicted in the calculated stop-band distribution shown in (d). (From Tamura et al.[16])

as shown in Figure 20(a). From a line scan across the center of this image [Figure 20(b)], the angular width of the dip is determined to be approximately 6°, corresponding to a detected frequency range of approximately 150 GHz, calculated with Eq. (7).

The effect of the superlattice in this experiment may be isolated by recording a corresponding phonon image of the substrate alone and subtracting it from the substrate + SL image. This difference image is shown in Figure 20(c). Not only does the $m = 1$ ring appear strongly, but an additional, noncircular transmission dip becomes apparent. A Monte Carlo calculation of this system shows both of these stop bands also, as displayed in Figure 20(d). The nonspherical stop band corresponds to a combination of L/FT (inner structure) and L/ST (outer structure) coupled-mode stop bands, merged by the experimental frequency resolution.

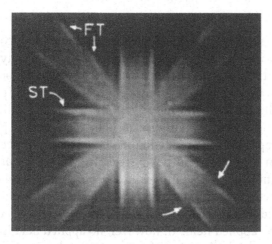

Figure 21 Phonon image of FT and ST phonons in the same system as in Figure 20 but with a 700-GHz detector. The arrows show an FT stopband. (From Tamura et al.[16])

By depositing a detector with a lower onset frequency (700 GHz), the same superlattice sample was used to observe the effect of intramode Bragg scattering in the transverse modes. The resulting phonon image, Figure 21, shows a distinct missing band in the FT caustic ridge (double arrows, lower right), as predicted in Figure 19(c). The ST stop bands at the experimental frequency are out of the experimental scan range, as indicated in Figure 19(b).

Phonon-spectroscopy experiments with amorphous Si superlattices (Si : H/Si : N) have been performed by Santos et al.[21] Using fixed tunnel junctions, which sample a range of incident angles (<20°), they observed extra dips in the phonon transmission that can be explained by intramode stop bands, as predicted by Tamura and Wolfe.[19]

Finally, we mention that extensive calculations (and a few phonon-imaging experiments) have been performed for a variety of nonperiodic super-lattices.[22-25] It is rather remarkable that these structures often show stop bands that are qualitatively similar to those of the periodic structures, even though there is no simple Bragg condition governing the phonon transmission. Moreover, Tamura has found some nonperiodic configurations which display narrow *passbands* as well as stop bands.[24]

A phonon-imaging experiment on a Fibonacci superlattice with 200 layers of alternating GaAs has shown distinct stop bands, as predicted for this non-periodic structure.[23] In addition, the ability of 700-GHz phonons to traverse so many interfaces without significant diffuse scattering is a testament to the quality of the epitaxial growing techniques. To appreciate this fact, recall the examples in Chapter 11 in which the scattering of high-frequency phonons at a single interface exposed to atmospheric conditions was primarily diffusive.

ii)   Ferroelectric domain boundaries – Another perfect match

Ferroelectricity is the property of some nonmetallic crystals to spontaneously develop an electric polarization in the absence of an externally applied field.[26] The phase transition from dielectric to ferroelectric behavior occurs at a critical temperature, T_c, the Curie temperature. In the ferroelectric phase the center of masses for the positive and negative ions in the crystalline unit cell are displaced, causing a net dipole moment. The ferroelectric phase transition is usually accompanied by a reduction in crystalline symmetry (a structural transition) from a nonpolar to polar structure. Only 10 of the 32 crystal classes have sufficiently low symmetry to display this polar behavior.

If the ferroelectric transition occurs in the presence of a sufficiently large applied electric field, the crystal forms a single domain with electric polarization aligned along the applied field. If, however, no applied field is present during the transition, multiple domains of opposing dipoles are usually formed so that the net electric moment of the crystal is near zero. The positive energy required to set up domain walls between regions of opposing polarizations is more than offset by the reduction in electrostatic energy surrounding a uniformly polarized sample.

Here we are interested in how domain walls affect the phonon propagation in such crystals. The abrupt breaking of lattice periodicity at the domain wall produces an acoustic mismatch, where phonon reflection and refraction can occur. Because the internal boundaries are free of contaminants, one should expect that this is an ideal situation to test acoustic-mismatch theory. To greatly simplify the analysis, we would like to have a ferroelectric crystal that contains a parallel series of planar domain walls.

Fortunately, nature has been kind to the experimenter: The common ferroelectric crystal KDP (potassium dihydrogen phosphate, KH_2PO_2) is readily prepared in the single-domain state by applying reasonable fields at the transition, and, if no field is applied, a uniform set of parallel domain walls spontaneously appear. Thus, by applying (or omitting) an electric field in the cool-down process one can prepare either single- or multiple-domain structures for a single sample and phonon detector. Of course, we hope that the scattering from domain walls is not masked by a stronger scattering from impurities or defects.

Such crystals are indeed available. Thermal-conductivity measurements[27,28] on KDP samples between 0.1 and 2 K reveal that the multiple-domain case has approximately 30% lower thermal conductivity than the single-domain case. And the reduction in conductivity is found to depend on the direction of thermal current. As in the case of dislocations in LiF (Chapter 9), the experiments suggest that certain subsets of phonons are contributing much more strongly to the thermal conductivity than others.

Above $T_c = 122$ K, KDP is paraelectric and has tetragonal symmetry.[29] This crystal has useful applications in optical harmonic generation and

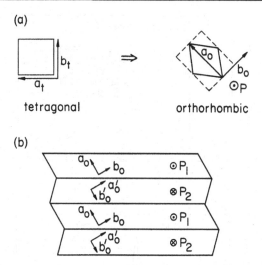

Figure 22 (a) Structural changes that accompany the ferroelectric phase transition in KDP. The electric dipole moment is shown as P. (b) Planar domains with thicknesses 3 to 10 μm and alternating polarization are formed upon cool down through the transition when no electric field is applied. (From Weilert et al.[28])

electro-optical devices.[30,31] At the ferroelectric transition temperature of 122 K, there is a structural change from tetragonal to orthorhombic, and domains of opposing electric polarization with 3–10-μm spacing spontaneously appear, as indicated in Figures 22(a) and 22(b). The structural distortion is greatly exaggerated in this figure; the actual changes in lattice constant are only ≃1%. Applying a field of 1–2 kV/cm along the tetragonal [001] axis while cooling through the transition causes a single orthorhombic domain to form.

The effect of domain-wall scattering on heat pulses has been measured by Weilert et al.[27,28,32] Time traces for phonons propagating along [001] are shown in Figure 23. As expected for the small orthorhombic distortions, the phonon velocities are not significantly affected, but the relative transmission of L and T phonons is markedly different for single- and multiple-domain cases.

The phonon-focusing patterns for the single- and multiple-domain cases are dramatically different, as shown in Figure 24. As expected, the single-domain crystal displays a pattern of caustics that deviates from fourfold symmetric, due to the orthorhombic structure. A mirror image of this pattern is obtained if the reverse poling field is applied during the phase transition. (The applied field is zero during the phonon measurement and it was not possible to reverse the polarization by applying electric fields at low temperature.) Remarkably, the FT features remain almost intact in the multiple-domain case although a more symmetric square is seen, as might be expected from averaging the

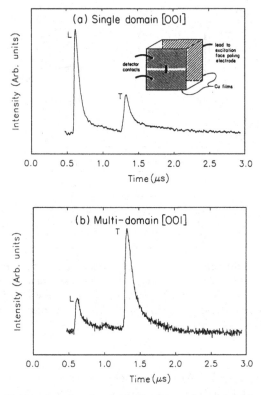

Figure 23 Heat pulses for the single-domain and multiple-domain cases of KDP, showing a differential scattering of longitudinal and transverse phonons from domain walls. The inset shows the experimental geometry used to apply the electric field during the cool down in order to form a single domain. (From Weilert et al.[32])

distorted orthorhombic patterns for opposite domains. The ST phonon-focusing structures, however, are nearly destroyed, and a new concentration of ST flux (identified as such by its late time of flight) appears as an intense horizontal line in the center of the picture. This new structure parallels the domain boundaries, suggesting that it is due to ST flux that is channeled between the boundaries.

The principal conclusion from these experiments is that phonons of different polarization are scattered differently by the domain boundaries. To analyze the scattering processes one must simulate both the bulk propagation and the acoustic mismatch at the interfaces.

In fact, the bulk problem alone is not trivial, due to the low crystal symmetry. From Chapter 5 (page 123) we see that there are six independent elastic constants for tetragonal crystals, and these are known for KDP in its paraelectric phase.[33] In Figure 22(a), the tetragonal a_t and b_t directions are

Figure 24 *Phonon images of KDP for (a) single-domain and (b) multiple-domain cases. The boxcar gate includes both L and T phonons, as indicated in Figure 23. (From Weilert et al.[28])*

equivalent, so $C_{11} = C_{22}$, $C_{13} = C_{23}$, and $C_{44} = C_{55}$. The orthorhombic distortion shown in this figure destroys this equivalence, thereby creating three new elastic constants, for a total of nine. Weilert et al.[28] parameterized this small distortion with a single parameter δ, giving rise to orthorhombic elastic constants such as,

$$C'_{11} = (1 + \delta)C_{11}, \qquad C'_{22} = (1 - \delta)C_{11}, \tag{10}$$

where the C_{IJ} are the room-temperature tetragonal values in the orthorhombic basis. Piezoelectric effects are ignored.[34] By adjusting δ, they found a good match between a Monte Carlo calculation and the experimental data of Figure 24(a). The best result (for $\delta = 0.05$) is shown in Figure 25(a). The excellent prediction of the caustic pattern indicates that the calculated polarization vectors, which are needed for the boundary scattering problem, will be adequately represented.

The next step is to calculate the reflection and transmission probabilities at the boundary between two sections of crystal with opposite polarization, as in Figure 22. For each phonon with wavevector $\mathbf{k}$ and polarization $\mathbf{e}$ one must

Figure 25 (a) Monte Carlo calculation of the phonon-focusing pattern in single-domain KDP, assuming δ = 5%. (b) Calculated heat flux for the multiple-domain case. (From Weilert et al.[32])

compute three reflection and three transmission coefficients, as discussed in Section C of Chapter 11. Fortunately all of the possible resulting phonons have a $k_{\parallel}$ that is identical to the initial phonon, and this $k_{\parallel}$ is conserved in future encounters with parallel interfaces. Therefore, for a given initial incident angle one need only calculate a 6×6 probability matrix for each of the two interface types (↑ to ↓ regions, or ↓ to ↑), and these two matricies can be used to compute the trajectory of a given phonon as it intersects many boundaries in the crystal.[35] As usual, the input to the Monte Carlo calculation is a uniform angular distribution of phonon wavevectors originating from the heater surface. When they cross the detector surface, the phonons are spatially binned without regard to their arrival times.

A Monte Carlo image based on a 3-mm-thick KDP crystal with a 10-μm domain wall spacing is shown in Figure 25(b).[28] This calculation reproduces the features of the experimental data [Figure 24(b)] very well. The bright horizontal line indeed arises from channeling of ST phonons, and the FT square corresponds to phonons that have hardly scattered or mode converted

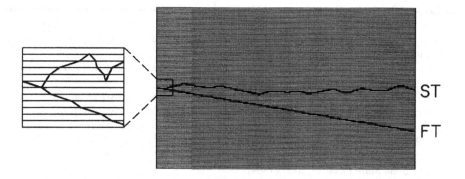

Phonons	Polarization	Reflection probability at a single wall	Net effect through crystal
Horizontal ST caustic	≈y	≈50% little mode conversion	channeled
Vertical ST caustic	≈x	≈1% transmitted to ST	near ballistic
FT caustic	≈z	<<1% reflection	near ballistic

Figure 26 Calculated trajectories of FT and ST phonons through a multiple-domain crystal of KDP. The FT phonons undergo only small angle scattering, whereas the ST phonons have a reflection probability of almost 50% at each interface, explaining the bright horizontal line in the multiple-domain images. The table summarizes the reflection probabilities for the phonons. (From Weilert et al.[32])

on their route to the detector surface. This situation is illustrated in Figure 26, which shows computed trajectories of an ST phonon and an FT phonon. The table lists the reflection probabilities for phonons with selected polarizations.

The detailed agreement between experiment and calculation validates the acoustic-mismatch theory for domain wall boundaries in KDP. This conclusion is underscored by the fact that the phonons encountered several hundred interfaces on their paths from source to detector. As discussed in Chapter 11, the basic tenets of this theory are the continuity across the interface of (a) the normal component of stress and (b) the displacement. The KDP computation reviewed here incorporates the full anisotropy of a low-symmetry crystal and deals with interfaces that are not symmetry planes. The experiments represent the only example so far of a phonon image that is grossly, but reversibly, modified by an applied field.

References

1. C. Höss, J. P. Wolfe, and H. Kinder, Total internal reflection of high-frequency phonons: a test of specular refraction, *Phys. Rev. Lett.* **64**, 1134 (1990).
2. G. L. Koos, A. G. Every, and J. P. Wolfe, Critical-cone channeling of thermal phonons at a sapphire-metal interface, *Phys. Rev. Lett.* **51**, 276 (1983).
3. A. G. Every, G. L. Koos, and J. P. Wolfe, Ballistic phonon imaging in sapphire: bulk focusing and critical-cone channeling effects, *Phys. Rev. B* **29**, 2190 (1984).
4. A. G. Every, Pseudosurface wave structures in phonon imaging, *Phys. Rev. B* **33**, 2719 (1986); Phonon focusing in reflection and transmission, *Phys. Rev. B* **45**, 5270 (1992).
5. D. C. Hurley, A. G. Every, and J. P. Wolfe, Ballistic phonon imaging of diamond, *J. Phys. C: Solid State Phys.* **17**, 3157 (1984).
6. G. A. Northrop and J. P. Wolfe, Phonon reflection imaging: a determination of specular versus diffuse boundary scattering, *Phys. Rev. Lett.* **52**, 2156 (1984).
7. P. Taborek and D. Goodstein, Phonon focusing catastrophes, *Solid State Commun.* **33**, 1191 (1980); Diffuse reflection of phonons and the anomalous Kapitza resistance, *Phys. Rev. B* **22**, 1550 (1980).
8. S. Burger, W. Eisenmenger, and K. Lassmann, Specular and diffusive scattering of high frequency phonons at sapphire surfaces related to surface treatment, in *LT-17 (Contributed Papers)*, U. Eckern, A. Schmid, W. Weber, and H. Wühl, eds. (Elsevier, Amsterdam, 1984).
9. E. Mok, S. Burger, S. Döttinger, K. Lassmann, and W. Eisenmenger, Effect of laser annealing on specular and diffuse scattering of 285 GHz phonons at polished silicon surfaces, *Phys Lett.* **114A**, 473 (1986).
10. D. Marx and W. Eisenmenger, Phonon scattering at silicon crystal surfaces, *Z. Phys. B* **48**, 277 (1982).
11. The details of interface scattering can be averaged out by choosing detectors with large areas. Quantitative reflection experiments and analysis with large-area fixed detectors have been performed by G. Müller and O. Weis, Quantitative investigation of thermal phonon pulses in sapphire, silicon, and quartz, *Z. Phys. B Condensed Matter* **80**, 25 (1990).
12. R. Wichard and W. Dietsche, Specular phonon reflection: refocusing and symmetry doubling, *Phys. Rev. B* **45**, 9705 (1992). See also Ref. 35.
13. V. Narayanamurti, Phonon optics, carrier relaxation and recombination in semiconductors, in *Phonon Scattering in Condensed Matter*, H. J. Maris, ed. (Plenum, New York, 1980), p. 385.
14. C. Colvard, T. A. Gant, M. V. Klein, and A. C. Gossard, Observation of folded acoustic phonons in a semiconductor superlattice, *Phys. Rev. Lett.* **45**, 298 (1980); M. V. Klein, Phonons in semiconductor superlattices, *IEEE J. Quantum Electron.* **22**, 1760 (1986); H. Brugger, G. Abstreiter, H. Jorke, H. J. Herzog, and E. Kasper, Folded acoustic phonons in Si-Si_xGe_{1-x} superlattices, *Phys. Rev. B* **33**, 5928 (1986).
15. J. Sapriel and B. Djafari Rouhani, Vibrations in superlattices, *Surface Sci. Rep.* **10**, 189 (1989); J. Menéndez, Phonons in GaAs-AlGaAs superlattices, *J. Lumin.* **44**, 285 (1989).
16. S. Tamura, D. C. Hurley, and J. P. Wolfe, Acoustic-phonon propagation in superlattices, *Phys. Rev. B* **38**, 1427 (1988).
17. V. Narayanamurti, H. L. Störmer, M. A. Chin, A. C. Gossard, and W. Wiegmann, Selective transmission of high frequency phonons by a superlattice: the "dielectric" phonon filter, *Phys. Rev. Lett.* **43**, 2012 (1979).
18. O. Koblinger, J. Mebert, E. Dittrich, S. Döttinger, W. Eisenmenger, V. Santos, and L. Ley, Phonon stopbands in amorphous superlattices, *Phys. Rev. B* **35**, 9375 (1987).
19. S. Tamura and J. P. Wolfe, Coupled-mode stop bands of acoustic phonons in semiconductor superlattices, *Phys. Rev. B* **35**, 2528 (1987). See also H. Kato, Acoustic SH phonons in a superlattice with (111) interfaces, *J. Acoust. Soc. Am.* **101**, 1380 (1997).

20. D. C. Hurley, S. Tamura, J. P. Wolfe, and H. Morkoc, Imaging of acoustic phonon stop bands in superlattices, *Phys. Rev. Lett.* **58**, 2446 (1987).

21. P. V. Santos, J. Mebert, O. Koblinger, and L. Ley, Experimental evidence for coupled-mode phonon gaps in superlattice structures, *Phys. Rev. B* **36**, 1306 (1987).

22. S. Tamura and J. P. Wolfe, Acoustic-phonon transmission in quasiperiodic superlattices, *Phys. Rev. B* **36**, 3491 (1987); D. C. Hurley, S. Tamura, J. P. Wolfe, K. Ploog, and J. Nagle, Angular dependence of phonon transmission through a Fibonacci superlattice, *Phys. Rev. B* **37**, 8829 (1988).

23. S. Tamura, Phonon transmission through periodic, quasiperiodic and random superlattices, in *Phonons 89*, S Hunklinger, W. Ludwig, and G. Weiss, eds. (World Scientific, Singapore, 1990), p. 703.

24. S.-I. Tamura, Resonant transmission of acoustic phonons in multisuperlattice structures, *Phys. Rev. B* **43**, 12646 (1991).

25. S. Mizuno and S.-I. Tamura, Transmission and reflection times of phonon packets propagating through superlattices, *Phys. Rev. B* **50**, 7708 (1994).

26. See, for example, C. Kittel, *Introduction to Solid State Physics*, 6th ed. (J. Wiley, New York, 1986), pp. 373–88.

27. M. Weilert, Phonon scattering by ferroelectric domain walls in potassium dihydrogen phosphate, Ph.D. Thesis, University of Illinois at Urbana-Champaign (1993).

28. M. A. Weilert, M. E. Msall, J. P. Wolfe, and A. C. Anderson, Mode dependent scattering of phonons by domain walls in ferroelectric KDP, *Z. Phys. B* **91**, 179 (1993).

29. R. J. Nelmes, Z. Tun, and W. F. Kuhs, A compilation of accurate structural parameters for KDP and DKDP, and a user's guide to their crystal structures, *Ferroelectrics* **71**, 125 (1987).

30. D. Eimerl, Electro-optic, linear, and nonlinear optical properties of KDP and its isomorphs, *Ferroelectrics* **72**, 95 (1987).

31. R. M. Hill and S. K. Ichiki, Optical behavior of domains in KH_2PO_4, *Phys. Rev.* **135**, A1640 (1964).

32. M. A. Weilert, M. E. Msall, A. C. Anderson, and J. P. Wolfe, Phonon scattering from ferroelectric domain walls: results of phonon imaging in KDP, *Phys. Rev. Lett.* **71**, 735 (1993).

33. O. Delekta and A. Opilski, Measurement of propagation velocity of ultrasonic wave at a frequency of 10 MHz in pure and deuterated KDP single crystals, *Arch. Acoust.* **5**, 181 (1982).

34. A. G. Every and V. I. Neiman, Reflection of electroacoustic waves in piezoelectric solids: mode conversion into four bulk waves, *J. Appl. Phys.* **71**, 6018 (1992).

35. Here the internal boundaries are normal to the detection surface. Transmission through multiple interfaces that are parallel to the detection surface has been generally considered by A. G. Every, Phonon focusing in reflection and transmission, *Phys. Rev. B* **45**, 5270 (1992).

13

Imaging ultrasound in solids

This book could have been entitled "Phonon Caustics," for it is these unusual concentrations of thermal energy resulting from *point* excitation of crystals that have provided a wealth of new perspectives on the physics of acoustic phonons. The concept of caustics, however, is not common to the parent field of physical acoustics. Why not? In large part, it is because acoustic experiments on solids usually employ *planar* sources of vibrational energy. Also, the *coherence* and *macroscopic wavelengths* of radio-frequency waves are not characteristics of nonequilibrium phonons. This chapter describes some recent ultrasonic-imaging experiments and calculations that help to establish the conceptual links between phonon imaging and physical acoustics. More importantly, we shall see that the extension of phonon-imaging techniques to ultrasound greatly expands the applicability of these methods.

Throughout the text we have used the word "acoustic" to designate the lowest branches of the vibrational spectrum – those waves which, in the limit of low frequency, have the same velocities and polarizations as sound waves in the solid. Indeed, many of the caustic patterns we have seen are predictable from continuum elasticity theory. Yet, we have implicitly assumed that the wavelengths of the phonons are smaller than the source and detector sizes, and certainly much smaller than the sample dimensions. This assumption is true even for the lowest frequencies observed in a typical heat-pulse experiment – approximately 100 GHz, corresponding to $\lambda \simeq 50$ nm. Thus, we have adopted the ray picture of geometrical optics. It stands to reason that the caustic patterns predicted for particle-like phonons should be modified at long wavelengths.

Maris[1] was the first to point out that phonon caustics cannot be perfectly sharp, even with vanishing source and detector sizes, due to the nonzero wavelengths of the phonons. In effect there is a kind of diffraction limit to the sharpness of the caustics. This intrinsic fuzziness is not easily measurable in a typical heat-pulse experiment because the phonon wavelengths are very short (tens of nanometers). Maris estimated that in order to observe the finite-wavelength broadening of the caustics in a perfect crystal of approximately 1-cm thickness, the source and detector sizes must be 0.1 μm or smaller. While not impossible, this resolution is approximately two orders of magnitude finer than that attained in any heat-pulse experiment to date. Subsequently, Novikov and Chernozatonskii[2] suggested the use of ultrasound generated at radio frequencies to verify Maris's predictions.

A. Phonons versus ultrasound

Radio-frequency acoustic waves[3] are usually generated by planar transducers, which produce a plane wave of well-defined frequency and wavevector. In contrast, a point disturbance in a crystal produces a nonspherical wavefront of acoustic energy with the shape of the wave surface, as indicated in the right side of Figure 20 on page 52. Also, with ultrasound a new aspect of the problem arises: *coherency*. Until now, we have dealt with thermally generated phonons, which may be regarded as an *incoherent* collection of vibrational waves. If a point source is coherently driven, interference effects must be considered.

   An interesting contrast between the usual ultrasound and phonon experiments is that the standard radio-frequency pulse-echo experiments measure *phase* velocity, whereas heat-pulse experiments measure *group* velocity. Figure 1 shows why. Acoustic power emanating from the planar source at position A propagates along the direction of group velocity, V, as demonstrated in the experiment of Figure 3 on page 26. In Chapter 2, we derived the relationship between V and the wavevector k, which is normal to the phase

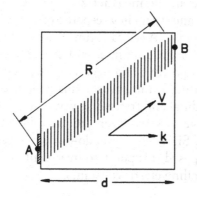

Figure 1 *Schematic diagram of an oblique acoustic wave emanating from a planar detector at point A and propagating to point B on the opposite parallel face. Group-velocity and wavevector directions are shown. The corresponding experiment is shown in Figure 3 on page 26.*

fronts. The result is simply

$$\mathbf{V} \cdot \mathbf{k} = \omega. \tag{1}$$

The phase velocity is a scalar quantity given by

$$\omega = vk. \tag{2}$$

The time of flight for a given phase front to travel a distance R from position A to position B (Figure 1) is

$$t_0 = R/V = d/v, \tag{3}$$

where d is the thickness of the crystal. The experimenter who divides crystal thickness, d, by the measured time of flight obtains the *phase* velocity, and the experimenter who divides the distance of energy propagation, R, by the time of flight obtains the *group* velocity.* Consequently, in the limit of point source and point detector (as in a heat-pulse experiment), one measures V.

Note also that the wave propagating from A to B in Figure 1 returns along the *same* path (i.e., from B to A) due to $k_\parallel$ conservation in the specular reflection. Amazingly, the acoustic echo returns to the source even for an oblique waves such as the one shown. By observing multiple-pass echoes in the radio-frequency experiment, extremely accurate measurements of the phase velocity can be made.[3]

A novel demonstration of the anisotropic propagation of radio-frequency waves emanating from a point source in a crystal has been reported by Every et al.[4] Their technique, performed at room temperature, is illustrated in Figure 2. An intense, focused laser beam of approximately 10-ns duration strikes the surface of a Si crystal and produces a thermoelastic deformation wave with components in the megahertz range. A piezoelectric film with a cross section of $1\ mm^2$ and frequency response centered around $4\ MHz$ detects the resulting acoustic pulse. A series of time traces are recorded for successive laser positions along a single scan line, as shown in the figure. The light and dark regions represent positive and negative displacements of the surface at the detector. The resulting $x-t$ image contains structures that resemble a cross section of the longitudinal and transverse wave surfaces, although this is not actually a spatial image. Clearly seen are the folds in the ST wave surface for the (100) plane of Si. Also, a pseudo–surface wave is identified. Every and Sachse[5] have proposed this point-source/point-receiver configuration as a way to extract crystalline elastic constants.

---

* Notice that R is not a measured quantity in the pulse-echo experiment, which employs the same transducer as generator and detector (Figure 1).

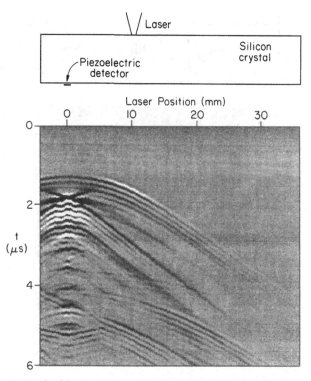

Figure 2 *A pulsed laser striking a Si crystal at room temperature causes thermoelastic expansion at the source, which produces an acoustic pulse propagating through the crystal. The piezoelectric detector on the reverse side detects the ultrasonic pulse. A series of vertically displayed time traces for a number of laser positions produces the x–t image shown. (From Every et al.[4])*

B.   Ultrasonic-flux imaging

Hauser et al.[6] have subsequently performed imaging experiments with coherent ultrasonic beams, as indicated at the top of Figure 3. Typically, 15-MHz ultrasound is produced with a conventional ultrasonic transducer incorporating a spherical lens that has a focal length $f = 1.9$ cm in water. A diffraction-limited beam of approximately 200-μm lateral dimension strikes the surface of a crystal, here Si. The large aperture of the acoustic lens ($f/1.5$), coupled with a large velocity ratio ($V_{water} = 1.4$ km/s; $V_{Si} = 5$–9 km/s), produces a beam with approximately 180° of angular dispersion in the crystal.

   On the opposing surface of the crystal is focused the receiving element, which is identical to the transmitter but connected to a radio-frequency amplifier, followed by a rectifying circuit and a boxcar integrator[7] for time selection. Either the transmitting or receiving transducer is scanned in a

Figure 3 Top: Schematic of the UFI experiment.   An ultrasonic beam, focused by an acoustic lens, propagates through the water and strikes the crystal. A similar receiver samples the ultrasound at a particular point on the opposite face. The detector position is scanned parallel to the surface by digitally controlled translation stages (see Figure 9 for electronics). (a) Resulting ultrasound image for a 2-cm cube of Si, revealing a combination of phonon-focusing effects and internal diffraction. (b) Phonon image of Silicon at 2 K, taken with a 1.5 × 1.5 cm² scan range and a tunnel-junction detector with an onset frequency of 450 GHz. (From Hauser et al.[6])

raster pattern, and the signal is collected for each pixel and displayed on a video monitor. (See Figure 9 below for further experimental details.) To emphasize that the point-source/point-detector configuration measures the flux of acoustic energy (associated with group velocity) in the crystal, the method has been named acoustic-flux imaging or ultrasonic-flux imaging (UFI), regardless of whether the acoustic amplitude, its absolute value, or its square is displayed.[8]

An ultrasonic-flux image of Si is shown in Figure 3(a). Immediately we see striking similarities and differences between this image and the corresponding heat-pulse image shown in Figure 3(b). First, the intense horizontal and vertical FT ridges present in the phonon image are absent in the ultrasound image. This is because the focused acoustic beam (which is a compression wave in the water) couples very weakly to the FT mode, which has a polarization nearly parallel to the surface. Second, the diagonal ST ramps bounded by sharp caustics in the phonon image are replaced in the ultrasound image by a series of parallel lines resembling diffraction fringes. Finally, the ST phonon-focusing structure in the middle of the phonon image (the ST box) becomes a square-shaped pattern of diffraction-like spots in the ultrasound image.

C. Internal diffraction

What is "diffracting" the ultrasonic beam? The answer is none other than the principal object of our attention in this book: the acoustic wave surface. Diffraction normally describes the interference of waves that are partially obstructed by solid objects. While the wave surface is not a solid object, it is a real-space structure associated with the wave itself. It is centered about the acoustic point source and clearly projected in a phonon image. As Maris predicted, the nonzero wavelengths limit the sharp resolution of the wave surface. And when a coherent source is employed, an interference pattern is observed [Figure 3(a)], an effect that Hauser et al.[6] called internal diffraction.

What is really happening? Let us try to understand the exact origin of this interference effect in the ultrasonic image. It takes (at least) two waves to interfere. What are the two waves involved?

Consider Figure 4. The focused ultrasound beam of the transmitter produces an oscillating stress at the surface of the crystal for a short time Δt – a so-called tone burst or wave train. Each oscillation of this local disturbance produces a Huygens wavelet radiating outwardly from the point source, creating a wavefront in the shape of the wave surface. Thus, we can consider the result of the tone burst to be a series of constant-t surfaces more closely spaced than the three illustrated in Figure 20 on page 52. In other words, the

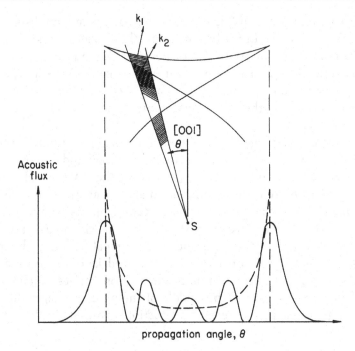

Figure 4 *A tone burst of ultrasound generated at a point on the crystal surface produces a series of expanding wavefronts in the shape of the wave surface. For a small angular section in the far field, the acoustic field is well approximated by a set of plane waves whose wavevectors are normal to sections of the wave surface. When these waves overlap, interference occurs, giving rise to the internal diffraction effect. (From Wolfe and Hauser.[14])*

group-velocity surface defines surfaces of constant phase in the expanding wave train.

If we concentrate on the small solid angle shown in Figure 4 (possibly the angle subtended by a detector), we see that the ultrasonic tone burst in this direction is approximated by the sum of three plane waves corresponding to three sections of the wave surface. As in Figure 1, the normals to the plane waves are not collinear with their radial propagation direction. Indeed the wavevectors are normal to the corresponding segments of wave surface, as argued in Chapter 2. If the tone burst is sufficiently long, we see that two of the acoustic fields overlap and have the potential to interfere. The degree of interference depends upon the directions of the polarization vectors, which should be close for waves with nearby k vectors. If the tone burst is lengthened such that the third portion of the wave surface is included, more complicated interference would occur, although this third segment has a much different wavevector and polarization from the other two. For simplicity, we ignore the third wave in the subsequent discussion and deal solely with $\mathbf{k}_1$ and $\mathbf{k}_2$. As the observation angle θ is rotated, the phase difference between the two partial

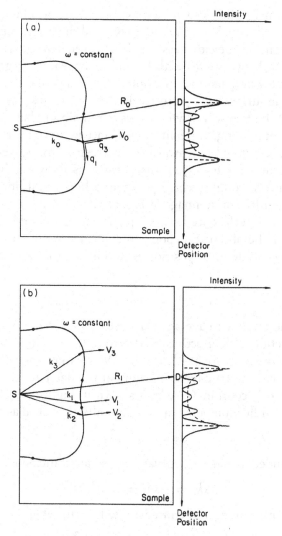

Figure 5 (a) Acoustic slowness surface superimposed on the Si crystal and centered on the point source, S. The early-arriving signal detected at D has a group velocity $\mathbf{V}_0$ corresponding to the inflection in the slowness surface and the fold in the wave surface. (b) When the detector is moved down, two early-arriving waves are detected, as also illustrated in Figure 4. Destructive interference between these two waves occurs when $(\mathbf{k}_1 - \mathbf{k}_2) \cdot \mathbf{R}_1 = \pi$. (From Hauser et al.[6] and Weaver et al.[9])

waves changes and a pattern of interference fringes are created, as shown schematically in the figure. This is internal diffraction.

A more quantitative measure of this self-interference effect can be gained by considering the corresponding acoustic slowness surface, superimposed on the crystal in Figure 5.[9] In Figure 5(a) we consider the source-to-detector direction, R_0, which coincides with a phonon caustic, or fold in the wave surface. The group velocity, V_0, is parallel to R_0, and the wavevector k_0 points to an inflection point on the slowness surface. As the detector is moved downward [Figure 5(b)], two waves with k_1 and k_2 appear with group velocities parallel to the new propagation direction, R_1. These are the waves discussed in Figure 4. If the surface were perfectly flat near k_0, then $k_1 \cdot R_1 = k_2 \cdot R_1$ (see the figure), and the two waves would always remain in phase at the detector, constructively interfering. In reality the surface is curved near the inflection point, so as R_1 is rotated downward the corresponding waves begin to destructively interfere, producing an Airy pattern, as shown in the figure.[1] This interference pattern roughly corresponds to a horizontal line scan through the center of the ultrasound image of Figure 3(b).

We plotted the interference between plane waves k_1 and k_2 in Figure 5(b) on page 28. It is helpful at this point to explicitly write out the superposition of the waves at detector position R_1 assuming that they have the same polarization,

$$e^{ik_1 \cdot R_1} + e^{ik_2 \cdot R_1} = 2\cos(\Delta k \cdot R_1/2)e^{ik_0 \cdot R_1}, \tag{4}$$

which is a plane wave with average wavevector $k_0 = (k_1 + k_2)/2$, modulated by a cosine function with effective wavevector $\Delta k/2 = (k_2 - k_1)/2$. The nulls in the envelope function are fixed in space. We see that when $\Delta k \cdot R_1 = \Delta k_\parallel \times R_1 = \pi$, the waves destructively interfere at the detector. To estimate $\Delta k_\parallel$, we choose the $q_1 q_3$ coordinate system shown in Figure 5(a). The slowness surface near the inflection point can be generally described as

$$q_3 = -aq_1^3/k_0^2, \tag{5}$$

where k_0 is included to make the curvature constant dimensionless. Therefore,

$$\Delta k_\parallel = -2q_3 = 2aq_1^3/k_0^2, \tag{6}$$

and so the first minimum in the intensity will occur when

$$q_1 = (\pi k_0^2/2aR_1)^{1/3} \propto R_1^{-1/3}\lambda^{-2/3}, \tag{7}$$

in terms of the average wavelength, $\lambda = 2\pi/k_0$. The real-space angle between the intensity maximum and the first minimum is[9]

$$\Delta\theta \simeq |V_1 - V_0|/|V_0| = 3a(\pi/2ak_0R_1)^{2/3} \propto (\lambda/R_1)^{2/3}. \tag{8}$$

This result may be contrasted to the standard result[10] for optical diffraction of a plane wave from a semi-infinite opaque plane, where the angular spacing between fringes scales as $(\lambda/R)^{1/2}$.

Up to this point, I have been deliberately vague about the actual stress fields induced at the crystal surface. In fact, the sound wave in the water can be modeled quite accurately, and the refraction of this incident field at the surface

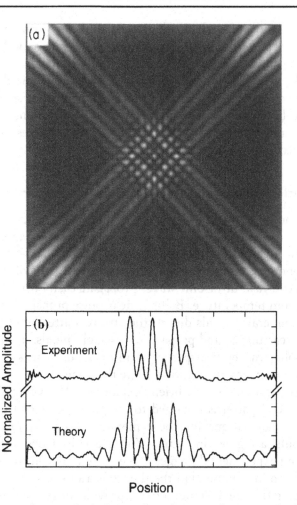

Figure 6 (a) Theoretical calculation of an ultrasonic-flux image of Si based on a continuum elastodynamic model for the received signal, assuming the experimental ultrasonic configuration of Figure 3. (b) Horizontal line scans across the theoretical and experimental images. (From Weaver et al.[9])

of an anisotropic medium can be modeled with acoustic-mismatch theory, as discussed in Chapter 11. The coupling of the incident pressure wave to various crystalline modes may be accounted for in detail. The rigorous calculation of the transmitted acoustic signal does require considerable computer time, but it has been performed for the present case by Weaver et al.,[9] and the results are shown in Figure 6(a). This image is virtually identical to the experimental image of Figure 3(a).

Horizontal scan lines through the experimental and theoretical images, analogous to the Airy pattern discussed above, are compared in Figure 6(b).

These traces (and, of course, the images) depend rather critically on the frequency and the crystal thickness, as systematically studied in Reference 9.

The bright features in the corners of the ultrasonic-flux images of Figures 3(a) and 6(a) deserve some attention. A broader scan range shows this effect to be part of a near-circular ring of high intensity, seen clearly in both experiment and theory.[11] It arises from head waves (pseudo–surface waves) as discussed in Chapter 12.

D.   Applications of ultrasonic-flux imaging

One of the most interesting aspects of the UFI technique is that it greatly enlarges the experimental domain for studying elastically anisotropic media. Imaging with heat pulses is generally limited to low temperatures because phonon–phonon scattering restricts the mean free path of high-frequency phonons. In contrast, ultrasonic waves propagate easily through many types of solids at room temperature. Ballistic heat-pulse propagation has also been limited to nonmetallic solids due to the strong scattering of phonons from carriers, whereas ultrasound propagates through metals. Finally, even non-crystalline solids can be transparent to ultrasound, so it is even possible to study elastic anisotropy in, say, composite materials.

All of these possibilities were briefly explored by Weaver et al.[9] Figures 7(a) and (b) show UFI patterns obtained for single crystals of Cu and Al. The corresponding calculations for these metals, with the known elastic constants as inputs, are given in Figures 7(c) and 7(d). General features of the theoretical and experimental images agree quite well. One of the factors that has not yet been incorporated in the theory is acoustic attenuation. It seems quite promising that the UFI method will provide unique new information on the anisotropy of acoustic attenuation. Conceivably the attenuation could arise from intrinsic sources, such as the carriers, or extrinsic sources, such as defects (e.g., dislocations).

One promising application of this technique is to alloys and phase transitions. Adapting his ultrasound-imaging apparatus for the transmission mode, Trivisonno[12] recently obtained the image shown in Figure 8 for a 50/50 alloy crystal of NiAl. The crystal is a cube of 1-cm dimension, and the frequency is 15 MHz. The ability to sensitively observe the elastic properties by internal diffraction suggests that this tool may be useful for observing phase transitions, such as the Martensitic transition, which has been reported for NiAl with a 63% atomic concentration of Ni at approximately 280 K.

Wesner et al.[13] have demonstrated a different approach to UFI; namely, recording both the *phase* and *amplitude* of the ultrasonic field simultaneously.

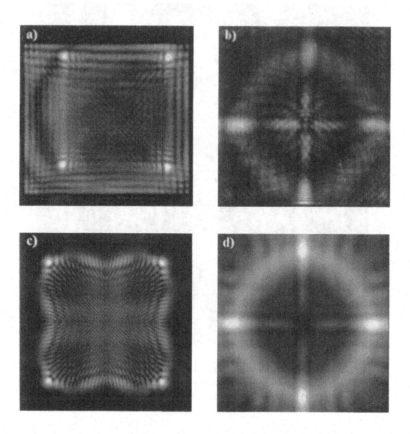

Figure 7 Experimental ultrasound images of single-crystal metals using 500-ns tone bursts at 15 MHz and (100) faces. (a) Cu, 16 mm thick, 17 × 17 mm² scan range. (b) Al, 7.35 mm thick, 12 × 12 mm² scan range. Calculated ultrasound images for (c) Cu and (d) Al, corresponding to the experimental images, but having somewhat different scan ranges. (From Weaver et al.[9] and Hauser.[16])

They developed a UFI system based on a commercial acoustic microscope. Their experiments utilize ultrasonic transducers with a frequency of 392 MHz and focal spots in the range of 1 μm. Phase-sensitive images of single crystals of LiTaO$_3$ and GaAs show extremely fine interference patterns in the form of a hologram. Of course, the acoustic phases of various interfering waves are very sensitive to crystal dimensions, scan range, surface irregularities, etc., so imperfections have large effects in the images. In the future, however, the use of near-gigahertz frequencies may permit the study of anisotropic acoustic propagation in small samples.

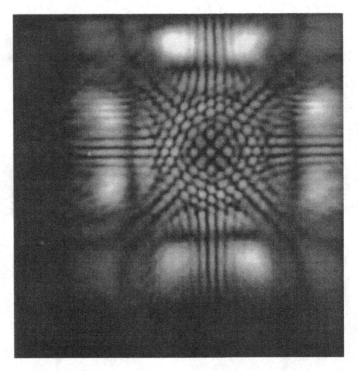

Figure 8 UFI of a 50/50 alloy crystal of NiAl at room temperature and 15-MHz frequency. Horizontal and vertical axes are {110}; the pattern is centered on (001). (From Trivisonno.[12])

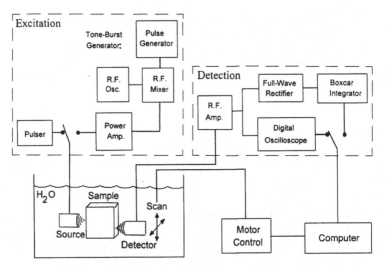

Figure 9 (a) Diagram of the UFI apparatus. (From Hauser.[16])

E. Acoustic wavefront imaging

The information accessible to a UFI experiment can be collected and analyzed in a variety of ways. A more detailed look at the apparatus is shown in Figure 9. Instead of utilizing the radio-frequency rectifier and boxcar integrator, one can choose to collect the entire time trace at a given spatial position, using a high-speed digital oscilloscope or a transient recorder. The signal acquired by the oscilloscope can be digitally rectified in the computer or processed in other ways.

To record a time trace for each detector position in an image takes a considerable amount of storage space. Nevertheless, Wolfe and Hauser[14] have collected such an $x–y–t$ data cube for several materials to examine the utility of this approach. The idea is illustrated in Figure 10, which shows (a) a typical time trace for ultrasound in Si, (b) an $x–t$ plane of a data cube for Si, giving the acoustic amplitude as a pseudo–three-dimensional plot, and (c) a graytone image of the same data (similar to Figure 2). Typical data cubes store up to 256 such $x–t$ images, each containing 256×256 data points. A series of $x–t$ images for several y planes in Si are reproduced in Figure 11. The complexity associated with the multiply folded ST wave surface near the cube axis is clearly displayed.

The data cube may also be played back as a series of $x–y$ images at successive times, producing a slow-motion view of the various acoustic waves emanating from a point source and striking the back side of the crystal. This type of "wavefront movie" allows the experimenter to clearly identify the origins of the various ballistic, reflected, and mode-converted components. Several frames of just such a movie taken with 15-MHz transducers are shown in Figure 12. The first three images show the L mode expansion, and the last three images display the expanding T mode. Waves reflected from the crystal surfaces are apparent in images d–f.

Figure 2 in the Preface of this book is a superposition of three snapshots like those in Figure 12 but taken with 50-MHz transducers. A similar data-collection procedure was used for high-frequency phonons, and the results were reproduced in Figure 4 on page 67. In the ultrasonic case, one sees the expanding wavefronts of transverse waves replete with the ubiquitous folds in the wave surface and internal interference effects. The phonon movie, on the other hand, contains the FT mode, which is absent from the ultrasound due to the polarization of the generating pressure wave.

All the data discussed so far in this section correspond to short-pulse excitation, which directly reveals the acoustic wavefronts. In contrast, Figure 13 shows one frame of a movie obtained when a tone burst is used. These data expose the internal diffraction effect from a different perspective: radial node lines are clearly seen in the folded region of the wave. An internal diffraction

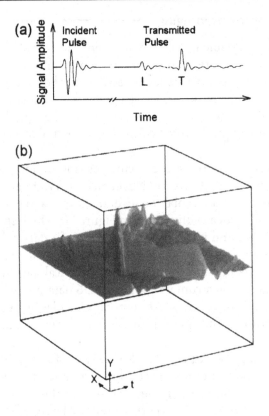

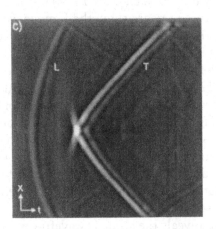

Figure 10 (a) Typical timetrace of a short ultrasound pulse traversing a Si crystal. (b) Schematic diagram of an acoustic data cube, showing a pseudo–three-dimensional plot of acoustic intensities in an x–t plane. 256 such arrays are contained in the data cube. (c) Graytone representation of the x–t plot in (b). (From Wolfe and Hauser.[14])

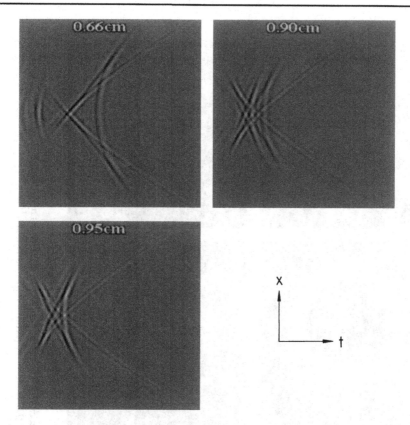

Figure 11 *Successive y slices of the data cube for Si showing x–t images of the ST wave at high resolution (50 MHz).*

image, such as the one in Figure 3(a) (rotated 45°), is effectively an average of the movie over all time.

Indeed, the internal diffraction effect is even contained in a data cube produced with a *short* excitation pulse and a *broad* receiver bandwidth. Consider the Si data cube that produced the high-resolution slices of ST wavefronts shown in Figure 14. If each of the time traces in this x–y–t data cube is Fourier transformed, the result is an x–y–v data cube! Now each x–y image corresponds to a particular frequency, as if a continuous-wave source were used.[15] Several of these images are shown in Figure 15. With this single data cube, the internal diffraction effect is reproduced for frequencies between 5 and 50 MHz. For comparison, a phonon image of Si, corresponding to acoustic waves of several hundred gigahertz, is also shown in the figure.

The series of images in Figure 14 brings us full circle to the initial predictions of Maris, as mentioned in the introduction of this chapter. We see that as the ultrasonic frequency of the waves is increased, the flux pattern

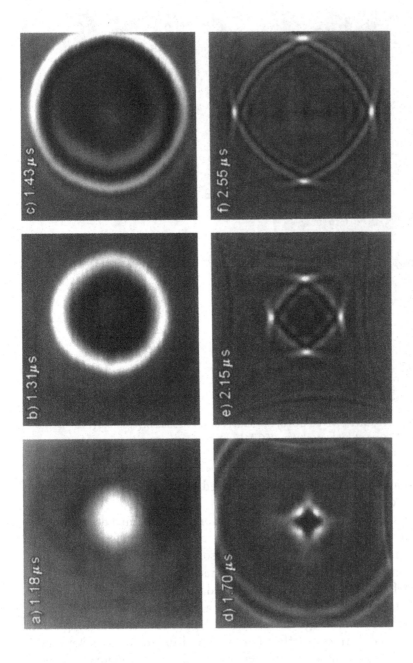

Figure 12 Successive time slices of the data cube for Si showing acoustic wavefronts of the longitudinal (a through c) and slow transverse (d through f) waves. 15-MHz resolution. (From Wolfe and Hauser.[14])

a) 1.18 μs

b) 1.31 μs

c) 1.43 μs

d) 1.70 μs

e) 2.15 μs

f) 2.55 μs

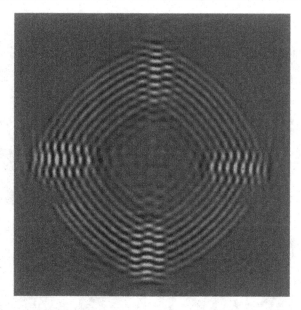

Figure 13 Waves produced by a 15-MHz tone burst traversing a 1-cm-thick Si crystal. The ST wave displays radial interference nodes along the fold directions. (From Wolfe and Hauser.[14])

approaches that of the high-frequency phonons. The ultrasound images are, in effect, diffraction-limited pictures of the phonon caustics. The frequencies in these experiments span the range of a few megahertz to nearly a terahertz – 10^6 to 10^{12} Hz.

F. Acoustic imaging in composite materials

We conclude this chapter on a practical note. The ability to observe vibrational wavefronts and internal diffraction emanating from a point source will no doubt be applicable to the analysis of new materials. By playing the data back as a video, the experimenter can sift through massive amounts of information, effectively using the human brain as an ultra-high-speed image processor to pick out the important details.

For example, carbon-fiber epoxy composites possess a combination of high strength and low weight that span a variety of uses – from tennis rackets to aircraft wings. The strength of these materials lies in the imbedding of carbon fibers in an epoxy base, combined with lamination of layers with different fiber orientation. These materials have an elastic anisotropy that lends itself naturally to the ultrasonic-imaging methods described above. Indeed, the

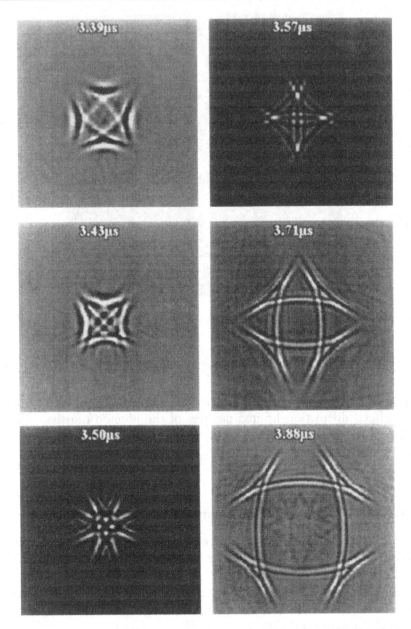

Figure 14 Series of x–y slices through the x–y–t data cube obtained with 50-MHz transducers, showing a high-resolution view of the evolving ST wavefront in Si. (From Wolfe and Hauser.[15]) This data cube is used to generate the Fourier-transformed data displayed in Figure 15.

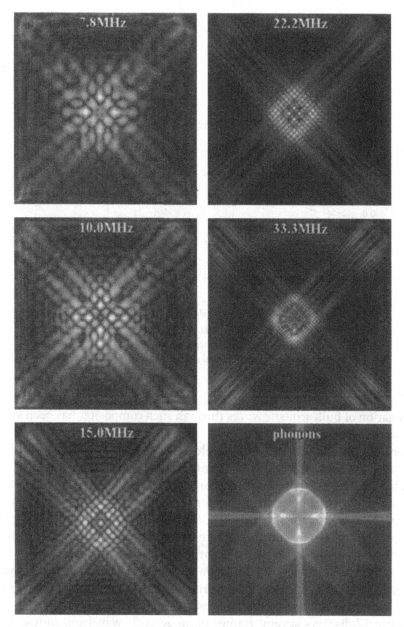

Figure 15 Series of x–y slices through the Fourier-transformed (x–y–v) data cube of Si. In effect, these experimental images are the wave patterns formed by a continuous wave at the specified frequency. (From Wolfe and Hauser.[15])

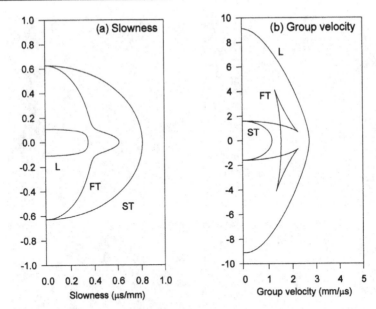

Figure 16 (a) Slowness curves calculated for a uniaxial composite with fibers oriented along the vertical axis. Elastic constants for this material are given in Ref. 20. (b) Group-velocity curves calculated with the same parameters. Note that the horizontal scale is twice the vertical scale.

propagation of bulk acoustic waves through such composites has been studied by several experimenters.[17-20]

Consider the simplest case of a sample in which all the laminates are bonded together with their carbon fibers oriented in the same direction. Ideally, the interfaces between laminates disappear, and the resulting sample has an axis of rotational symmetry along the fiber direction. This is the same acoustic symmetry as a hexagonal lattice, where the slowness and wave surfaces are invarient to rotations about the symmetry axis (see Chapter 5). Calculated cross sections of these two surfaces for this uniaxial composite are shown in Figure 16.[20] Folds appear in the wave surface, just as for crystals, and the rather large elastic anisotropy gives rise to an obvious interaction between the L and FT modes, similar to that observed in TeO_2 (Chapter 5).

Figure 17 displays several frames of an acoustic wavefront movie of this material.[20] The elliptic-like L wave appears first, and the FT wave with its prominant folds is observed at late times [Figure 17(c)]. The simple acoustic model described above appears to be well justified. It should be noted, however, that these composite structures are quite lossy: waves with frequencies above a few megahertz are strongly attenuated. In Chapter 14, we shall consider further ultrasound experiments on this and other composite samples from the perspective of surface acoustic waves.

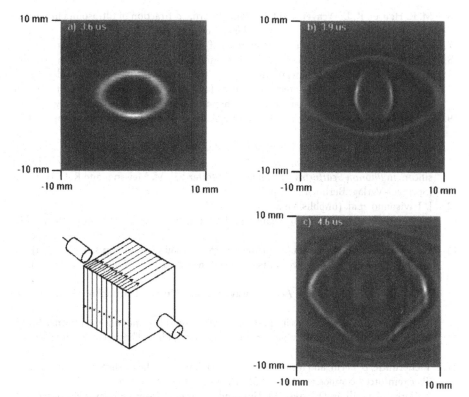

Figure 17 Images of wavefront arrivals from a 1-cm-thick uniaxial composite sample with fibers oriented horizontally (i.e., rotated 90° from Figure 16). These x–y scans are from Wolfe and Vines.[20] Comparing to Figure 16, one can identify the elliptic-like L wavefront and the folded FT wavefront. The ST wavefront is not directly excited by the incident pressure wave in the water.

References

1. H. J. Maris, Effect of finite wavelength on phonon focusing, *Phys. Rev. B* **28**, 7033 (1983).
2. V. V. Novikov and L. A. Chernozatonskii, Feasibility of observing the focusing of ultrasound energy flux, *Sov. Phys. Acoust.* **34**, 215 (1988).
3. B. A. Auld, *Acoustic Fields and Waves in Solids* (Krieger, Malabar, 1990).
4. A. G. Every, W. Sachse, K. Y. Kim, and M. O. Thompson, Phonon focusing and mode conversion in silicon at ultrasonic frequencies, *Phys. Rev. Lett.* **65**, 1446 (1990); A. G. Every and W. Sachse, Imaging of laser generated ultrasonic waves in Si, *Phys. Rev. B* **44**, 6689 (1991); *J. Acoust. Soc. Am.* **95**, 1942 (1994). See also G. Rümpker and C. J. Thomson, Seismic-waveform effects of conical points in gradually varying anisotropic media, *Geophys. J. Int.* **118**, 759 (1994).
5. A. G. Every and W. Sachse, Determination of the elastic constants of anisotropic solids from acoustic-wave group-velocity measurements, *Phys. Rev. B* **42**, 8196 (1990). See also L. Wang, Determination of the ray surface and recovery of elastic constants of anisotropic elastic media: a direct and inverse approach, *J. Phys. Condensed Matter* **7**, 3863 (1995).

6. M. R. Hauser, R. L. Weaver, and J. P. Wolfe, Internal diffraction of ultrasound in crystals: phonon focusing at long wavelengths, *Phys. Rev. Lett.* **68**, 2604 (1992).

7. See Section A of Chapter 3 for a description of the function of this instrument.

8. The terminology also emphasizes the distinction between this method, which measures the *angular variations* of ultrasonic flux in elastically anisotropic media, and the more common ultrasonic-imaging techniques used for flaw location and nondestructive evaluation. A similar contrast occurs between phonon imaging and thermal imaging.

9. R. L. Weaver, M. R. Hauser, and J. P. Wolfe, Acoustic flux imaging in anisotropic media, *Z. Phys. B* **90**, 27 (1993).

10. E. Hecht, *Optics*, 2nd ed. (Addison-Wesley, Reading, MA, 1987), p. 456.

11. M. R. Hauser, R. L. Weaver, and J. P. Wolfe, Internal diffraction of acoustic waves in silicon, in *Phonon Scattering in Condensed Matter VII*, M. Meissner and R. O. Pohl, eds. (Springer-Verlag, Berlin, 1993), p. 77.

12. J. Trivisonno et al. (unpblished data).

13. J. Wesner, K. U. Wurz, K. Hillmann, and W. Grill, Imaging of coherent phonons, *ibid.*, p. 68.

14. J. P. Wolfe and M. R. Hauser, Acoustic wavefront imaging, *Ann. Phys.* **4**, 99 (1995). See also J. P. Wolfe, Acoustic wavefronts in crystalline solids, *Phys. Today* **48**, 34 (September 1995).

15. J. P. Wolfe and M. R. Hauser, Acoustic wavefronts and internal diffraction, *Physica B* **219 & 220**, 702 (1996).

16. M. R. Hauser, Acoustic waves in crystals: I. Ultrasonic flux imaging and internal diffraction, II. Imaging phonons in superconducting niobium, Ph.D. Thesis, University of Illinois at Urbana-Champaign (1995).

17. T. Lhermitte, B. Perin, and M. Fink, Dispersion relations of elastic shear waves in cross-ply fiber reinforced composites, *Proc. IEEE Ultrason. Symp.* **89**, 1175.

18. C. Corbel, F. Guillois, D. Royer, M. Fink, and R. De Mol, Laser-generated elastic waves in carbon-epoxy composite, *IEEE Trans. Ultrason. Ferroelectr. Freq. Control* **40**, 710 (1993).

19. W. Sachse, B. Castagnede, I. Grabec, K. Y. Kim, and R. L. Weaver, Recent developments in quantitative ultrasonic NDE of composites, *Ultrasonics* **28**, 97 (1990).

20. J. P. Wolfe and R. E. Vines, Acoustic wavefront imaging of carbon-fiber/epoxy composites, in *Proceedings of 1996 IEEE Ultrasonics Symposium* (IEEE, Piscataway, NJ, 1997), p. 607.

14

Imaging surface acoustic waves

A. Context

With few exceptions this book has concerned itself with the propagation of vibrational waves in the *bulk* of a crystal. As indicated in Section B of Chapter 12, there are also vibrational modes associated with the surface of the medium. They are known as Rayleigh surface waves (RSWs) and pseudo–surface waves (PSWs). As with bulk waves, the velocities of these surface waves depend on the propagation direction, so one might reasonably expect phonon-focusing effects. Indeed, Tamura and Honjo[1] and Camley and Maradudin[2] have predicted phonon-focusing patterns for Rayleigh waves in a variety of media. An example of the latter results[2] is shown in Figure 1: (a) is the slowness curve for Rayleigh waves on the (111) surface of Ge and (b) shows the corresponding singular directions of energy flux emanating from a point source in this plane. Phonon caustics arise from the zero-curvature regions of the slowness surface, just as discussed with bulk waves in Figure 8 of Chapter 2.

Unfortunately, high-frequency surface phonons are likely to have much shorter mean free paths than bulk phonons, due to the sensitivity of these waves to surface irregularities. The scattering of surface phonons from such defects, plus the potentially huge background of scattered bulk phonons, makes it extremely difficult to observe the ballistic propagation and phonon focusing of surface phonons. A recent study by Höss and Kinder,[3] however, has reported experimental evidence for the focusing of surface phonons on a laser-annealed Si surface.

At frequencies in the megahertz range, the focusing of surface acoustic waves has been clearly demonstrated. For an acoustic source, Kolomenskii and Maznev[4] have employed a giant laser pulse (Nd:YAG, 10 mJ, 10 ns) focused to a 7-μm spot on the surface of a polished Ge crystal at room temperature. The incident energy is sufficient to melt the Ge in the small excitation region

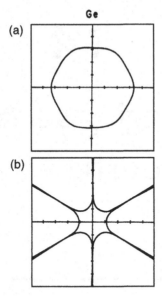

Ge

(a)

(b)

Figure 1 (a) Slowness curve for the Rayleigh surface wave on the (111) surface of Ge. Inflection points give rise to singular flux. (b) Focusing pattern showing caustic lines along the surface. Caustic directions correspond to the group velocities (normals) at inflection points in (a). (From Camley and Maradudin.[2])

and, by the thermoelastic effect, generate an intense surface wave propagating outward from the excitation point. Prior to the pulse, the crystal is covered with a dust of 1–2-μm Al_2O_3 particles. These microparticles are ejected from the surface in regions where the surface wave exceeds a certain intensity. Figure 2 shows a photograph of the dusted surface after the laser pulse. The dark spokes are where the microparticles have been removed and thus represent directions of strong focusing of acoustic energy in the (111) plane. The results agree nicely with the calculations for Ge in Figure 1.

Kolomenskii and Maznev have also performed time-of-flight measurements of the intense surface waves by using the deflection of a focused HeNe laser beam as a contactless point detector. With this high-resolution detection method and an angular scanning technique described below, Maznev et al.[5] have directly observed the folds in the Rayleigh wave surface of GaAs, as shown in Figure 3. By tracing the arrival times of the pulses in Figure 3(b), one can identify the folded wave structure plotted at the top of Figure 3(a).

B.   Anisotropic propagation of coherent surface waves ──────

By adapting the ultrasonic methods discussed in Chapter 13, Vines et al.[6] have developed a quantitative means of characterizing the propagation of

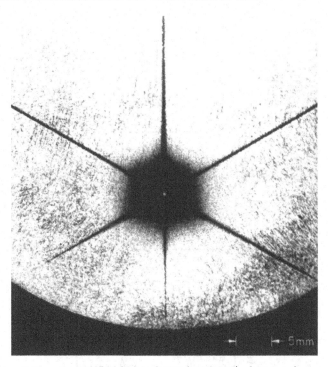

Figure 2 Experiment of Kolomenskii and Maznev[4] showing the powder pattern of a Ge surface after point excitation with a giant laser pulse. The dark lines are where the powder is removed due to intense surface waves radiating from the excitation point, corresponding to the caustic lines in Figure 1(b).

coherent surface acoustic waves in anisotropic media. Not surprisingly, the introduction of this approach has provided some new perspectives on surface acoustic waves. In this section, the basic methods and a sampling of the experimental measurements are described. Detailed explanations of the results require an understanding of the complex topologies of both RSWs and PSWs in anisotropic media, which are introduced in the following section.

The experimental arrangement is simple, as depicted schematically in Figure 4. The sample is mounted on a rotating platform and immersed in water. Ultrasonic transducers – transmitter and receiver – are focused to points on the sample surface, separated by a distance d. By rotating the crystal about an axis normal to this surface, the propagation of surface waves along any propagation direction in the surface can be recorded. As in the bulk experiment described in Chapter 13, a broad angular distribution of wavevectors is generated and detected by these transducers, so the measurement reveals the acoustic energy flux (a superposition of plane waves) along a given direction. Consequently, we expect to see evidence of both phonon focusing and internal diffraction.

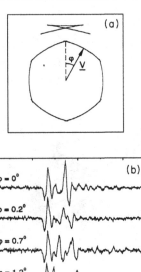

*Figure 3 (a) Wave surface for Rayleigh waves in the (111) plane of GaAs.
The magnification at the top (near the (112) direction) shows the small fold
in this group-velocity surface. (b) Experimental angle-time plot showing
the predicted fold structure for GaAs. (From Maznev et al.[5])*

Figure 5(a) shows the data obtained for a tone burst of 15-MHz ultrasound
propagating across a (001) surface of Si.* Each vertical line of this image
represents a time trace of the pulse. Light and dark regions correspond to
positive and negative amplitudes, respectively. A series of these time traces
are recorded as the sample is rotated, producing the angle – time (or θ–t)
image shown.[7]

To determine the energy propagating along a certain direction, the absolute
value of the signal is integrated over an entire time trace. This integrated signal

---

* The surface of this 2-cm cube of Si is highly polished; however, identical data are obtained for lapped (i.e.,
  highly damaged) surfaces because the ultrasonic waves have wavelengths much larger than the damaged
  layer – a condition not valid for high-frequency phonons.

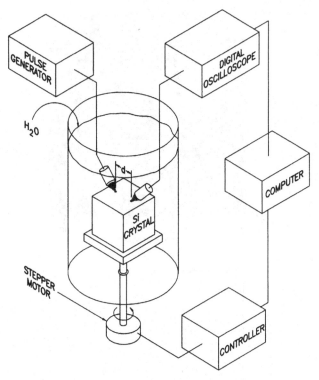

Figure 4 Schematic diagram of the ultrasound imaging experiment of Vines et al.[6] The sample is immersed in water. The transmitter and receiver transducers are focused to fixed points on the surface, and the crystal is rotated about an axis normal to the surface. A time trace of ultrasound is recorded for each angular position, producing an angle-time image.

is plotted in Figure 5(b). As expected, the ultrasonic flux varies significantly as a function of angle. Below, these data will be compared to a calculation of the acoustic flux (Figure 12). It turns out that the broad structures near 20° and 70° in Figure 5 are associated with phonon focusing of the RSW. The oscillations are due to internal diffraction.

Additional insights into the surface wave propagation are obtained by using a short pulse rather than a tone burst. Figure 6 shows an assortment of experimental results for different crystals. These experiments are analogous to the wavefront imaging experiments described in Chapter 13 and shown in Figure 2 of the Preface, which was obtained for bulk wave propagation in Si. A rich variety of wavefront structures are observed in the surface wave experiments, and not all the structures are associated with Rayleigh waves; portions originate from PSWs. Now we take a look at the theoretical basis for all these possibilities.

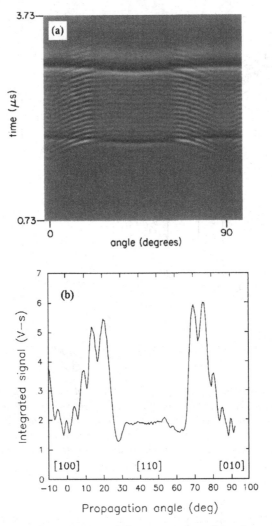

Figure 5 (a) Angle-time image of the (001) surface of Si, obtained with a tone burst of 15-MHz ultrasound containing approximately 16 oscillations, and d = 10.0 mm. (b) Plot of time-integrated amplitude of the ultrasound recorded in (a). (From Vines et al.[6])

C. Surface acoustic waves on a free surface

What is a surface wave? It is composed of a linear combination of solutions to the wave equation for an infinite medium, constructed to satisfy the boundary conditions of a free or loaded surface.[8] In this context the solutions to the bulk wave equation are called partial waves. To qualify as a true surface wave (an RSW), all partial-wave components must be localized near the surface;

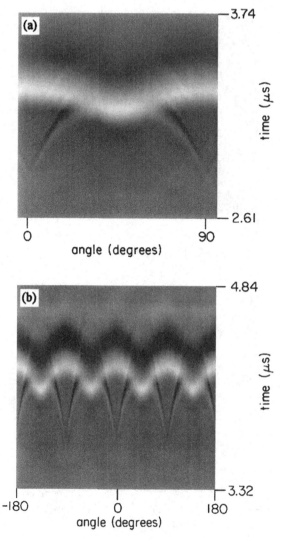

Figure 6 *Angle-time images for short-pulse excitation. (a) (001) surface of Si, d = 16.13 mm. (b) (001) surface of Ge, d = 7.057 mm. (c) Natural crystal of c-cut calcite (CaCO₃), d = 5.00 mm. (d) (110) surface of calcium fluoride, d = 8.88 mm. (From Vines et al.[6])*

that is, they must be evanescent waves that decay exponentially with distance into the solid. In addition to Rayleigh waves, it is often possible to satisfy the boundary conditions with a combination of evanescent and nondecaying waves, resulting in a PSW.

First we consider what partial waves are available. Recalling the relation between wavevector and slowness, $\mathbf{k} = \omega\mathbf{s}$, we write a solution to the wave

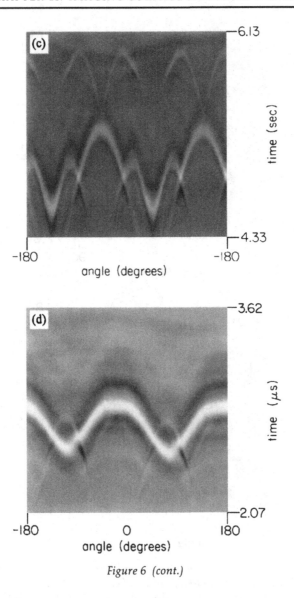

Figure 6 (cont.)

equation of the infinite medium [Eq. (15) on page 42] as

$$u_l = e_l e^{i\omega(\mathbf{s}\cdot\mathbf{r}-t)}. \tag{1}$$

Plugging this into the wave equation, we obtain the three linear equations for the polarization components,

$$(C_{ijlm}s_j s_m - \rho\delta_{il})e_l = 0, \qquad i = 1, 2, \text{ and } 3. \tag{2}$$

The existence of nontrivial solutions to these equations requires that the

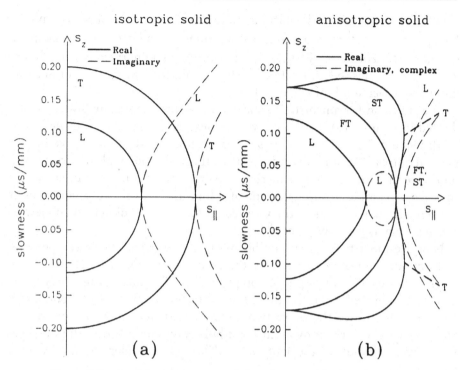

Figure 7 Real and imaginary parts of the solutions, $s_z^{(n)}$, to Eq. (3) for a given surface slowness, $s_{\parallel}$. There are always six solutions for each $s_{\parallel}$, but the solutions may be real, imaginary, or complex. (a) Elastically isotropic medium. (b) (010) plane of an anisotropic solid similar to Si. Dashed curves show the real (T) and imaginary (FT, ST) parts of the complex mode.

determinant of the above matrix must vanish:

$$|C_{ijlm}s_j s_m - \rho\delta_{il}| = 0. \tag{3}$$

This is a sixth-order equation in s_i. Because we will be dealing with a planar surface of the medium, we define the cartesian coordinates x, y in the plane of the surface, making the z axis normal to the surface. For any $\mathbf{s}_{\parallel} = s_x\hat{x} + s_y\hat{y}$, Eq. (3) can be solved for s_z, yielding six solutions, $s_z^{(n)}$, where $n = 1, 2, \ldots 6$. In Chapters 2 and 4 we considered only real solutions for s_z, which generate three slowness sheets – L, FT, and ST – corresponding to nondecaying bulk waves. Now, in order to satisfy the boundary conditions, we must consider also the complex solutions, which allow waves that exponentially increase or decrease with distance into the sample.

For simplicity we first consider an isotropic solid. Taking $v_L/v_T = (C_{11}/C_{44})^{1/2} = 1.7$, we solve Eq. (3) over a range of $s_{\parallel}$ and plot the solutions $s_z^{(n)}$ in Figure 7(a). The heavy solid lines are the purely real solutions – a single L mode with velocity v_L and two degenerate T modes with velocity v_T. For small $s_\parallel$ there are six real solutions, counting both positive and negative s_z. For $v_L^{-1} < s_\parallel < v_T^{-1}$ there are four real solutions (T) and two purely imaginary solutions (L). For $s_\parallel > v_T^{-1}$ there are six purely imaginary solutions, shown as the dashed lines in Figure 7(a). The hyperbolic form of these solutions was discussed in Section B of Chapter 12.

The situation for anisotropic solids is somewhat more complicated, as indicated in Figure 7(b), which shows slowness curves similar to those in the (010) plane of Si ($s_y = 0$, $s_x = s_\parallel$). The concavity of the ST curve in this plane requires that, for some values of $s_\parallel$, there are four real solutions for s_z on this branch alone. As $s_\parallel$ is increased, pairs of real solutions merge and are replaced by complex-conjugate pairs of solutions. Notice that for any chosen $s_\parallel$ there are six solutions to the wave equation. A three-dimensional view of this topology is given by Every.[9]

Now we must take linear combinations of the partial wave components to form surface waves that satisfy the boundary conditions. The reader is referred to Eqs. (32) through (39) on pages 287–88, which set up the formalism for this procedure in the context of refraction at a solid/solid boundary. For the free surface there are two distinct cases that must be considered: (1) surface waves driven by an incident bulk wave, and (2) free-running surface waves without an incident wave. Then we consider the fluid loaded surface.

i) Free surface driven by an incident bulk wave

We imagine first a free surface driven by an incident bulk wave in the solid, as discussed in Chapter 12. In this case we consider the following linear combination of solutions to the wave equation (i.e., a sum of partial waves):

$$\mathbf{u} = \mathbf{e}^{(0)} e^{i\omega[\mathbf{s}^{(0)} \cdot \mathbf{r} - t]} + \sum_{n=1}^{3} \Gamma_n \mathbf{e}^{(n)} e^{i\omega[\mathbf{s}^{(n)} \cdot \mathbf{r} - t]}, \tag{4}$$

where the first term represents the incident bulk wave and the remaining terms are reflected waves. The reflected waves are solutions to the wave equation that are phase matched to the incident wave (i.e., they have the same $\mathbf{s}_\parallel$). Only three of the six possible phase-matched waves are required in the sum: the bulk waves with group velocities pointing into the solid and the evanescent waves with intensities that exponentially decrease (not increase) into the solid. For the free-surface boundary conditions, $\sigma_{xz} = \sigma_{yz} = \sigma_{zz} = 0$ at $z = 0$, the coefficients of the partial waves must satisfy the three linear equations:

$$A_{in} \Gamma_n = B_i, \tag{5}$$

where A_{in} and B_i are given in terms of the elastic constants, polarization vector, and slowness vector [Eq. (38) on page 288]. The B_i represent the contribution from the incident wave. The solution to these equations is given by[5]

$$\Gamma_n = \bar{A}_{mn} B_m / |A_{pq}|, \tag{6}$$

where $|A_{pq}|$ is the boundary condition determinant and $\bar{A}_{mn}$ is the cofactor of A_{mn}.

For any given value of $s_{\parallel}$ defined by the incident wave, it is possible to find a combination of waves that satisfy the free-surface boundary conditions. If $s_{\parallel} < v_L^{-1}$, the solution will involve up to three reflected bulk waves, L, FT, and ST (i.e., there exists no surface wave). As $s_{\parallel}$ exceeds v_L^{-1}, a solution will be found that involves the L evanescent wave [Figure 7(a)] and one or more bulk T waves. For $s_{\parallel}$ slightly larger than v_L^{-1} the amplitude of the evanescent wave can be very large; this is the driven PSW discussed in Chapter 12 (e.g., page 306). Values of $s_{\parallel}$ beyond all the real slowness curves are not considered here because there is no bulk driving wave in that region.

ii) Free-running surface waves

The presence of the incident bulk wave driving the surface helped greatly in satisfying the free-surface boundary conditions. Notice that in Eq. (4) there are three parameters (Γ_n) to adjust, and a solution [Eq. (6)] can always be found. If the driving term (i.e., the incident wave) in Eq. (4) is removed, then there are only two degrees of freedom, namely, the ratios of the three Γ_n. Another way of saying this is that $B_i = 0$ in Eqs. (5) and (6). This implies that a nontrivial solution for Γ_n exists only when

$$|A_{pq}| = 0. \tag{7}$$

This condition cannot be satisfied for an arbitrary choice of $s_{\parallel}$, but it is satisfied for certain values of $s_{\parallel}$. Figure 8(a) shows the value of this boundary condition determinant for a free surface of the isotropic solid depicted in Figure 8(b). For this choice of $v_L/v_T = 1.7$, the determinant vanishes at a velocity $v_S = 1.08 v_T$. This Rayleigh wave is composed of two evanescent waves – one L and one T. The T partial wave is polarized normal to the surface, so the particle motion is ellipsoidal in the saggital plane containing the surface normal and the wavevector. The RSW propagates without attenuation.

Now let us take a look at what happens for an anisotropic solid. Figure 8(c) shows the value of the boundary condition determinant for a (001) surface of Si and a wavevector pointing 15 degrees away from the [110] axis. The slowness curves for the saggital plane are plotted in Figure 8(d). The determinant vanishes at only one point, just beyond the slowness curve for the ST bulk wave. The RSW is composed of three evanescent partial waves, and it has a velocity slightly slower than that of the ST bulk wave with a wavevector along the surface. Because of this additional complexity, the RSW in an anisotropic medium is sometimes called a generalized Rayleigh wave.

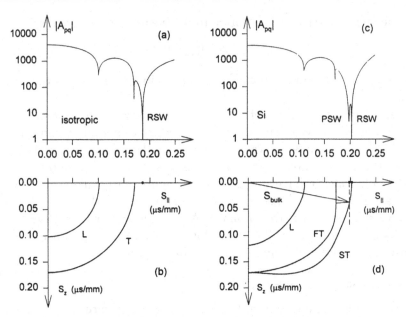

Figure 8 (a) Absolute value of the boundary condition determinant as a function of slowness for the free surface of an isotropic solid. The value of this determinant equals zero when the boundary conditions (zero stress) are met, as is the case for the RSW. (b) Corresponding slowness curves for the isotropic solid. (c) Absolute value of the boundary condition determinant as a function of slowness for the free (001) surface of Si, calculated at a propagation direction 30° from the [100] axis. (d) Corresponding slowness curves for Si. For this anisotropic solid a PSW appears that is phase matched to a bulk ST wave.

For the anisotropic case depicted in Figures 8(c) and 8(d), the determinant has a pronounced minimum between the bulk FT and ST slownesses along the surface. Lim and Farnell[10] pointed out that this type of behavior signals a new type of surface wave – the PSW. Up to this point we have considered only real values of $s_{\parallel}$. However, by adding a small amount of imaginary component to the *parallel* slowness – i.e., $s_{\parallel} = s_{\parallel\mathrm{real}}(1 + i\varepsilon)$ – it is possible to drive the local minimum in the determinant to zero. In this case, the boundary conditions are satisfied by a choice of two evanescent waves and a bulk ST wave, whose wavevector points into the solid, as shown in Figure 8(d). (See Ref. 11 for a more complete description of these partial waves.) Physically, an imaginary component of $s_{\parallel}$ implies that the PSW decays as it propagates along the surface. It is radiating its energy into the bulk via the phase-matched ST wave.

Figure 9(a) is a polar plot of the allowed values of $s_{\parallel}$ for the free-running RSWs and PSWs on the (001) surface of Si. The RSW traveling along [100] is elliptically polarized in the saggital plane containing the surface normal and the wavevector. As $s_{\parallel}$ is rotated away from the [100] direction the RSW

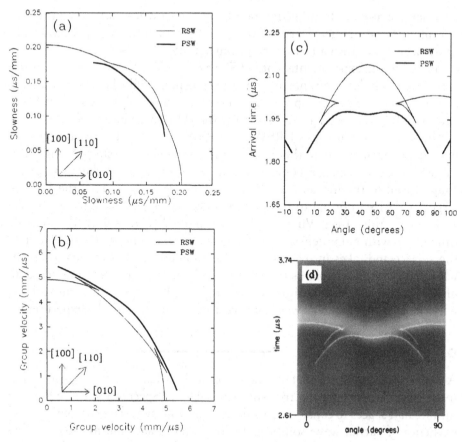

Figure 9 (a) Calculated slowness curves for RSWs and PSWs on the (100) surface of Si. For the PSW the real part of the slowness is plotted. (b) Calculated group-velocity curves corresponding to (a). (c) Calculated arrival times of RSWs and PSWs. (d) Superimposed on an experimental angle-time image of Si are the arrival-time curves as in (c), but with a dot density proportional to the predicted wave intensity. (From Vines et al.[6])

gains a component of transverse polarization parallel to the surface. The polarization of the RSW propagating along [110] is completely transverse and parallel to the surface, and its velocity is degenerate with the ST bulk wave (also known as the shear horizontal (SH) wave because its polarization is parallel to the crystal surface). In contrast, the PSW traveling along the [110] direction is polarized in the saggital plane due to L and FT partial-wave components and is uncoupled to the bulk wave. As $s_\parallel$ is rotated away from the [110] direction, the PSW couples to the bulk wave and gains a transverse component of polarization in the plane of the surface.

The corresponding group-velocity curves for these surface waves are plotted in Figure 9(b). The RSW displays folds in the group-velocity curve which, as usual, correspond to inflection points on the slowness curve. Figure 9(c) is a plot of the predicted arrival time, d/V, as a function of propagation angle, where $d = 1$ cm and V is the group velocity. This computation may be compared to the experimental data of Figure 6(a). The temporal width of the experimental pulse is not narrow enough to exhibit folds; however, one may conclude that the broad experimental pulse is composed mainly of an RSW near the {100} direction and a PSW near the [110] direction ($\theta = 45°$).

The relative intensities of RSWs and PSWs in the data are governed in part by the polarization of these waves and by phonon focusing. Along [100] the RSW has a large component of polarization normal to the surface and couples well to the incident water wave. Likewise, along [110] the PSW is largely polarized normal to the surface and couples well to the water wave. To quantify this effect, Vines et al.[6] have plotted [Figure 9(d)] the arrival-time curve with a dot density that increases with the normal component to the surface and takes into account the focusing effect. The calculation (thin white line) roughly explains the relative intensities in the experimental data of Figure 6, which is superimposed in Figure 9(d). In particular, it explains why the folds in the RSW wave surface are not observed in the experiment.

D. Fluid loaded surfaces

A proper treatment of surface waves on a fluid loaded solid is somewhat more complicated than the free surface problem. Specifically, there are two infinite half spaces to consider, each described by its own elasticity tensor. The existence of partial wave solutions for the fluid in addition to those for the solid leads to new types of surface waves, as well as modifications of the free-surface waves. For example, there exist surface waves composed of evanescent partial waves in both media, known as Scholte waves. Also, with fluid loading the RSW becomes a leaky Rayleigh wave, which radiates its energy into the fluid as it travels.

The problem is illustrated in Figure 10 for a water loaded Si surface. Figure 10(a) indicates how a surface wave with transverse polarization normal to the surface generates a longitudinal pressure wave in the liquid. For Si the phase velocity of the RSW along [100] is $v_S = 4.928 \times 10^5$ cm/s and the speed of sound in water is $v_\ell = 1.483 \times 10^5$ cm/s (there are no transverse modes in a fluid). As shown by the slowness curves plotted in Figure 10(b), conservation of $s_\parallel$ implies that the water wave is emitted in the direction $\theta_\ell = \sin^{-1}(v_\ell/v_S) = 17.5°$. Because the acoustic impedances, $Z = \rho v$, are nearly an order of magnitude different for water and Si, the RSW velocity is hardly changed from that on the free surface, although the RSW pulse is broadened in time due to the water loading, as seen in the data of Figure 9(d).

Vines et al.[6] and Tamura et al.[11] took a logical approach to this problem: they computed free-running wave solutions to the fluid loaded surface,

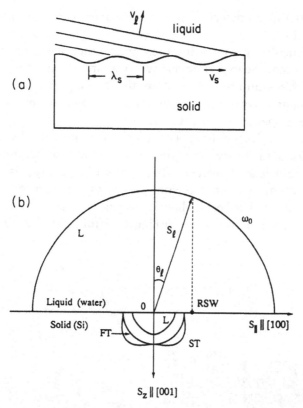

Figure 10 (a) Diagram showing the coupling of a transverse surface wave on a solid with the (longitudinal) sound wave in a liquid. (b) Slowness curves corresponding to the water/Si(001) interface. The RSW frequency is not greatly affected by the fluid loading due to the disparity in acoustic impedances ($Z_L = 1.48 \times 10^5$ g cm^{-2} s^{-1} for water and $Z_s = 1.36 \times 10^6$ g cm^{-2} s^{-1} for the transverse mode along [100] in Si); however, its energy leaks into the fluid due to phase matching to the water wave with the slowness vector shown.

allowing the parallel slowness to be complex, as discussed above. Their calculations showed interesting extensions in the surface wave branches that were induced by the liquid loading. For example, fluid loading of the Si surface causes the calculated PSW branch to extend all the way to the {100} axes. These calculations showed new folds in the wave surfaces that were not apparent in the experiments, possibly because fine details are washed out by the liquid damping. Ikehata et al.[12] subsequently studied the surface modes on the (110) faces of CaF$_2$ and Ge and also predicted extensions of the RSW and PSW branches induced by the liquid loading, in partial agreement with experiments. Branches that were not observed in the ultrasound images have large energy emissivity into the solid rather than into the liquid, suggesting that one should attempt to detect these surface waves from the solid side rather than the liquid side.

Another difficulty in comparing these modal theories to the transmission experiments is that the experiments involve an inhomogeneous combination of driven and free-running waves. Also, the imaging experiments generally employ transient waves produced by a concentrated source of ultrasound, rather than homogeneous plane waves.

Many of these difficulties have been recently circumvented by Every et al.,[13] who have calculated the dynamic response produced by an impulsive point source on a fluid loaded surface. Their algorithm computes the dynamic Green's function of two joined elastic half-spaces: the anisotropic solid and the contacting fluid. The results have been plotted exactly in the form of the experimental data and are shown in Figure 11 for several different crystals.

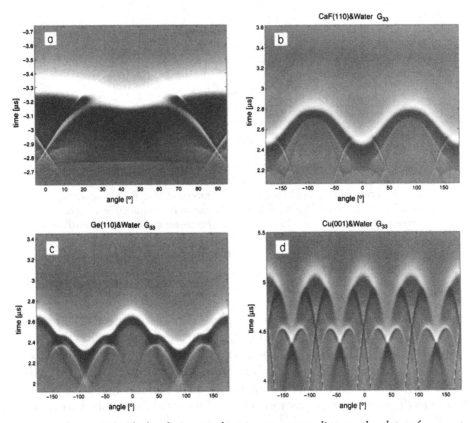

Figure 11 Calculated Green's functions corresponding to the data of Figure 6 and Reference 6. (a) Water-loaded Si(100) surface with $d = 16.31$ mm, (b) water-loaded $CaF_2(110)$ surface, (c) water-loaded Ge(110) surface, (d) water-loaded Cu(001) surface. (From Every et al.[13])

The images agree beautifully with the data in Figure 6 and in Ref. 6 and show the detailed information available from this type of experiment and analysis. This area of investigation is rapidly developing at the time of this writing. It will be interesting to see how the Green's function and modal theories converge.

E. Calculation of focusing and internal diffraction

We now return briefly to the tone-burst experiments of Figure 5. These experiments can be simulated by considering an angular distribution of free surface waves, using the calculated slowness and wave surfaces in Figures 9(a) and 9(b).[14] In the experiment, the ultrasonic transducer excites a 67° cone of wavevectors in the plane of the surface. Vines et al.[6] calculated the intensity at the detector point for a coherent superposition of plane waves with this distribution, and the result is shown as the upper curve in Figure 12.

The oscillations occurring in this calculation are due to the internal diffraction effect associated with the folded group-velocity surface of the Rayleigh wave [Figure 9(b)]. The physics behind this effect is illustrated in Figures 4 and 5 on pages 338–39. The calculated flux in Figure 12 is not identical to the experimental data of Figure 5(b), in part because the calculation assumes a pure 15-MHz source, whereas the experiment possesses a broader bandwidth. Nevertheless, the internal diffraction effect is clearly seen as oscillations in the detected flux.

To isolate the effect of phonon focusing, a second calculation that assumes an incoherent distribution of high-frequency phonons was performed. The

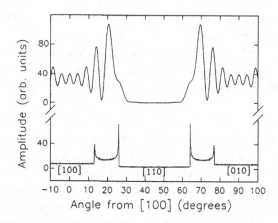

Figure 12 Top curve: Calculated amplitude of a 15-MHz RSW on a (001) surface of Si for d = 10 mm. Bottom curve: Predicted RSW intensity for high-frequency phonons in Si. (From Vines et al.[6])

result is plotted as the lower curve in Figure 12. We see that the broad increases in acoustic flux centered around 20° and 70° are due to phonon focusing, observed in Figure 5(b) by ultrasonic imaging.

F. Application of surface-wave imaging to composite materials

As discussed in the last section of Chapter 13, a technologically important material with elastic anisotropy is the carbon-fiber/epoxy composite. Wavefront images through the bulk of one of these materials showed the same type of wave-surface folding that is observed in crystals. Not surprisingly, the surface-wave imaging methods described in this chapter can be applied to such composites, and the results are sensitive to the particular structure fabricated.

Figures 13(a) and 13(c) show a comparison of $\theta-t$ images in two such samples.[15] In both cases, 15-MHz line-focus transducers are used to produce ultrasonic waves with a well-defined wavevector. The sample producing Figure 13(a) is a uniaxial composite similar to the one used in the transmission experiments of Figure 17 on page 353. The laminates are pressed together so that all the fibers are aligned along one axis, as shown in the sketch at the bottom of Figure 13. The second sample was made by combining laminates with their carbon-fiber directions alternating between 0° and 90°, as indicated in the sketch at the right. The purpose of this latter structure, of course, is to produce a high-strength material.

One of the common features of the two images is the lack of ultrasound pulses along the 0° and 180° propagation directions, where the wavevector is parallel to the carbon fibers. The longitudinal sound velocity is very fast in these directions, and the angles of the incident waves (determined by the orientation and numerical aperature of the transducer) are adjusted for good acoustic coupling between the water and the transverse waves, not the L waves. Near $\theta = 90°$, however, several waves are detected. Moreover, the $\theta-t$ image of the 0–90° composite is quite different from that of the uniaxial composite, simply because the elastic properties of these two structures are very different.

The speed of the ST bulk wave for most of the angles in Figures 13(b) and 13(d) is slower than the sound speed in water, which corresponds to an arrival time of 6.7 μs. Recall that the velocity of a Rayleigh wave along a given direction is slower than that of the slowest bulk wave. Therefore, it is not possible to phase match a Rayleigh wave to the incident water wave, so no Rayleigh wave is generated in the water-immersion geometry. Instead the data of Figure 13 shows only surface-skimming bulk waves, as can be seen by a comparison of the images to the calculations of bulk-wave arrival times in Figures 13(b) and 13(d). Theoretical simulations of these transient pulses using the dynamic Green's function method has been accomplished by Every

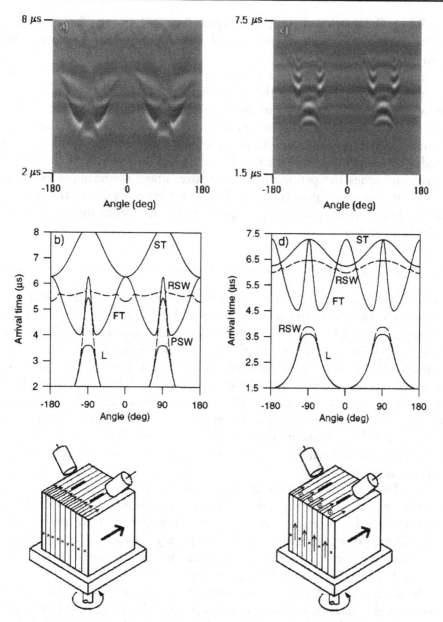

Figure 13 (a) and (c) Angle-time images of ultrasound propagating along the surface of two carbon-fiber/epoxy composites, as shown schematically for θ = 90° at the bottom of the figure. The ultrasound is generated and detected by line-focus transducers (i.e., with cylindrical lenses) in order to select wavevector rather than group velocity. (b) and (d) Calculations of arrival times for bulk and surface waves, as reported by Wolfe and Vines.[14]

and Briggs.[16] Theory and experiment agree very well, and the theory predicts additional effects (Scholte waves) beyond the reach of the present experiments with immersion transducers.

Another curious result can be seen in the data: Despite the use of a single broadband incident pulse, the transmitted signals in the 0–90° composite show multiple oscillations, some continuing for more than 2 μs [Figure 11(c)]. The selective transmission of waves with certain frequencies probably arises from dispersive propagation. As we have seen from the discussion of semiconductor superlattices in Chapter 12, when the wavelength of an acoustic wave approaches the period of the structure, Bragg scattering is expected. Ultrasonic frequency gaps (i.e., stop bands) are expected due to the periodic lamination of these composites, suggesting new directions for research and technology in the future.

G.   A phononic lattice

In fact, Bragg scattering of surface acoustic waves from *two-dimensional* periodic structures has recently been observed by the k-vector scanning technique described in the last section. Consider the experiment shown schematically in Figure 14(a). The surface of a solid block is drilled with a hexagonal array of holes, which are filled with a material of different elastic properties. An ultrasonic wave is focused to a line (labeled T) using a cylindrical acoustic lens. A receiving transducer is cylindrically focused to a parallel line (labeled R) some distance away on the surface. By rotating the sample, the experimenter is able to continuously scan the direction of the wavevector with respect to the axes of the two-dimensional lattice and thus measure the acoustic transmission as a function of this angle.

In order to observe Bragg scattering from this lattice, one must choose a range of wavelengths comparable to the lattice spacing, a. If the incident wavevector lies on one of the Brillouin-zone boundaries of the array, a band gap will appear at the corresponding frequency, just as for electrons in a crystalline medium. By scanning the direction and frequency of the wave, one can hope to map out the band structure.

The scattering of classical waves, such as light and sound, from a periodic medium has been a popular topic in recent years. A decade ago, Yablonovitch[17] proposed that such a "photonic crystal" for electromagnetic waves could have significant practical value. For example, if a three-dimensional lattice composed of two materials with differing dielectric constants could be constructed with a band gap that encompassed all propagation directions, then spontaneous emission from an excited electronic state in that energy gap would be greatly inhibited. This could be particularly useful for reducing radiative losses in semiconductor lasers. Also, if a medium could be constructed that did not support microwaves of certain frequencies, one could imagine

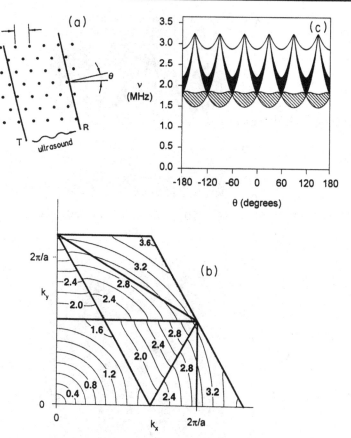

Figure 14 (a) Experiment for observing the transmission of ultrasonic waves through a periodic medium. The dots represent cylindrical plugs of one medium embedded in another. Line-focus transducers transmit (T) and receive (R) the surface waves. (b) Calculation of the contours of constant frequency (labeled in MHz) based on the theory of Ref. 20, modified for the measured surface wave velocities in Al and a polymer. Frequency gaps appear at the Brillouin-zone boundaries, shown as the heavy lines. (c) Shaded area is the frequency gap between the first and second zones; black area is the frequency gap between the second and third zones.

constructing waveguides and other optical devices for this frequency range.[18] Since Yablonovich's paper, several configurations of photonic lattices have been extensively studied and constructed.[19]

The study of long-wavelength *elastic* waves in a periodic medium (a so-called phononic lattice) is still in its infancy. The three-dimensional elastic problem is mathematically more complicated than the photon problem because longitudinal polarization is also possible (and the elastic wave equation is more complicated than the electromagnetic wave equation), so theorists have concentrated on two-dimensional periodic arrays, such as the one shown in

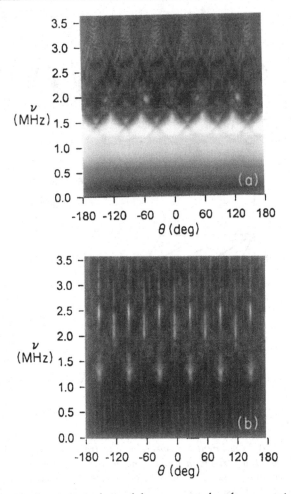

Figure 15 (a) Fourier transform of the transmitted surface wave signal for the polymer/Al lattice with a 26% filling fraction. Dark areas indicate the frequency gaps as a function of wavevector direction, θ, with respect to the x axis of the lattice [Figure 14(a)]. (b) Fourier transform of the reflected signal from the hexagonal lattice, using a single transducer for source and detector. Bright spots indicate backscattering of surface waves from the lattice when the wavevector lands on a zone boundary.

Figure 14. The interest in finding structures that show a complete band gap for all propagation directions (like a semiconductor) has carried over from the photon case, although this is not a necessary condition for studying the physics of such systems.

Vines and I recently performed the experiment shown in Figure 14(a). Surface waves are particularly suitable for studying a two-dimensional lattice because a well-defined wavevector direction can be generated and scanned. The lattice parameter and frequency range were chosen with the help of a theoretical analysis of Kushwaha et al.[20] Their calculation was designed for transverse bulk waves propagating in a plane normal to a hexagonal array of long rods, so we took the velocities to be those of the surface waves in each isotropic medium.

The constant-frequency curves in the first three Brillouin zones are plotted in Figure 14(b) for a commercial polymer (Crystalbond) filling the holes in an Al substrate. The lattice constant is $a = 1$ mm, and the filling fraction is 26%. The heavy lines in this figure are the Brillouin-zone boundaries, where the band gaps occur. Figure 14(c) is a plot of the first two frequency gaps (shaded regions) as a function of the propagation direction. The structure of the band gaps depends on the particular densities and elastic constants of the two media. For example, if the two media are reversed (Al posts surrounded by the polymer), a complete energy gap between the first and second zones is predicted for all angles.

The experimental results for a phononic lattice consisting of polymer-filled holes in Al with a filling fraction of 26% are shown in Figure 15. A short pulse is applied to the generator, and the Fourier transform of the transmitted signal is recorded as a function of propagation angle. Figure 15(a) shows the occurrence of band gaps (dark regions) in the transmission spectrum, qualitatively similar to those predicted for this structure. The experiment, however, corresponds to a water-loaded surface, and the transducer bandwidths are limited to the 0.5–2.0-MHz region. Also, the theoretical predictions may be modified from those of Figure 14 when the elastic half-space is properly accounted for. Nevertheless, it is clear from these initial experiments and calculations that such phononic lattices provide interesting systems for future studies.

The diffraction of surface waves from a periodic medium is further illustrated in Figure 15(b). This is a reflection image for the same lattice as described above. A single transducer [producing wavefront T in Figure 14(a)] is used as both transmitter and receiver. Only waves that are reflected from the lattice with a wavevector directly opposing the incident wavevector are detected, giving rise to the diffraction spots along the directions of the reciprocal lattice vectors. Phononic lattices of this sort reveal the classical diffraction of coherent elastic waves.

References

1. S. Tamura and K. Honjo, Focusing effects of surface acoustic waves in crystalline solids, *Jpn. J. Appl. Phys.* **20** (Suppl. 3) 17 (1980).

2. R. E. Camley and A. A. Maradudin, Phonon focusing at surfaces, *Phys. Rev. B* **27**, 1959 (1983).

3. C. Höss and H. Kinder, Propagation and focusing of high-frequency Rayleigh phonons, in *Proceedings of Fourth International Conference on Phonon Physics*, Sapporo, Japan (1995), *Physica B* **219, 220**, 706 (1996).

4. A. A. Kolomenskii and A. A. Maznev, Observation of phonon focusing with pulsed laser excitation of surface acoustic waves in silicon, *JETP Lett.* **53**, 423 (1991); Phonon focusing effect with laser generated ultrasonic surface waves, *Phys. Rev. B* **48**, 14502 (1993).

5. A. A. Maznev, A. A. Kolomenskii, and P. Hess, Time-resolved cuspidal structure in the wavefront of surface acoustic pulses on (111) GaAs, *Phys. Rev. Lett.* **75**, 3332 (1995). See also D. K. Banerjee and J.-H. Pao, Thermoelastic waves in anisotropic solids, *J. Acoust. Soc. Am.* **56**, 1444 (1974).

6. R. E. Vines, M. R. Hauser, and J. P. Wolfe, Imaging of surface acoustic waves, *Z. Phys. B* **98**, 255 (1995); R. E. Vines, S. I. Tamura, and J. P. Wolfe, Surface acoustic wave focusing and induced Rayleigh waves, *Phys. Rev. Lett.* **74**, 2729 (1995).

7. Similar results can be obtained with a single line-focus transducer in a reflection geometry: N. N. Hsu, D. Xiang, S. E. Fick, and G. V. Blessing, Time and polarization resolved ultrasonic measurements using a lensless line-focus transducer, in *Proceedings of the 1995 IEEE Ultrasonics Symposium* (1996), p. 867. The use of single point-focus transducers have been employed extensively for the study of surface waves and materials characterization. See, A. Briggs, *Acoustic Microscopy* (Clarendon Press, Oxford, 1992); Z. Yu and S. Boseck, Scanning acoustic microscopy and its applications to material characterization, *Reviews of Modern Physics* **67**, 863 (1995).

8. See, for example, A. A. Maradudin, Surface acoustic waves, in *Nonequilibrium Phonon Dynamics*, W. E. Bron, ed. (Plenum, New York, 1985), Chapter 10. A practical variety of surface wave phenomena is contained in A. H. Nayfeh, *Wave Propagation in Layered Anisotropic Media* (North Holland, Amsterdam, 1995).

9. A. G. Every, Pseudosurface wave structures in phonon imaging, *Phys. Rev. B* **33**, 2719 (1986).

10. T. C. Lim and G. W. Farnell, Search for forbidden directions of elastic surface-wave propagation in anisotropic crystals, *J. Appl. Phys.* **39**, 4319 (1968); F. R. Rollins, Jr., T. C. Lim, and G. W. Farnell, Ultrasonic reflectivity and surface wave phenomena on surfaces of copper single crystals, *Appl. Phys. Lett.* **12**, 236 (1968).

11. S. Tamura, R. E. Vines, and J. P. Wolfe, Effect of liquid loading on the surface acoustic waves in solids, *Phys. Rev. B* **54**, 5151 (1996).

12. A. Ikehata, R. E. Vines, S. Tamura, and J. P. Wolfe, Images of surface acoustic waves on the (110) face of cubic crystals with liquid loading, *Physica B* **219 & 220**, 710 (1996).

13. A. G. Every, A. A. Maznev, and G. A. D. Briggs, Surface response of a fluid loaded anisotropic solid to an impulsive point force: application to scanning acoustic microscopy, preprint (1997); A. G. Every, K. Y. Kim, and A. A. Maznev, The elastodynamic response of a semi-infinite anisotropic solid to sudden surface loading, *J. Acoust. Soc. Am.* **102**, 1346 (1997).

14. See also, S. I. Tamura and M. Yagi, Finite wavelength effects on the ballistic propagation of surface acoustic waves, *Phys. Rev. B* **49**, 17387 (1994).

15. J. P. Wolfe and R. E. Vines, Acoustic wavefront imaging of carbon-fiber/epoxy composites, in *Proceedings of 1996 IEEE Ultrasonics Symposium* 1997.

16. A. G. Every and G. A. D. Briggs, Surface response of a fluid loaded solid to impulsive line and point forces: application to scanning acoustic microscopy, submitted for publication (1997).

17. E. Yablonovitch, Inhibited spontaneous emission in solid-state physics and electronics, *Phys. Rev. Lett.* **58**, 2059 (1987).

18. P. F. Goldsmith, Quasi-optical techniques, *Proc. of the IEEE* **80**, 1729 (1992).

19. See the review by E. Yablonovitch, Photonic band-gap crystals, *J. Phys. Condensed Matter* **5**, 2443 (1993).

20. M. S. Kushwaha and P. Halevi, Band-gap engineering in periodic elastic composites, *Appl. Phys. Lett.* **64**, 1086 (1994); M. S. Kushwaha, P. Halevi, G. Martinez, L. Dobrzynski, and B. Djafari-Rouhani, Theory of acoustic band structure of periodic elastic composites, *Phys. Rev. B* **49**, 2313 (1994).

15

Interactions of ballistic phonons with electrons

The phonon is an elementary excitation of a crystal. How it couples to other elementary excitations is critical to many areas of condensed-matter physics. Foremost are the interactions of phonons with electrons. In this final chapter, I will touch upon a few emerging technologies and applications of phonon imaging that mainly involve the interactions of electrons and phonons. The topics chosen include the optical properties of semiconductors and heterostructures, mechanisms of superconductivity, and development high-energy particle detectors.

Of course, we have already encountered the interaction of nonequilibrium phonons with electrons. Chapter 10 dealt with the spatial and frequency distributions of phonons *created* by photoexcited carriers in semiconductors. In contrast, an increasing variety of experiments employ ballistic phonons to *probe* both intrinsic electronic states and those associated with impurities. In addition, it is possible in some cases to *detect* nonequilibrium phonons by their effect on the optical and electronic properties of crystals. We begin by considering a few crystalline systems that are suitable for time- and space-resolving phonon detectors, while keeping in mind the broader goal of elucidating the physics of new materials or devices.

A. Optical detection of nonequilibrium phonons

For the most part, the optical properties of crystals in the visible or near-infrared region are associated with transitions between electronic states. Phonons of sufficient energy can induce transitions between low-lying electronic states, thereby affecting the optical absorption or emission of the solid. A considerable variety of crystalline systems undergo changes in their optical properties in the presence of nonequilibrium phonons, and, therefore, these systems may be useful as heat-pulse detectors, often giving frequency

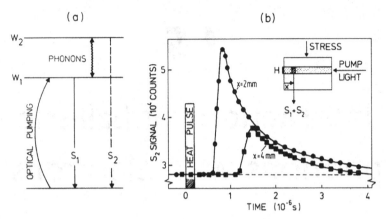

Figure 1 (a) Stress splitting of the excited states (W_1 and W_2) of Eu^{2+} in CaF_2 allows for optical detection of phonons with energy equal to this splitting. The S_2 luminescence is proporational to the phonon population at this energy. (b) Time-of-flight experiment with 600-GHz phonons. Two spatially selected regions at different phonon path lengths yield the heat pulses shown. (From Renk.[2])

information about the phonons. Reviews by Bron,[1] Renk,[2] and Wybourne and Wigner[3] should be consulted for references on this subject. Here, we mention just a few types of experiments and crystalline systems to expose the wide-ranging possibilities.

In pioneering experiments of this sort, Anderson and Sabisky[4] applied a magnetic field to split the Zeeman levels of Tm^{2+} ions in SrF_2. Nonequilibrium phonons with energy equal to the Zeeman splitting cause a change in the spin population, which can be detected by the absorption of circularly polarized light. This provides a phonon detector with frequency tunability up to approximately 500 GHz. Similarly, phonon-induced fluorescence has been observed for magnetically split impurity levels[5] in LaF_3 and stress-split impurity levels[6] in CaF_2 and SrF_2. Spectroscopic information about heat-pulse phonons has also been gained by inducing energy-level crossings of magnetic ions in similar insulator crystals.[7]

An example of optical detection of phonons with time, space, and frequency resolution is shown in Figure 1. Figure 1(a) shows the electronic energy levels of Eu^{2+} in CaF_2. The splitting between the W_1 and W_2 levels is directly proportional to the uniaxial stress applied to the crystal. The crystal is cooled to liquid-He temperature, and the W_1 level is populated by resonant optical pumping with a tunable laser. In the absence of phonons, the W_2 level is not populated and no S_2 fluorescence is observed. Nonequilibrium phonons with energy equal to the level splitting, however, can excite electrons from W_1 to W_2, which fluoresces at energy S_2. Figure 1(b) shows the heat pulses generated by an electrical pulse at the crystal surface and detected by the S_2 fluorescence.

By changing the crystal stress (and, thus, the frequency of detected phonons), it is possible to observe the changeover from ballistic to diffusive propagation due to elastic scattering from defects. In addition, by optically exciting a large region of the crystal with high-energy photons, the anharmonic decay of phonons can be measured. In this way, Baumgartner et al.[8] and Happek et al.[9] have observed the theoretically predicted ν^5 dependence of this inelastic scattering rate. (See Section F in Chapter 8, for a discussion of this process.)

A number of authors (for example, Refs. 10–13) have studied phonon trapping and diffusion of phonons in ruby (Al_2O_3 doped with Cr^{3+}) by optically exciting the ions and observing phonon-induced changes in the fluorescence of crystal-field-split levels. By spatially selecting the fluorescence, Akimov et al.[14] observed effects of phonon focusing in Al_2O_3. Several workers[15] have observed the transient frequency distribution of nonequilibrium phonons by observing the vibronic sidebands in the photoluminescence of sharp impurity lines in diamond, CdS, SrF_2, and CaF_2. Bron et al.[16] have used the vibronic sideband method to study the quasidiffusion process (discussed in Chapter 10).

While a number of these optical methods can detect nonequilibrium phonons with space, time, and frequency resolution, their potential for phonon imaging has hardly been exploited. Also, most experiments to date have been limited to specific crystalline systems with appropriate dopants. No one has yet demonstrated a universal material that could be deposited on many materials (analogous to metal-film bolometers and tunnel-junction detectors) and provide an optical probe of the acoustic-phonon population. With the advent of molecular beam and other epitaxial growth techniques, however, a variety of heterostructures and quantum wells with unique optical properties are possible, generally with compound semiconductors such as GaAs. Along these lines, Akimov and Shofman[17] observed phonon-induced changes in the impurity luminescence of GaAs.

One might ask if there are *intrinsic* electronic states in semiconductors whose optical signature can be modulated by nonequilibrium phonons. The basic excited electronic state of a semiconductor is the exciton, which is an electron and a hole bound together by their coulomb attraction. Excitons are produced by optical excitation and they are observed by the luminescence emitted when the electron and the hole recombine. Furthermore, high densities of excitons in the elemental semiconductors Ge and Si exhibit a remarkable range of excitonic matter, which includes biexcitons (the analog of H_2) and droplets of electron-hole liquid (see also Chapter 10). Several experiments have examined the interaction of nonequilibrium phonons with these excitonic species. From the perspective of phonon detection, Akimov et al.[18] have observed a modulation of electron-hole liquid luminescence in Ge due to a flux of nonequilibrium phonons.

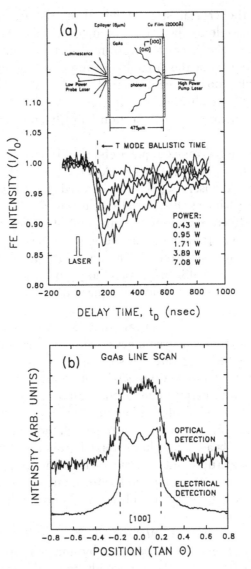

Figure 2 (a) Optical detection of phonons in GaAs by observing the reduction in luminescence from photoexcited excitons in an epilayer. A slight
heating of the epilayer by the phonons reduces the population of the excitons at low thermal energy, which give rise to the recombination signal.
(b) By scanning the detection laser a spatial distribution of the ballistic
phonons is obtained. The signal is compared to the conventional line scan
across the ST box, as obtained with a superconducting tunnel junction.
(From Ramsbey et al.[19])

Ramsbey et al.[19] have shown that the exciton luminescence in a relatively pure epitaxial layer of GaAs is modulated by nonequilibrium phonons.[20] In their experiment, a sharply focused laser beam creates a local population of excitons in the epilayer. If nonequilibrium phonons enter this probe region from another heat source, the recombination luminescence from the excitons (both free and impurity bound) decreases in intensity. A simple interpretation of this effect is that the phonons heat the free-exciton gas, reducing the number of excitons that can recombine radiatively.*

Because the lifetime of the excitons created by the probe laser is only approximately a nanosecond, they can be used as a time- and space-resolved phonon detector. Basically, the probe beam is pulsed at a specified time after a heat pulse is created (with another laser) on the opposite side of the crystal. By scanning the time difference between probe pulse and source pulse, one gains a temporal trace, as shown in Figure 2(a) for several excitation powers. By fixing the time delay and scanning the probe beam, one observes the angular distribution of phonon flux in the heat pulse. Figure 2(b) shows a profile of a phonon-focusing structure obtained in this way, compared to one taken in the standard manner with a superconducting tunnel-junction detector. At present, the sensitivity of this optical technique is considerably lower than that obtained with superconducting detectors, but the sensitivity is expected to vary greatly with different semiconductors and dopants.

A useful type of spatially resolving phonon detector has been developed by Kent and coworkers.[21,22] It consists of a 1–10-μm-thick CdS film evaporated on the sample and covered with a Cu film in an interdigital pattern. When the sample is cooled to low temperature, this semiconducting film has megaohm resistance. A laser beam is briefly focused on one portion of the exposed CdS film, locally promoting electrons to shallow traps. With the laser off, this small portion of the film is sensitized to nonequilibrium phonons, which can liberate the trapped electrons and cause a local reduction in conductivity. After the nonequilibrium phonon signal is recorded, the film is reset to its high-resistance state by passing a high current through it. As in the scannable tunnel-junction detector described in Chapter 3, the activating laser beam can be scanned in a raster pattern to produce an image. The novel aspect of CdS bolometer is that, unlike superconducting detectors, it works in magnetic fields up to at least 11 T. Kent has utilized this device in studies of two-dimensional electron systems.[22]

B.   Phonon detection in semiconductor heterostructures

We also mention two novel phonon-detection techniques involving technologically important semiconductor systems. First is the phonoconductivity

---

* Because GaAs is a direct-gap semiconductor, only those excitons with the (very small) wavevector of an emitted photon can radiate.

method of Lassmann and coworkers.[23,24] Epitaxial semiconductor films containing impurity atoms are grown on compatable substrates. The essential idea is that nonequilibrium phonons of sufficient energy can ionize the donors or acceptors, producing free electrons or holes in the conduction or valence bands that can be detected by a change in the conductivity of the film. This is the thermal analog of the *photo*conductivity effect. By using a tunable tunnel-junction source, a rich variety of excitation spectra are obtained. In practice, it is found that the sensitivity of the conducting film is enhanced by illumination with light, which elevates the carriers into metastable states that are more easily ionized by the phonons. In principle, the phonoconduction effect could be used for phonon imaging by using the usual scanned heat source. Frequency selectivity could be controlled by the choice of semiconductors and impurities.

The second system consists of a heterostructure fabricated on a GaAs crystal in the form of a field-effect transistor. Electrical bias of this device produces a region of mobile electrons that are confined to a very thin layer in the GaAs. This is a means of producing a two-dimensional electron gas (2DEG) whose areal density (number of carriers/area) may be continuously varied merely by changing the electrical bias. Nonequilibrium phonons incident on the device can be absorbed by the electrons, imparting their momentum to the 2DEG. If the phonon is obliquely incident on the surface, the component of transferred momentum along the conducting plane produces a voltage across the device. Thus, the 2DEG becomes a phonon detector whose voltage response (+ or −) depends on the direction of the incident phonon wavevectors.

This effect has been artfully demonstrated using phonon-imaging techniques. Karl et al.[25] raster scanned a laser beam across the GaAs surface opposite a 2DEG. The image in Figure 3(a) is the calculated focusing pattern for GaAs. The experimental image in Figure 3(b) is obtained with a 2DEG detector. The data show a phonon-focusing pattern that is multiplied by a positive or negative factor that represents the direction and magnitude of the phonon wavevectors in the conducting plane. The central square is the ST focusing structure near [100], and the outwardly radiating spires are due to FT phonons.

Experiments of this sort not only underscore the potential for using semiconducting devices for phonon detection, they provide new sources of information about the electron–phonon interactions. For example, the results of Figure 3 argue for the dominance of a piezoelectric coupling between the acoustic phonons and electrons rather than a deformation-potential mechanism. This conclusion is reached by comparing Figures 3(b) to the simulations for these two cases, displayed in Figures 3(c) and 3(d). Related experiments have examined the frequency dependence of the coupling as well

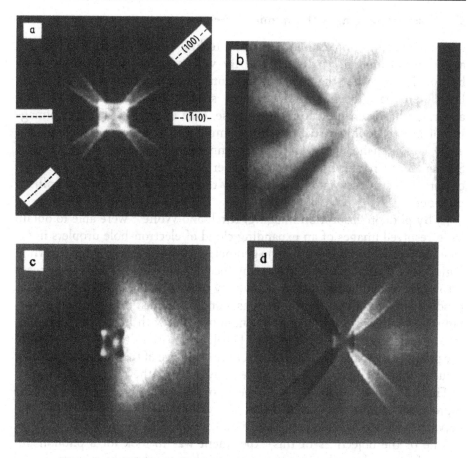

Figure 3 (a) Calculated phonon-focusing pattern for GaAs. (b) Measured pattern obtained by using a 2DEG (phonon-drag) detector. Bright and dark areas correspond to positive and negative voltages, determined by the component of the phonon momentum in the detection plane. (c) Theoretical phonon-drag pattern assuming deformation potential coupling. (d) Theoretical phonon-drag pattern assuming piezoelectric coupling. The latter calculation coincides with the data in (b). (Data and calculations by Karl et al.[25])

as the directional emission of phonons from a field-effect transistor.[26] These and other experiments on the 2-dimensional electron gas[27,28] have stirred up considerable theoretical interest.[29] In a novel spin-off experiment, Knott et al.[30] have examined the spatial distribution of currents in the 2DEG structure by imaging He bubbles generated in the superfluid He bath, analogous to Eisenmenger's first fountain-pressure experiment (Figure 5 on page 18).

C. Shadow imaging with phonons

Phonon imaging, unlike the microscopy techniques of acoustical imaging[31] and thermal imaging,[32] deals primarily with angular distributions of vibrational flux rather than the observation of spatial structures. Nevertheless, there have been a few instances so far where a phonon-imaging apparatus has been used for this purpose. One example is the microimaging of buried doping structures in a commercial Si wafer, which Schreyer et al.[33] accomplished with a large-area scannable tunnel-junction detector (see Section B of Chapter 3). Below, two other examples are described in which objects inside the crystal are observed by their shadow in a beam of ballistic phonons.

(i) By phonon absorption imaging, Kirch and Wolfe[34] were able to obtain time-resolved images of an expanding cloud of electron-hole droplets in Ge (see Section G of Chapter 10). The geometry for this experiment is shown in Figure 4(a). The droplets are created by focusing a pulsed Nd:YAG laser beam into the bottom of a slot. A synchronously timed pulse of ballistic phonons is produced on one surface by another laser and detected at the opposite surface. Figure 4(b) shows the transmitted phonon signal with no droplets present; it is a section of the FT ridge near [110]. When the droplets are present, the phonons are partially absorbed. A time sequence of the expanding droplet cloud is obtained by changing the time delay between the two lasers, as shown in Figures 4(c)–4(f). This proved to be a novel method for characterizing the rapidly expanding cloud of droplets and calibrating the coupling strength between phonons and electron-hole liquid.

One of the objectives of this experiment was to look for a phonon hot spot, which was hypothesized to arise from phonon–phonon interactions and extend nearly a millimeter from the excitation point. In fact, no phonon hot spot was resolvable in these experiments, consistent with the discussion of Chapter 10. Experiments using a similar shadow geometry but with fixed tunnel-junction source and detector in a phonon-spectroscopy mode yielded the absorption spectrum of the electron-hole liquid in Ge.[35] Those experiments demonstrated the predicted transparency of this Fermi fluid when the phonon wavevectors exceed the Fermi wavevector of the electrons.

(ii) The second example of shadow imaging does not directly involve the interaction of ballistic phonons with electronic states. Huebener et al.[36] wanted to test the ability of an electron-beam-scanned phonon-imaging setup to observe defects in the interior of a crystal. They bored small holes in a sapphire crystal with a high-power focused laser beam. These induced defects were easily imaged by the ballistic phonons. Metzger et al.[37] have shown that by using a small number of identical detectors and the standard electron-beam scanning, one can use the phonon-imaging technique to determine the xyz position of microscopic (>10-μm) defects within a crystal.

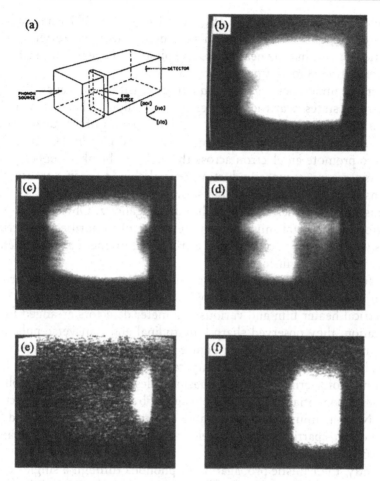

Figure 4 (a) Schematic of the phonon absorption imaging experiment of Kirch and Wolfe.[34] The crystal is Ge at 1.8 K. (b) Phonon signal with no electron-hole droplets present. This FT ridge is cutoff by the slot at the right, where the droplets will be introduced. (c) Phonon signal at $t \simeq 0$, just after the droplet cloud has been created. The dark region adjacent to the slot is a shadow image of the droplet cloud. (d) Phonon image at $t = 1$ μs, showing the expansion of the droplet cloud under the influence of a phonon wind from the excitation region. (e) and (f) are image (b) with images (c) and (d) subtracted, respectively, showing more directly the profile of the electron-hole droplet cloud as it expands into the crystal.

D. Ballistic phonons in superconductors

An extreme example of electrons hindering the ballistic motion of phonons is in a metal. The high density of free carriers in a normal metal limits the mean free path of thermal phonons to microscopic distances, even at low temperatures. Of course, quantum theory tells us that only a small fraction of the electrons (of the order $k_B T / E_{\text{Fermi}} \simeq 10^{-4}$ at $T = 10$ K) are available for scattering phonons; nevertheless, this is a huge density of scattering centers. It is little wonder that the heat pulses and phonon images shown so far in this book are restricted to nonmetallic crystals.

If a metal undergoes a superconducting transition, however, the density of electronic states near the Fermi energy is radically changed, as indicated schematically in Figure 8 on page 74. With the opening of the superconducting gap at temperatures below T_c, phonons with energies $h\nu$ less than 2Δ are unable to promote an electron across the gap (i.e., break a Cooper pair). At zero temperature, a superconducting crystal should become transparent to phonons with $h\nu < 2\Delta$, which (as shown in Table 2 on page 74) includes phonons with frequencies of several hundred gigahertz. Obviously, to observe this effect, the crystal must be relatively free of impurities and structural defects which scatter phonons – not a trivial requirement as most metals are highly susceptible to dislocations.

The first demonstration of ballistic heat pulses in a superconductor was accomplished with single crystals of lead by Narayanamurti et al.[38] Using an electrical heater film and various bolometer detectors arranged for [111] propagation, they observed sharp longitudinal and transverse pulses below approximately 2 K, as shown in Figure 5. As expected, the mean free path of the phonons is reduced as the temperature of the crystal is raised, because the number of thermally excited quasiparticles (which can scatter phonons) increases exponentially. In addition, using a Pb-oxide-Pb tunnel-junction detector, Narayanamurti et al. were able to directly detect the rapid diffusive motion of quasiparticles (unpaired electrons) at temperatures below approximately 3 K.

Recently, the ballistic propagation of phonons through a single crystal of Nb was been studied by Hauser.[39] A calculation of the FT and ST slowness surfaces for Nb – a cubic crystal with positive anisotropy factor – is shown in Figure 6. Monte Carlo simulations of the expected phonon-focusing patterns for ballistic phonons are shown in Figure 7.

In his experiments, Hauser generated phonons by laser excitation of a Cu film deposited on a near-[110] face of Nb. The 1.8-mm-thick crystal was cooled to a lattice temperature of 1.8 K, which is well below $T_c = 9.5$ K. The phonons were detected with a broadband Al bolometer of dimension $5 \times 10 \, \mu\text{m}^2$. The image in Figure 8(a) was obtained at low excitation power. Corresponding time traces of the heat pulses are plotted in Figure 8(b). The phonon-focusing pattern observed for this broad time gate is in good agreement with that predicted in Figure 7(d). The sharp ballistic pulses and well-defined caustics imply that most of the phonons generated at low excitation levels have frequencies below $2\Delta / h = 730$ GHz.

Further informative observations were made in these experiments: (1) As the temperature of the crystal is raised from roughly 1.6 to 2.1 K, the

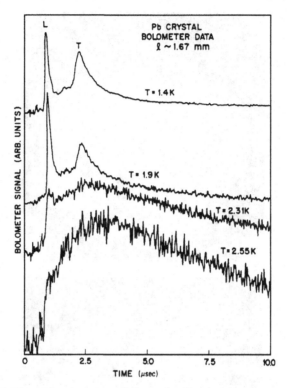

*Figure 5 Heat pulses transmitted through a Pb crystal at various temper-
atures below the superconducting critical temperature of 7.2 K. Ballistic
pulses from longitudinal and transverse phonons are observed below approximately 2 K. Above that temperature, diffusive phonon propagation
is observed due to scattering from thermally excited quasiparticles (electrons). (From Narayanamurti et al.[38])*

ballistic signal decreases and a pronounced tail develops, indicative of phonon
scattering from thermal quasiparticles (unpaired electrons) in the bulk of the
superconductor. (2) At low temperature (1.6 K), the ballistic signal tends to
saturate as the excitation density is increased. This effect is quantitatively
understood by modeling the excited metal film as a Planckian source and
computing the fraction of phonons that have frequencies above the superconducting gap. The high-frequency phonons are scattered out of the ballistic
pulse as they break Cooper pairs. One can see from Figure 9(a) that for a film
temperature of 12 K, approximately half of the generated phonons will be scattered, and at 20 K a much larger fraction is lost. The solid line in Figure 9(b)
is a calculation of the ballistically transmitted flux as the excitation power is
increased. This calculation compares favorably to the measured LO phonon
signals shown in the figure. (3) The lifetime and spatial extent of the phonon
source produced by sharply focused laser excitation of the metal film increase
with laser power. These effects have not yet been quantitatively explained,

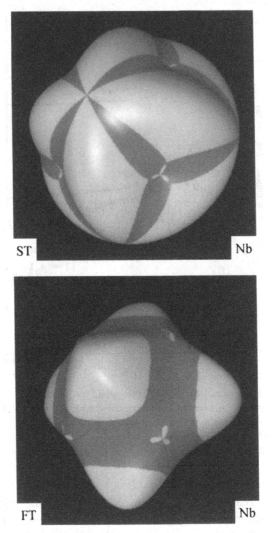

Figure 6 Calculation of the ST and FT slowness surfaces in Nb. Light and dark regions are convex (or concave) and saddle regions, respectively. (From Hauser.[99])

but they are likely to be associated with the interaction of quasiparticles and phonons near the source.

The increase in the phonon source size with increasing excitation density is dramatically displayed in Figure 10. Images (a) through (c) are the images obtained as the excitation power is increased from that in Figure 8(a). The signal gain is reduced as the power is increased in order to keep the brightness the same in all three photos. One readily observes that the caustics broaden at higher power, indicating that the source of the ballistic phonons is spreading

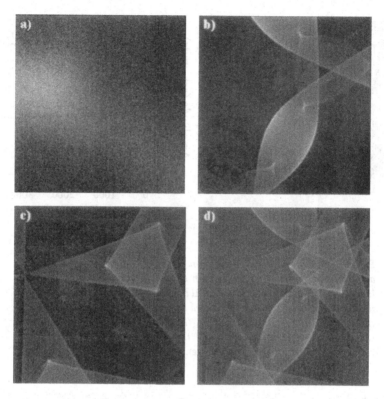

*Figure 7 Calculation of the phonon-focusing patterns for (a) longitudinal,
(b) fast transverse, and (c) slow transverse modes in Nb. (d) Superposition
of all modes. A 2-mm-thick crystal and 3.27 × 3.27 mm² scan range are assumed, and elastic constants are from Simmons and Wang.[47] (Calculation
by Hauser.[39])*

out. In addition, a remarkable pattern of dark caustics appears precisely where
the low-power caustics occur. In fact, the so-called high-power image actually
represents a subtraction of a low-power image from a high-power image.*
The bright caustics in the high-power image are broadened due to the larger
phonon source size and they form a light background for the sharper caustics
at low power to be observed. The dark caustics observed in the experimental
images of Figures 10(b) and 10(c) are simulated by the calculation in Figure 10(d), which subtracts a narrow-source Monte Carlo calculation from a
broad-source calculation.

---

* These data are actually a single image recorded with a small leakage of continuous-wave power superimposed on the the intense laser pulses. When the laser cavity is dumped, creating the 10-ns pulse, the
continuous-wave leakage is destroyed and requires a recovery time of $\simeq 1\,\mu$s. Thus, a negative low-power
image is superimposed. The effect was not intentional and led to much speculation before its technical
origin was ascertained. Generally, the leakage of continuous-wave power in a cavity-dumped mode is
much lower than in this experiment.

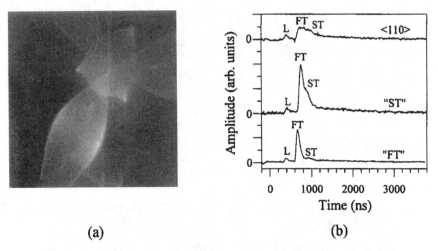

(a)   (b)

Figure 8 (a) Experimental phonon image of a 1.8-mm-thick crystal of Nb at 1.8 K. A 15-ns Ar+ laser pulse with 14.6-nJ energy was used (relatively low power). The Cu heater film and Al bolometer film were separated from the crystal by an oxide layer produced by anodization. (b) Heat pulses for three different crystalline directions. The trace labeled "ST" is on an ST caustic near the ⟨111⟩ direction and the trace labeled "FT" is on an FT caustic near ⟨110⟩. (Data from Hauser.[39])

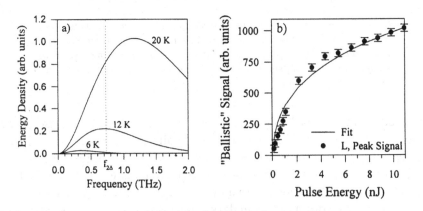

Figure 9 (a) Calculated Planckian distribution for several source temperatures. For frequencies above $f_{2\Delta}$, phonons have sufficient energy to break Cooper pairs in the superconductor, and consequently their mean free paths will be greatly reduced. (b) Dots are experimental measurements of the peak signal for the longitudinal ballistic signal. The nonlinear increase in the intensity with excitation energy is well explained by a model (solid curve) that takes into account the increase in temperature of the heater film. (From Hauser.[39])

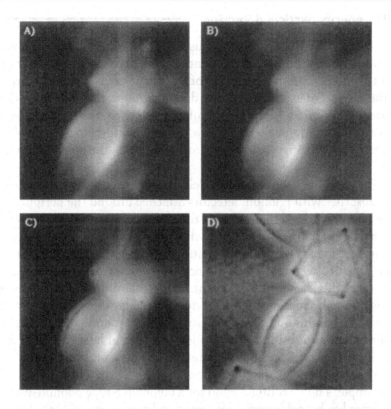

Figure 10 Experimental phonon images of Nb at increasing excitation energy: (a) 52 nJ, (b) 99 nJ, and (c) 156 nJ. Compared to the low-energy image in Figure 8a, the bright caustics increase in width due to an increase in phonon source size. Superimposed on (b) and (c) are dark caustics, which arise from a subtraction of a low-power image, demonstrating the significant change in source size with power due to interactions with quasi-particles (see Figure 9). (d) Simulation of image (c) obtained by subtracting a calulated image with a 50-μm source from one with a 180-μm source. (From Hauser.[99])

The imaging of ballistic heat flux through superconductors may become a useful probe of the superconducting state. In particular, the acoustic attenuation of high-frequency phonons is predicted to be anisotropic in many superconductors, and the ability to scan the phonon wavevectors over a wide range of angles could yield detailed information about the electron–phonon coupling and the superconducting gap. This approach would seem to be particularly valuable for high-T_c superconductors, for which there is increasing evidence of anisotropic energy gaps.[40] Currently only small single crystals of these ceramic superconductors are available, requiring the development of microimaging techniques.

E. High-energy particle detection

Most elementary-particle detectors in use today collect the charge produced by an ionizing event, although a significant fraction of the deposited energy is in the form of phonons. A measurement of this portion may contain information about the event. For example, massive dark-matter particles (if they exist) are postulated to produce a high fraction of thermal excitations. Based on these and other considerations, there are now concerted efforts to develop phonon-based particle detectors. A survey of some of these efforts may be found in Ref. 41.

A typical phonon-based detector for elementary particles consists of a large-volume absorber with phonon detectors attached around the periphery. Ideally, the observer would like to know the energy deposited in the event, the type of particle, and the direction in which the particle was traveling. Practically, only a subset of this information is gained by detecting the pulse heights. One of the potentials of phonon-based detectors is providing information about the location of the event. Positional information is gained by comparing signals from multiple detectors placed on the absorbing medium. Relative pulse heights and times of arrival of the phonon pulses contain spatial information as a result of the phonon focusing of heat flux from a point source. In such devices an understanding of the phonon source and the bulk scattering processes (e.g., quasidiffusion) is crucial.[42]

For example, a novel configuration for a bolometer-type phonon-based particle detector has been developed by Young and coworkers.[43,44] Two large-area meander lines of Ti metal film (evaporated on Si) are biased in temperature just below their sharp transition temperature. Phonons produced in the Si crystal as the final products of an α-particle absorption event cause the Ti to go normal where the local phonon flux is sufficiently intense. The heights of the voltage pulses across the detectors indicate the total lengths of the superconducting lines that have been driven normal and, thus, contain information about the anisotropy of the phonon flux. The relative amounts of ballistic and quasidiffusive phonons caused by the absorption of a high-energy particle are the subject of much interest.[45] The photoexcitation experiments described in Chapter 10 lend some clues as to the complexity, and perhaps resolution, of these problems.[46]

References

1. W. E. Bron, Spectroscopy of high-frequency phonons, *Rep. Prog. Phys.* **43**, 301 (1980).
2. K. F. Renk, Studies on nonequilibrium phonons by optical techniques, in *Nonequilibrium Phonon Dynamics*, W. E. Bron, ed. (Plenum, New York, 1985).
3. M. N. Wybourne and J. K. Wigmore, Phonon spectroscopy, *Rep. Prog. Phys.* **51**, 923 (1988).

4. C. H. Anderson and E. S. Sabisky, Spin-phonon spectrometer, in *Physical Acoustics*, *Vol. 8*, W. P. Mason and R. N. Thurston, eds. (Academic, New York, 1971), p. 1.

5. J. M. Will, W. Eisfeld, and K. F. Renk, *Appl. Phys. A* **31**, 191 (1983); R. S. Meltzer, J. E. Rives, D. J. Sox, and G. S. Dixon, in *Phonon Scattering in Condensed Matter*, W. Eisenmenger, K. Lassmann, and S. Dottinger, eds. (Springer, Berlin, 1984), p. 115.

6. W. Eisfeld and K. F. Renk, Tunable optical detection and generation of terahertz phonons in CaF_2 and SrF_2, *Appl. Phys. Lett.* **34**, 481 (1979).

7. L. J. Challis, An introduction to crossing effects in phonon scattering, in *Nonequilibrium Phonon Dynamics*, W. E. Bron, ed. (Plenum, New York, 1985).

8. R. Baumgartner, M. Engelhardt, and K. F. Renk, Spontaneous decay of high frequency acoustic phonons in CaF_2, *Phys. Rev. Lett.* **47**, 140 (1981).

9. U. Happek, W. W. Fischer, and K. F. Renk, Decay of terahertz phonons in $Ca_{1-x}Sr_xF_2$ mixed crystals, in *Phonons 89*, S. Hunklinger, W. Ludwig, and G Weiss, eds. (World Scientific, Singapore, 1990), p. 1248.

10. For example, K. F. Renk and J. Deisenhofer, Imprisonment of resonant phonons observed with a new technique for the detection of 10^{12}-Hz phonons, *Phys. Rev. Lett.* **26**, 764 (1971).

11. For example, A. A. Kaplyanskii, S. A. Basun, V. A. Rachin, and R. A. Titov, Anisotropy of resonant absorption of high frequency 0.87×10^{12} Hz phonons in an excited state of Cr^{3+} ions in ruby, *JETP Lett.* **21**, 200 (1975).

12. For example, R. S. Meltzer, and J. E. Rives, New high energy monoenergetic source for nanosecond phonon spectroscopy, *Phys. Rev. Lett.* **38**, 421 (1977); R. S. Melzer, J. E. Rives, and W. C. Egbert, Resonant phonon-assisted energy transport in ruby from 29-cm^{-1}-phonon dynamics, *Phys. Rev. B* **25**, 3026 (1982).

13. J. I. Dijkhuis, A. van der Pal, and H. W. de Wijn, Spectral width of optically generated bottlenecked 29 cm^{-1} phonons in ruby, *Phys. Rev. Lett.* **37**, 1554 (1976).

14. A. V. Akimov, S. A. Basun, A. A. Kaplyanskii, V. A. Rachin, and R. A. Titov, Direct observation of the focusing of acoustic phonons in a crystal of ruby, *JETP Lett.* **25**, 461 (1977).

15. For example, J. Shah, R. F. Leheny, and A. H. Dayem, I_1 emission line in CdS as a phonon spectrometer, *Phys Rev. Lett.* **33**, 818 (1974); W. E. Bron and W. Grill, Phonon spectroscopy: I. Spectral distribution of a phonon pulse, II. Spectral, spatial and temporal evolution of a phonons pulse, *Phys. Rev. B* **16**, 5303, 5315 (1977).

16. W. E. Bron, Y. B. Levinson, and J. M. O'Conner, Phonon propagation by quasidiffusion, *Phys. Rev. Lett.* **49**, 209 (1982).

17. A. V. Akimov and V. G. Shofman, Recombination of hot electrons with holes at trapping centers in epitaxial *n*-type GaAs, *Sov. Phys. Semicond.* **25**, 684 (1991).

18. A. V. Akimov, A. A. Kaplyanskii, and E. S. Moskalenko, Phonon wind on excitons in Si, *Physica B* **169**, 382 (1991).

19. M. T. Ramsbey, I. Szafranek, G. Stillman, and J. P. Wolfe, Optical detection and imaging of nonequilibrium phonons in GaAs using excitonic photoluminescence, *Phys. Rev. B* **49**, 16427 (1994).

20. See also A. Yu. Blank, E. N. Zinov'ev, L. P. Ivanov, E. I. Kovalev, and I. D. Yaroshetskii, Exciton recombination in GaAs induced by nonequilibrium acoustic phonons, *Sov. Phys. Semicond.* **25**, 39 (1991). Further references on related subjects may be found in Ref. 19.

21. D. C. Hurley, G. A. Hardy, P. Hawker, and A. J. Kent, Spatially resolving detector for heat-pulse experiments in magnetic fields, *J. Phys. E. Sci. Instrum.* **22**, 824 (1989).

22. A. J. Kent, Phonon imaging in two-dimensional electron systems, in *Phonon Scattering in Condensed Matter VII*, Vol. 112 of Springer Series in Solid State Sciences (Springer-Verlag, Berlin, 1993), p. 351.

23. W. Burger and K. Lassmann, Energy-resolved measurment of the phonon-ionization of D⁻ and A⁺ centers in Si with superconducting-Al tunnel junctions, *Phys. Rev. Lett.* **53**, 2035 (1984); Continuous high-resolution phonon spectroscopy up to 12 meV: measurement of the A⁺ binding energies in Si, *Phys. Rev. B* **33**, 5868 (1986).

24. K. Lassmann and W. Burger, Phonon-induced electrical conductance in semiconductors, in *Phonon Scattering in Condensed Matter V*, Vol. 68 of Springer Series in Solid State Physics, A. C. Anderson and J. P. Wolfe, eds. (Springer-Verlag, Berlin, 1986), p. 116.

25. H. Karl, W. Dietsche, A. Fischer, and K. Ploog, Imaging of the phonon drag effect in GaAs-AlGaAs-heterostructures, *Phys. Rev. Lett.* **61**, 2360 (1988); F. Dietzel, W. Dietsche, and K. Ploog, Electron–phonon interaction in the quantum Hall effect regime, *Phys. Rev. B* **48**, 4713 (1993); Y. M. Kershaw, S. J. Bending, W. Dietsche, and K. Eberl, Quantitative model of phonon-drag imaging experiments, *Semicond. Sci. Technol.* **11**, 1036 (1996).

26. M. Rothenfusser, L. Köster, and W. Dietsche, Phonon emission spectroscopy of a two-dimensional electron gas, *Phys. Rev. B* **34**, 5518 (1986).

27. See, for example, W. Dietsche, F. Dietzel, U. Klass, and R. Knott, Phonon absorption and emission experiments in quantum-Hall-effect devices, in *Phonon Scattering in Condensed Matter VII*, Vol. 112 of Springer Series in Solid State Physics, M. Meissner and R. O. Pohl, eds. (Springer-Verlag, Berlin, 1993), p. 345.

28. A. J. Kent, K. R. Strickland, and M. Henini, Phonon emission from a hot two-dimensional hole gas, *ibid.*, p. 359.

29. Several theoretical studies on this subject are reported in M. Meissner and R. O. Pohl, eds., *Phonon Scattering in Condensed Matter VII*, Vol. 112 of Springer Series in Solid State Physics (Springer-Verlag, 1993), pp. 341–382.

30. R. Knott, W. Dietsche, K. von Klitzing, K. Eberl, and K. Ploog, Inside a 2D electron system, images of potential and dissipation, *Solid State Electron.* (UK) **37**, 689 (1994).

31. A review of acoustic microscopy is given in *IEEE Trans. Sonics Ultrasound* **SU-32(2)** (1985). A chronicle of imaging with acoustic waves may be found in the International Conference Series, the volumes are entitled *Acoustical Imaging*. They are the *Proceedings of International Symposia on Acoustical Imaging* (1978–1995 are volumes 8–25) *Acoustical Imaging* (Plenum, New York).

32. See, for example, A. Rosencwaig, Thermal-wave imaging, *Science* **218**, 223 (1982).

33. H. Schreyer, W. Dietsche, and H. Kinder, Micro phonography of buried doping structures in Si, in *Phonon Scattering in Condensed Matter V*, Ref. 24, p. 315.

34. S. J. Kirch and J. P. Wolfe, Phonon-absorption imaging of the electron-hole liquid in Ge, *Phys. Rev. B* **29**, 3382 (1984).

35. W. Dietsche, S. J. Kirch, and J. P. Wolfe, Phonon spectroscopy of the electron-hole liquid in germanium, *Phys. Rev. B* **26**, 780 (1982).

36. R. P. Huebener, E. Held, W. Klein, and W. Metzger, Imaging of spatial structures with ballistic phonons, in *Phonon Scattering in Condensed Matter V*, Ref. 24, p. 305; W. Klein, E. Held, and R. P. Huebener, Imaging of spatial structures in crystals with ballistic phonons, *Z. Phys. B* **69**, 69 (1987).

37. W. Metzger, R. P. Huebener, R. J. Haug, and H.-U. Habermeier, Imaging of oxide precipitates with ballistic phonons, *Appl. Phys. Lett.* **47**, 1051 (1985).

38. V. Narayanamurti, R. C. Dynes, P. Hu, H. Smith, and W. F. Brinkman, Quasiparticle and phonon propagation in bulk, superconducting lead, *Phys. Rev. B* **18**, 6041 (1978).

39. M. R. Hauser, Acoustic waves in crystals: I. Ultrasonic flux imaging and internal diffraction, II. Imaging phonons in superconducting Nb, Ph.D. Thesis, University of Illinois at Urbana-Champaign (1995).

40. D. J. Van Harlingen, Phase-sensitive tests of the symmetry of the pairing state in the high-temperature superconductors – evidence for $d_{x^2-y^2}$ symmetry, *Rev. Mod. Phys.* **67**, 515 (1995).

41. *Fifth International Workshop on Low Temperature Detectors, J. Low Temp Phys.* **93**, 185–858 (1993). See also the review by N. E. Booth and D. J. Goldie, Superconducting particle detectors, *Supercond. Sci. Technol.* **9**, 493 (1996).

42. Bolometers and tunnel junctions for high-energy particle detection are considered in the following sampling of early articles: H. J. Maris, Design of phonon detectors for neutrinos, Ref. 24, p. 404; F. von Feilitzsch, T. Hertrich, H. Kraus, Th. Peterreins, F. Probst, and W. Seidel, Low temperature calorimeters, *Nucl. Instrum. Methods. Phys. Res. A* **271**, 332 (1988); B. Cabrera, J. Martoff, and B. Neuhauser, Acoustic detection of single particles, *Nucl. Instrum. Methods. Phys. Res. A* **275**, 97 (1989); Th. Petereins, F. Pröbst, F. von Feilitzsch, R. L. Mössbauer, and H. Kraus, A new detector of nuclear radiation based on ballistic phonon propagation in single crystals at low temperatures, *Phys. Lett. B* **202**, 161 (1988).

43. B. A. Young, B. Cabrera, and A. T. Lee, Observation of ballistic phonons in silicon crystals induced by α-particles, *Phys. Rev. Lett.* **64**, 2795 (1990).

44. B. A. Young, B. Cabrera, A. T. Lee, and B. L. Dougherty, Detection of elementary particles using silicon crystal acoustic detectors with titanium transition edge phonon sensors, *Nucl. Instrum. Methods. Phys. Res. A* **311**, 195 (1992).

45. B. Cabrera, B. L. Dougherty, A. T. Lee, M. J. Penn, J. G. Pronko, and B. A. Young, Prompt phonon signals from particle interactions in Si crystals, Ref. 41, p. 365; H. J. Maris, Phonon physics and low temperature detectors of dark matter, Ref. 41, p. 355.

46. S. E. Esipov, M. E. Msall, B. Cabrera, and J. P. Wolfe, Ballistic phonon emission from electron-hole droplets: applied to the nuclear recoil problem, Ref. 41, p. 377.

47. G. Simmons and H. Wang, *Single Crystal Elastic Constants and Calculated Aggregate Properties Handbook* (MIT Press, Cambridge, MA, 1971).

Appendix I

Algebraic solution
to the wave equation

As discussed in Chapters 2 and 4, the wave equation for an anisotropic medium leads to the characteristic equation

$$|D_{il} - v^2\delta_{il}| = 0, \tag{1}$$

where D_{il} is the 3×3 Christoffel matrix that depends on the crystal density ρ, the elastic constants C_{ijlm}, and the direction cosines n_i of the wavevector,

$$D_{il} = C_{ijlm}n_j n_m/\rho. \tag{2}$$

The three eigenvalues v_α^2 and eigenvectors $e(\alpha)$ of the Christoffel matrix are readily found by diagonalizing it numerically for a given wavevector direction. However, there may be times when an analytic solution to Eq. (1) is desired. Every[1] has found a closed-form solution for the phase velocities and polarization vectors that contains some useful insights.[2] We summarize his results here. Defining $v^2 = X$, Eq. (1) may be written

$$X^3 + q X^2 + r X + s = 0, \tag{3}$$

where q, r, and s are defined in terms of the C_{ijlm}'s and direction cosines n_i of the wavevector. [See, for example, Eq. (6) on page 92.] The three roots of this equation have the form

$$v_\alpha^2 = M + N\cos[\psi + (2/3)\pi\alpha], \qquad \alpha = 0, 1, 2, \tag{4}$$

where M, N, and ψ depend on the C_{ijlm}'s and n_i's, as given in Every's paper. Thus the square velocities may be represented as the phasor diagram shown in Figure 1. The angles between the three phasors are 120 degree. The center of the circle is at M, and its radius is N. For cubic crystals, the mean square velocity is simply $M = (C_{11} - C_{44})/3\rho$, but N and ψ depend on wavevector direction $\mathbf{n}$. So, as $\mathbf{n}$ changes in a given crystal, both the phasor angle and the radius of the circle change.

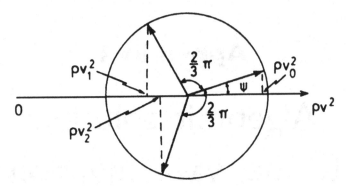

Figure 1 Phasor diagram representing the three eigenvalues, ρv_α. (From Every.[1])

In all but a few highly anisotropic crystals, ψ is less than $30°$ for all $\mathbf{n}$; therefore, the L velocity, v_0 (denoted v_3 in Chapter 4), and the T velocities, v_1 and v_2, are well separated. That is, the L and T slowness surfaces do not touch. (Recall that L, FT, and ST are defined here as the inner, middle, and outer surfaces.)

Every points out that for cubic crystals N and ψ depend only on the n_j's, ρ, and the combination of elastic constants given by

$$C_3 = (C_{11} - C_{22} - 2C_{44})/(C_{11} - C_{44}). \qquad (5)$$

So, the relative variation of v_α with direction depends on this single parameter. Does this mean that the shape of the slowness surface depends only on a single parameter? Unfortunately not, because the constant $M = (C_{11} - C_{44})/3\rho$ is added to this angular-dependent term, so the average radius of the surface (hence, the curvature) depends on M, which cannot be expressed solely in terms of C_3. Two parameters (e.g., $a = C_{11}/C_{44}$ and $b = C_{12}/C_{44}$) are required to specify the shape of the slowness surface and, thus, the phonon-focusing patterns.

Analytic expressions for the group velocity involve the phase-space derivatives, $\partial v/\partial n_i$, which can be found by implicit differentiation of Eq. (3). Expressions for the Jacobian further involve derivatives of the group velocity, $\partial V_i/\partial n_j$, which are second-order derivatives of v. Equations for these phase-velocity derivatives have been reported by Northrop and Wolfe[3] and are available in software form.[4] As previously mentioned, however, group velocity and Jacobian for a given $\mathbf{k}$ and α are conveniently calculated numerically by matrix manipulations on a computer.

References

1. A. G. Every, General, closed-form expressions for acoustic waves in cubic crystals, *Phys. Rev. B* **22**, 1746 (1980).

2. See also D. M. Barnett, J. Lothe, K. Nishioka, and R. J. Asaro, Elastic surface waves in anisotropic crystals: a simplified method for calculating Rayleigh velocities using dislocation theory, *J. Phys.* **F3**, 1083 (1973).

3. G. A. Northrop and J. P. Wolfe, Ballistic phonon imaging in germanium, *Phys. Rev. B* **22**, 6196 (1980).

4. G. A. Northrop, Acoustic phonon anisotropy: phonon focusing, *Comp. Phys. Commun.* **28**, 103 (1982).

Appendix II

Abbreviated tensor notation and group velocity

It is sometimes useful to use the abbreviated notation,

$$C_{IJ} = C_{ijlm},\tag{1}$$

for the fourth-rank elasticity tensor, as defined by Eqs. (9) and (10) on page 41. In this convention, $\mathbf{C}$ is a 6×6 matrix and stress, σ, and strain, ε, are six-component column vectors,

$$\sigma = \begin{bmatrix} \sigma_1 \\ \sigma_2 \\ \sigma_3 \\ \sigma_4 \\ \sigma_5 \\ \sigma_6 \end{bmatrix}, \qquad \varepsilon = \begin{bmatrix} \varepsilon_1 \\ \varepsilon_2 \\ \varepsilon_3 \\ \varepsilon_4 \\ \varepsilon_5 \\ \varepsilon_6 \end{bmatrix},\tag{2}$$

with elements defined by the conventions

$$\begin{bmatrix} \sigma_{xx} & \sigma_{xy} & \sigma_{xz} \\ \sigma_{xy} & \sigma_{yy} & \sigma_{yz} \\ \sigma_{xz} & \sigma_{yz} & \sigma_{zz} \end{bmatrix} = \begin{bmatrix} \sigma_1 & \sigma_6 & \sigma_5 \\ \sigma_6 & \sigma_2 & \sigma_4 \\ \sigma_5 & \sigma_4 & \sigma_3 \end{bmatrix},\tag{3a}$$

$$\begin{bmatrix} \varepsilon_{xx} & \varepsilon_{xy} & \varepsilon_{xz} \\ \varepsilon_{xy} & \varepsilon_{yy} & \varepsilon_{yz} \\ \varepsilon_{xz} & \varepsilon_{yz} & \varepsilon_{zz} \end{bmatrix} = \begin{bmatrix} \varepsilon_1 & \frac{1}{2}\varepsilon_6 & \frac{1}{2}\varepsilon_5 \\ \frac{1}{2}\varepsilon_6 & \varepsilon_2 & \frac{1}{2}\varepsilon_4 \\ \frac{1}{2}\varepsilon_5 & \frac{1}{2}\varepsilon_4 & \varepsilon_3 \end{bmatrix}.\tag{3b}$$

In standard notation, the dot product between a vector $\mathbf{e}$ and a symmetric second rank tensor A is

$$A \cdot \mathbf{e} = \begin{bmatrix} A_{xx} & A_{xy} & A_{xz} \\ A_{xy} & A_{yy} & A_{yz} \\ A_{xz} & A_{yz} & A_{zz} \end{bmatrix} \begin{bmatrix} e_x \\ e_y \\ e_z \end{bmatrix} = \begin{bmatrix} A_{xx}e_x + A_{xy}e_y + A_{xz}e_z \\ A_{xy}e_x + A_{yy}e_y + A_{yz}e_z \\ A_{xz}e_x + A_{yz}e_y + A_{zz}e_z \end{bmatrix}.\tag{4}$$

To obtain the same result in abbreviated notation, the matrix multiplication must take the form

$$
e \cdot A =
\begin{bmatrix}
e_x & 0 & 0 & 0 & e_z & e_y \\
0 & e_y & 0 & e_z & 0 & e_x \\
0 & 0 & e_z & e_y & e_x & 0
\end{bmatrix}
\begin{bmatrix}
A_1 \\
A_2 \\
A_3 \\
A_4 \\
A_6
\end{bmatrix},
\tag{5}
$$

where the subscript correspondences of Eq. (3a) are assumed for A.

In abbreviated notation, Hooke's law is written

$$
\sigma = C \cdot \varepsilon,
\tag{6}
$$

where the single dot implies a summation over one abbreviated subscript. Newton's second law is written

$$
\rho(d^2 \mathbf{u}/dt^2) = \nabla \cdot \sigma,
\tag{7}
$$

with the definition [similar to Eq. (5)]

$$
\nabla \cdot =
\begin{bmatrix}
\dfrac{\partial}{\partial x} & 0 & 0 & 0 & \dfrac{\partial}{\partial z} & \dfrac{\partial}{\partial y} \\
0 & \dfrac{\partial}{\partial y} & 0 & \dfrac{\partial}{\partial z} & 0 & \dfrac{\partial}{\partial x} \\
0 & 0 & \dfrac{\partial}{\partial z} & \dfrac{\partial}{\partial y} & \dfrac{\partial}{\partial x} & 0
\end{bmatrix}.
\tag{8}
$$

This is equivalent to Eq. (14) of Chapter 2, which is written in component form. Likewise, the form of the strain tensor [Eq. (11) on page 41] is

$$
\varepsilon = \nabla \mathbf{u} =
\begin{bmatrix}
\dfrac{\partial}{\partial x} & 0 & 0 \\
0 & \dfrac{\partial}{\partial y} & 0 \\
0 & 0 & \dfrac{\partial}{\partial z} \\
0 & \dfrac{\partial}{\partial z} & \dfrac{\partial}{\partial y} \\
\dfrac{\partial}{\partial z} & 0 & \dfrac{\partial}{\partial x} \\
\dfrac{\partial}{\partial y} & \dfrac{\partial}{\partial x} & 0
\end{bmatrix}
\begin{bmatrix}
u_x \\
u_y \\
u_z
\end{bmatrix}.
\tag{9}
$$

For the plane-wave displacement

$$
\mathbf{u} = u_0 e\, e^{i(\mathbf{k} \cdot \mathbf{r} - \omega t)}
\tag{8}
$$

the strain becomes

$$\varepsilon = i u_0 \mathbf{k} \mathbf{e}\, e^{i(\mathbf{k}\cdot\mathbf{r} - \omega t)}, \tag{9}$$

with the abbreviated form

$$\mathbf{k}\mathbf{e} = \begin{bmatrix} k_x e_x \\ k_y e_y \\ k_z e_z \\ k_z e_y + k_y e_z \\ k_z e_x + k_x e_z \\ k_y e_x + k_x e_y \end{bmatrix}. \tag{10}$$

This 1×6 matrix appears in Eqs. (35) and (37) on pages 49–50 for the power flow and group velocity, respectively. For a cubic crystal, the product is [using Eq. (10)]

$$C \cdot \mathbf{k}\mathbf{e} = \begin{bmatrix} C_{11}k_x e_x + C_{12}(k_y e_y + k_z e_z) \\ C_{11}k_y e_y + C_{12}(k_x e_x + k_z e_z) \\ C_{11}k_z e_z + C_{12}(k_x e_x + k_y e_y) \\ C_{44}(k_z e_y + k_y e_z) \\ C_{44}(k_z e_x + k_x e_z) \\ C_{44}(k_y e_x + k_x e_y) \end{bmatrix} \equiv \begin{bmatrix} B_1 \\ B_2 \\ B_3 \\ B_4 \\ B_5 \\ B_6 \end{bmatrix}, \tag{11}$$

which, with Eq. (5), leads directly to the group velocity,

$$\mathbf{V} = (1/\rho\omega)\mathbf{e} \cdot (C \cdot \mathbf{k}\mathbf{e}) = \frac{1}{\rho\omega} \begin{bmatrix} e_x B_1 + e_y B_6 + e_z B_5 \\ e_x B_6 + e_y B_2 + e_z B_4 \\ e_x B_5 + e_y B_4 + e_z B_3 \end{bmatrix}. \tag{12}$$

This expression gives $\mathbf{V}$ in terms of the chosen $\mathbf{k}$ vector and the polarization vector $\mathbf{e}$, which is an eigenvector of the Christoffel equation with input $\mathbf{k}$ and C. Notice that in this continuum theory the magnitude of $\mathbf{k}$ is proportional to ω. The group velocity in Eq. (12) may be written in terms of the slowness vector $\mathbf{s} = \mathbf{k}/\omega$, whose components are the solution of Eq. (22) on page 44. In other words, one may replace $\mathbf{k}/\omega$ by $\mathbf{n}/v$, where $\mathbf{n} = \mathbf{k}/k$ is the wave normal, and the phase velocity v is found by solving Eq. (18) on page 43.

Appendix III

Survey of phonon focusing in cubic crystals

As a supplement to Chapter 4, this Appendix contains two theoretical studies that may be useful to those who wish to know the qualitative features of phonon focusing in a particular crystal. These unpublished calculations were produced by Donna Hurley at the University of Illinois.

Figure 1 shows slices of the slowness surface in the (110) planes for the four major regions of elastic-parameter space. The small letters, ℓ and t, indicate the polarization of the waves at various portions of the surfaces.

Figure 2 shows the calculated phonon-focusing patterns for increasing elastic anisotropy in the negative-Δ regime. Both the fourfold [100] axis and the threefold [111] axis are readily identified in these wide-angle views.

Figure 3 shows the calculated phonon-focusing patterns for increasing elastic anisotropy in the positive-Δ regime.

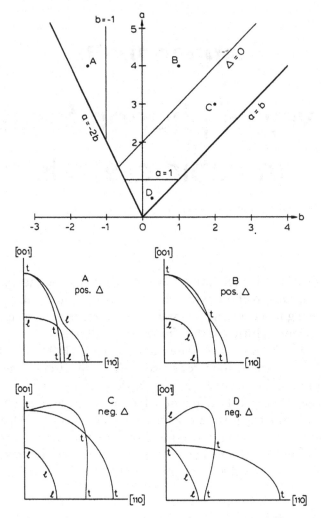

Figure 1 Top: Elastic parameter space with $a = C_{11}/C_{44}$ and $b = C_{12}/C_{44}$. Bottom: Slowness curves in the (110) plane for the four points labeled A, B, C, and D.

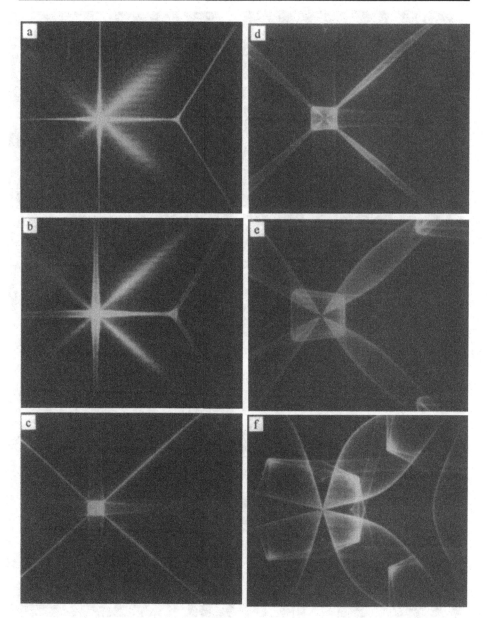

Figure 2 *Phonon-focusing patterns for crystals in the negative-Δ regime, showing the effect of increasing anisotropy. (a) Diamond, (b) Cr_2FeO_4, (c) MgO, (d) Ge, (e) CdTe, (f) CoO.*

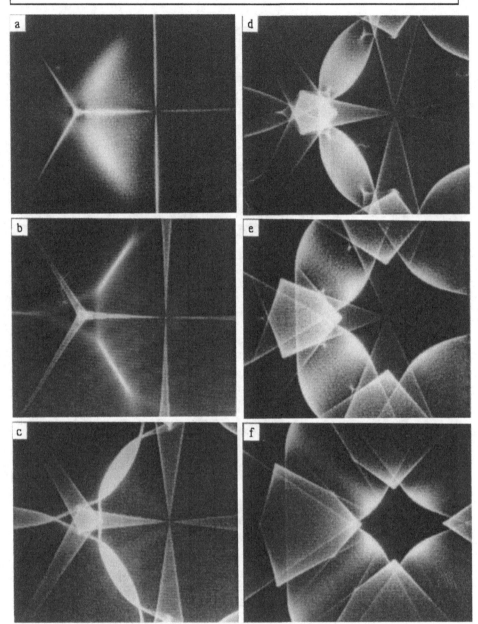

Figure 3 Phonon-focusing patterns for crystals in the positive-Δ regime.
(a) CsBr, (b) TlBr, (c) NaF, (d) NaCl, (e) KF, (f) KCl.

Index